# Principles of Agronomy

# Principles of Agronomy

Edited by **Jamie Hanks**

R CALLISTO REFERENCE

New York

Published by Callisto Reference,
106 Park Avenue, Suite 200,
New York, NY 10016, USA
www.callistoreference.com

**Principles of Agronomy**
Edited by Jamie Hanks

International Standard Book Number: 978-1-63239-640-2 (Hardback)

Printed in the United States of America.

# Contents

# Preface

I am honored to present to you this unique book which encompasses the most up-to-date data in the field. I was extremely pleased to get this opportunity of editing the work of experts from across the globe. I have also written papers in this field and researched the various aspects revolving around the progress of the discipline. I have tried to unify my knowledge along with that of stalwarts from every corner of the world, to produce a text which not only benefits the readers but also facilitates the growth of the field.

This book brings forth some of the most innovative concepts and elucidates the unexplored aspects of agronomy. It tries to understand the properties and interactions of soil and plants and is a combination of ecology, earth science, genetics, biology, chemistry, and economics. Plant physiology, soil classification, soil fertility, crop rotation, irrigation and drainage, plant breeding, pest control, insect and weed control are among the list of areas covered under this field. This discipline plays a crucial role in ensuring food security, soil health, sustainability of agricultural resources and improving nutritional value of already existing food crops. The various sub-fields of agronomy along with the technological progress that have future implications are glanced at in this book. It collated the researches of veterans from across the globe to unravel the unexplored aspects of this subject. As this field is emerging at a rapid pace, the contents of this book will help the readers understand the modern concepts and applications of the subject. It will prove to be a beneficial source of knowledge for agronomists, agriculturists, environmentalists, researchers and students alike.

Finally, I would like to thank all the contributing authors for their valuable time and contributions. This book would not have been possible without their efforts. I would also like to thank my friends and family for their constant support.

**Editor**

# Trace Element Management in Rice

**Abin Sebastian * and Majeti Narasimha Vara Prasad**

Department of Plant Sciences, University of Hyderabad, Hyderabad-500046, Telangana, India;
E-Mail: mnvsl@uohyd.ernet.in

\* Author to whom correspondence should be addressed; E-Mail: abinhcu@gmail.com

Academic Editor: Gareth J. Norton

---

**Abstract:** Trace elements (TEs) are vital for the operation of metabolic pathways that promote growth and structural integrity. Paddy soils are often prone to TE limitation due to intensive cultivation and irrigation practices. Apart from this, rice paddies are potentially contaminated with transition metals such as Cd, which are often referred to as toxic TEs. Deficiency of TEs in the soil not only delays plant growth but also causes exposure of plant roots to toxic TEs. Fine-tuning of nutrient cycling in the rice field is a practical solution to cope with TEs deficiency. Adjustment of soil physicochemical properties, biological process such as microbial activities, and fertilization helps to control TEs mobilization in soil. Modifications in root architecture, metal transporters activity, and physiological processes are also promising approaches to enhance TEs accumulation in grains. Through genetic manipulation, these modifications help to increase TE mining capacity of rice plants as well as transport and trafficking of TEs into the grains. The present review summarizes that regulation of TE mobilization in soil, and the genetic improvement of TE acquisition traits help to boost essential TE content in rice grain.

**Keywords:** trace elements; bioavailability; mobilization; fertilizer; bio-fortification; root phenology; molecular physiology; root biology; metal transporters; photosynthesis

---

## 1. Introduction

Rice is the main dietary source of trace elements (TEs) in Asian populations. However, rice grain possesses relatively low TE content compared to other cereal grains [1–3]. Hence enhancement of TEs in rice could be a practical solution to avoid TE deficiency disorders prevalent among populations which consume rice as staple food (Table 1). Apart from this, rice grain contaminated with toxic TEs such as Cd evokes the need of food safety measures that help to avoid toxic TE accumulation in rice. It is well known that nutrient cycles influence the mineral composition of the crop produce [4,5]. More than 90% of rice is cultivated in low land conditions where the soils are subjected to intermittent wet and dry periods. The field is subjected to water either by rain or irrigation. Occurrence of intermittent flooding especially in low land rain fed fields causes temporal changes in mobility of TEs [6–8]. However, irrigation is favorable to control the flooding periods and opens ways to manage TEs in the field through fertilization. The low land fields are also characterized by adequate water supply at the time of flowering, buffering of soil temperature and low risks of weed competition. All these features of low land fields help to prevent crop failure, and hence low land cultivation of rice is the most common cultivation practice.

**Table 1.**Trace elements (TEs) and human health.

| TEs | Health Disorder |
|---|---|
| Se | Keshena (China), bone, arthritis, cardio vascular and Cancer |
| Zn | Dwarf ness (Iran & Egypt) infertility, impaired taste and smell |
| Cu | Anemia, Skeletal defects |
| Mn | Tremors, Stiff muscles |
| I | Goiter |
| Fe | Anemia |
| Mg | Depression, Nervous system disorders |
| Mo | Mouth/esophageal cancer, Neurologic damage |
| Co | Anemia |
| Na | Coma |
| Toxic TEs | Toxicity |
| Hg | Mina-Mata, Neurological disorders |
| Cd | Itai-Itai, Bone crippling, Cancer, Heart problems |
| F | Yellowing of teeth, Skeletal deformities, Dental caries |
| Pb | Lung cancer, Neurological disorders |
| As | Skin, Lung cancer |
| Cr | Lung cancer, Kidney dysfunction, Skin ulcer |

Adaptability of rice plants to varying fertilization strategies promises TE management in low land rice paddies. Most often the rice paddies are fertilized with elements such as nitrogen, phosphorous, and potassium that enhance productivity. Zn, Fe, B, Mn, Cu, Mo and Si are the critical TEs for rice growth. Wetland rice fields are prone to Zn deficiency because of increased availability of macronutrient cations during the course of submergence [9]. But Fe and Mn deficiency is common in upland paddy fields due to formation of metal oxides as a consequence of higher aeration [10]. Moisture stress is well known to cause B deficiency in rice paddies whereas deficiency of Si is often related to mineral reserve deficits in soil rich in organic matter. Paddy soils also face Cu deficiency because of the application of lime to correct the acidic soil pH. TEs not only serve as enzyme cofactors but also play an important role in inhibition of plant uptake of toxic TEs such as Cd [11]. Thus one of the major drawbacks of micronutrient deficiency is the increase in availability of toxic TEs to crops.

Rice paddies can also be contaminated with toxic TEs such as Cd because of the usage of Cd contaminated phosphate fertilizer [12–14]. Apart from this, irrigation using mine water leachate also contaminates paddy soils with toxic TEs [15]. Thus rice plants are prone to toxic TE accumulation, which ultimately leads to a diet rich in toxic TEs. Since toxic TEs are detrimental to human health, accumulation of these elements needs to be reduced in rice grain. The present review focuses on the role of soil factors and molecular physiological processes on TE management in rice grain. The regulatory roles of nutrient cycles deserve attention as these cycles determine the TE availability to plants [16].The importance of soil physiochemical characteristics and biological activities responsible for TE mobilization in paddy soils is also discussed. It is well known that nutrient acquisition is the key physiological process, which has a direct role in plant uptake of TEs. This process depends on multiple factors such as root architecture, photosynthetic efficiency and activity of metal transporters. Hence the scope of manipulation of these factors is also explored. Advent of fertilization strategies, availability of rice germplasm and knowledge about genes responsible for TE accumulation can help to formulate approaches for TE fortification in rice.

## 2. Nutrient Cycle and Plant Uptake of TEs

Plant uptake of both toxic and non-toxic TEs depends on nutrient cycling in the soil [17,18]. Hence keeping balance of nutrient cycling is important to ensure crop yield together with restriction of toxic TE accumulation in rice grain. However, nutrient cycling in rice paddy agro ecosystems is often unbalanced due to anthropogenic intervention (Figure 1). Seasonal variations in soil physicochemical processes, especially in lowland rain fed fields where intermittent wet and dry periods occur, cause fluctuations in nutrient cycling. The flooding periods in rice fields cause leaching of cations [19]. Flooding also causes shifts in soil physicochemistry and results in precipitation or co-precipitation reactions of TEs. Soil microbial activity and incorporation of organic matter also affects solubility as well as plant uptake of TEs [20,21]. The difference between nutrient input into the soil through fertilization and nutrient removal during harvest also causes disturbances of nutrient cycles in paddy soils. Most often the fields receive fertilizer input in significantly higher amounts than needed by the rice plants. This could lead to an accumulation of mineral nutrients, especially ammonium, in the field that will reduce soil pH, and hence increase the mobility of soil TEs.

A nutrient cycle comprises the in and out flow of nutrients of the paddy field. Crop harvest is the major route of nutrients outflow in a paddy field. However, fertilizer application often replenishes the nutrient pool. It has been reported that TE availability in the soil depends on the parent material from which the soil is formed [22]. Even though paddy soils around the world are different because of variation in parent material from which they are derived, repeated agricultural practices such as fertilization have rendered these soils uniform, especially with regard to macronutrient content.

**Figure 1.** Field management practices that influence TE availability in rice fields (**a**); Irrigation (**b**); Puddling (**c**); Dry land cultivation (**d**);Mixed or co-cropping (**e**); *Azolla*cultivation (**f**); System of rice intensification (**g**); Fertilizer application (**h**); Burning of crop residues (**i**); Rice cultivation near mine tailings.

TE concentrations in most paddy soils are higher than in non-paddy soils because of anthropogenic input. Decay of organic matter also contributes to the TE pool. Crop residue, especially roots and culms account for a major portion of the organic matter in paddy soil. Since the concentration of many of the TEs in rice plants are—in the decreasing order—root, shoot and grain, retention and decay of crop residues in the soil significantly contribute to replenishment of the TEs pool in the paddy soil. Occurrence of TEs in the soil solution is limited as they bind with organic matter or parent material [23]. Thus a majority of TEs are associated with the solid phase of soil where they occur as precipitate or co-precipitate of mineral salts. Co-precipitation usually occurs with newly formed chemical species such as carbonates, phosphates, oxides and hydroxides [24,25]. These kinds of co-precipitation help to limit exposure of toxic TEs to biota, and also slow down the release of TEs into the soil solution. As seen among soil types of the world, when the concentration of a particular mineral nutrient exceeds a certain threshold value, mineral ions precipitate in the paddy soil. Similarly when the concentration of the nutrient ion is lower than the chemical solubility equilibrium, dissolution of that particular element occurs. Nutrient ions in the paddy soil solution are found in association with organic acids, inorganic ion pairs and free ions [26]. All these contribute to the dynamics of nutrient cycling in paddy soils that influence plant availability of TEs (Figure 2).

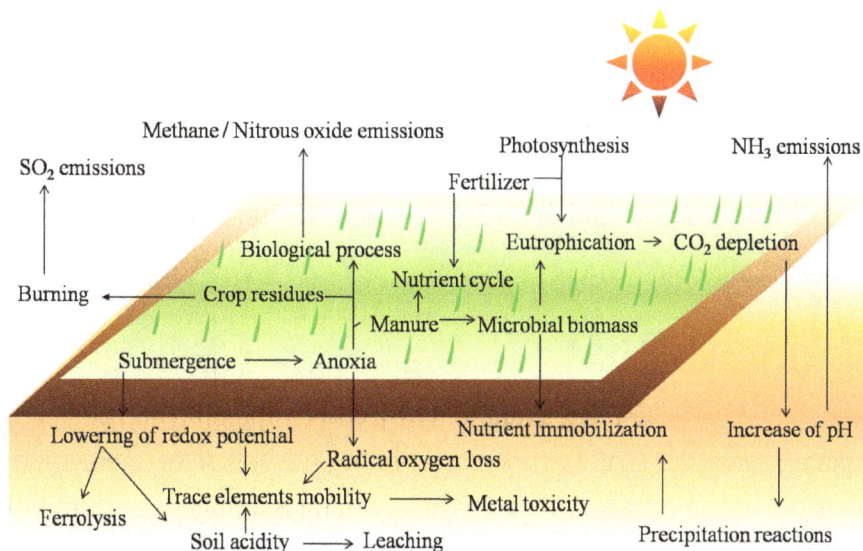

**Figure 2.** Nutrient cycling and mobilization of TEs in rice paddies.

## 2.1. Regulatory Role of Macronutrients

Macronutrients are the primary nutrients required for plant growth. Macronutrients affect TEs availability by shifting soil pH and precipitation reactions. Application of ammonium-based fertilizers is common in paddy fields for replenishing soil nitrogen. The applied fertilizer stimulates algal growth in paddy fields with enhanced utilization of $CO_2$. This reduces the carbonic acid content in submerged paddy fields with a subsequent rise of soil pH (even up to a pH of 9). But increase of pH leads ammonia evolution as well as precipitation of TEs. Apart from this, nitrogen loss occurs through leaching as well as denitrification [27]. Ammonia is also produced during the breakdown of organic materials such as manure, and volatilization of this gas further increases soil pH, which causes precipitation of TEs. It must be noted that a significant quantity of ammonia can be adsorbed into soil colloids, and hence plants growing in soils with a lower colloid contents, such as clay and humus, are more prone to suffer from nitrogen deficiency [28]. $SO_2$ held with soil particles causes a lowering of soil pH during submergence, and hence increases mobility of TEs. Microbial degradation of S-containing compounds also releases various S-containing gases, such as hydrogen sulfide ($H_2S$), carbon sulfide ($CS_2$), carbonyl sulfide (COS) and methyl mercaptan ($CH_3SH$) in submerged paddy soils which can react with TEs, which leads to precipitation of TEs [29,30].

Phosphorous (P) in paddy fields often meets the crop's demand. However, tillage practices often remove particulate P in the paddy soil through erosion. A significant amount of P loss also occurs during flooding related runoff in submerged rice paddies. In strongly acidic soils, P undergoes a precipitation reaction with elements such as Al, Mn and Fe. These precipitates are sparingly soluble, and hence rice plants growing in such soils face P deficiency as well as a deficit of Mn or Fe [31]. On the other hand, when the field is submerged, TEs undergo a reduction reaction that enables the release of P as well as TEs into the soil solution. Potassium (K) is prone to leaching, and this leads to a shift of soil pH towards the acidic phase, which mobilizes both TEs and toxic TEs. Apart from this, acidic soils tend to mask plant available K, and hence result in K deficiency, which accelerates TE toxicity [32]. It is noteworthy that poultry manure and wood ashes are good sources of K and have the capacity to immobilize toxic TEs [33]. Temporally sequential wetting and drying patterns in the rice field may also lead to the release of fixed K together with mobilization of TEs in the solid phase. Both Ca and Mg are also prone to leaching, especially in acidic paddy soils that cause more availability of toxic TEs for plant uptake. However, application of lime and dolomite in acidic paddy soils not only increases soil pH but also acts as potential source of Ca and Mg that can decrease plant uptake of toxic TEs.

## 2.2. Regulatory Role of Micronutrients

TEs act on each other with regard to plant availability by co-precipitation and competition for metal binding sites or transporters. Solubility of TEs increases during the course of submergence [34]. Mass flow of elements into roots especially that of Fe and Mn, accelerates at this stage. Similarly, movement of TEs towards the subsoil also increases in speed. Plaque formation in rice roots with Fe and Mn compounds is a noticeable feature of paddy soils during periods of submergence. This process occurs due to the development of an oxygen rich network of aerenchyma tissue in rice roots. When reduced forms of Fe/Mn reach the surface of this oxygen rich zone, these elements undergo oxidation and precipitate with silica or phosphate. This kind of precipitation also occurs in oxygen rich plow sole and surface soil. Thus, during a course of submergence, the movement and precipitation of Fe or Mn causes characteristic paddy soil development where a zone of reddish-brown streaks of the root channels is sandwiched between oxygenated surface layers and the Fe or Mn illuviated zone [35]. It is noteworthy that the downward movement of soluble Fe or Mn towards comparatively oxygen rich zones followed by precipitation reactions cause depletion of these elements from the top soil. The cycling of TEs in the solid phase of soil to the solution phase generally occurs either by adsorption-desorption reactions or precipitation-dissolution processes [36]. TEs in the solid phase often form stable bonds with surface functional groups, and hence the availability of the element to the biota depends on the strength of the bond formed between the particular nutrient element and the functional groups present in the solid phase [37]. Soil microbial processes play a significant role in the breakage of these bonds, and thus microbial action controls TEs availability. The biogeochemical transformation of Fe or Mn not only affects solubility and speciation of Fe or Mn but also the availability of toxic TEs such as Cd. Ability of Fe to form a complex with S reduces the chance of precipitation of Cd in the form of CdS in submerged soils. This situation also decreases the chance of adsorption of Cd to Fe or Mn oxides. Thus reductive dissolutions of Fe or Mn oxides in submerged soils lead to mobilization of Cd from the solid phase of the soil. However, the higher reactivity of $FeS_2$ may favor sorption of some parts of Cd in the solution. Similarly, Fe in the soil protects the rice plants from other toxic TEs.

Periodical changes of submergence and drainage sequences also cause variation in occurrence of forms of TEs in paddy soils, which influence TEs availability to plants (Table 2). For example Zn occurs in at least seven kinds of ionic forms or mineral salts in paddy soils ($Zn^{2+}$, $ZnCl^+$, $ZnSO_4$, $ZnOH^+$, $ZnCO_3$, $ZnS$, $ZnSiO_3$). It is well known that availability of nutrient ion to biota depends on the chemical speciation. Most often TEs undergo speciation with $OH^-$, $Cl^-$, $SO_4^{2-}$, $HCO_3^-$ and $F^-$. Humic substances and fulvic acids rich in poly carboxylic acids are also potential speciation agents of TEs in paddy soil [38]. Amino acids and tricarboxylic acids also play a significant role in the speciation of mineral nutrient ions [39,40].TEs in the solid phase are mobilized during suspension of organic matter or soil particles. Apart from these, puddling of rice paddies also contributes to mobilization of TEs in the soil sediments. However, flood related anoxia in the field in turn favors formation of hydroxide salts of TEs that make them unavailable for plant uptake [41,42]. Irrigation with chlorinated water also contributes to mobilization of TEs through the formation of water-soluble TEs salts with chlorine. Environmental factors are also involved in TE cycling in paddy soils [43,44]. Small quantities of TEs are incorporated into the field during rain fall and are deposited by the wind. TEs undergo leaching in the presence of runoff water, and hence rain fed rice paddies often face threat of macronutrient leaching [45]. Leaching of macronutrient cations not only causes nutrient deficiency in the soil but also contributes to soil acidification. On the other hand, acidification of soil accelerates the weathering process which in turn increases the TE pool in the paddy soil. However, this event often favors an increase of Al/Fe/Mn ions and causes Al/Fe/Mn toxicity in rice plants rather than promoting plant growth.

**Table 2.** Ionic forms of major TEs and toxic TEs in the soil solution.

| TEs | Occurrence |
| --- | --- |
| Mn | $Mn^{2+}$ |
| Fe | $Fe^{2+}$, $Fe(OH)_2^+$, $Fe(OH)^{2+}$, $Fe^{3+}$ |
| Zn | $Zn^{2+}$, $Zn(OH)^+$ |
| Mo | $MoO_4^{2-}$, $HMoO_4^-$ |
| B | $H_3BO_3^-$ |
| Cu | $Cu^{2+}$, $Cu(OH)^+$ |
| Co | $Co^{2+}$ |
| Toxic TEs | Occurence |
| Pb | $Pb^{2+}$ |
| Hg | $Hg^{2+}$ |
| Cd | $Cd^{2+}$, $Cd^{1+}$ |
| As | As(III), As(V) |
| Cr | $Cr^{3+}$, $Cr^{6+}$ |

## 3. Soil Dynamics Act on Mobilization of TEs

The dynamics of the paddy soil properties occurs mainly due to changes in physiochemical properties, biological activities and fertilization. Physiochemical characters of soils such as redox potential, pH, alkalinity, cation exchange capacity and texture significantly influence TEs mobilization in paddy soil [46–48]. An important feature of paddy fields consists of the progressive changes in oxidation and reduction states upon irrigation. Management of water supply of rice in the field leads to temporal changes in the water table called hydroperiods as well as changes in redox status of paddy soils. Submerged rice fields are also characterized by a low reduction potential because of poor aeration. In submerged conditions, TEs in the paddy soil undergo reduction reactions that reduce ionic forms of these elements in the soil solution. Biological processes and fertilization also play a crucial role in mobilization of TEs and hence help to control the mobility of TEs in paddy soils.

## 3.1. Redox Process

Reducing periods in the paddy soil causes redox depletion of elements such as Fe [49]. Reducing events cause formation of more $Fe^{2+}$ from the $Fe^{3+}$ pool, which migrate away from reduced zones. The migrated $Fe^{2+}$ later precipitates as $Fe^{3+}$ in oxygen rich zones. The process causes depletion of iron coatings from soil mineral grains. This causes the redoximorphic appearance of paddy soils with contrasting colors of redox depletions (gray color) and zones of oxidized Fe (blue-green). TEs posses multiple oxidation states, and the reducing surroundings favor solubility of TEs [50,51]. Most often metal ions such as Cu and Zn occur in association with oxides of other elements such as Mn. Hence the solubility of the oxides of TEs also favors the solubility of other TEs by releasing adsorbed TE ions. Submergence induced anoxia in the field favors the formation of TE sulfides and this makes many of TE cations immobile in soil. The organic matter of soil as well as carbonates also contribute to the above-mentioned anoxia induced TE immobilization in paddy soils.

Paddy soils differ in their degree of depletion of oxygen because of differences in soil texture, bulk density, aggregate stability, organic matter content, bio-pore formation, thermal properties and seasonal variability. The puddled paddy soil, which is saturated with water, undergoes alternate frost action during the winter season that breaks up the large soil masses and hence increases aeration events [52]. When the soil oxygen is depleted (Eh $\sim$ 0.34), N in the $NO_3^-$ undergoes reduction reactions resulting in the formation of $NO_2^-$, $N_2$, etc., and causesa drop in redox potential (Eh $\sim$ 0.24). At this point, Mn reduction followed by a sequence of reduction reactions of other TEs occurs with a drop of the reduction potential of the soil. This indicates transformation of macronutrients and TEs in sequential order. The reduction reactions during submergence follow the sequence N > Mn > Fe > S > C. Many of the TEs other than Mn and Fe undergo reduction reactions preferentially in the order between Fe and S. When the soil is drained, oxidation reactions occur in the reverse sequence. Another characteristic feature of paddy soils is the anoxia mediated methane production [53]. This occurs at $E_h$ below −0.2 V. Thus monitoring of nitrous oxide and methane helps to determine the reducing conditions in paddy soils. It can be concluded that dissolution or precipitation of TEsin paddy soils depend to a large extent on the dynamics in the anoxia mediated shifts in redox potential. Hence control of irrigation, which helps to regulate anoxic conditions in the field, could be a practical solution for the management of TE availability to rice plants.

## 3.2. Soil pH

Soil pH influences TE solubility and hence TE mobility [54,55]. Lowering of soil pH increases the solubility of TEs. Changes in redox potential are associated with shifts in soil pH. Decrease in reduction potential of soil causes a phenomenon called ferrolysis that accelerates soil acidity mediated weathering of the parent rock material [56,57]. When paddy soils are submerged, cations displaced by $Fe^{2+}$ migrate out ofthe reduced zone resulting in a deficiency of cations. When such a paddy soil is drained, the $Fe^{2+}$ in the soil becomes $Fe^{3+}$, and results in the precipitation of ferric oxide, which adsorbs TEs [58].This often results in the occurrence of a $H^+$ ion as the major cation in the paddy soil and causes lowering of the soil pH. Thus even though solubility of TEs could be higher in paddy soil, plant available TEs are limited.

Rice plants growing in submerged acidic soils often experience TEs or toxic TEs toxicity [59,60]. Al, Mn and Fe are the chief elements that cause metal toxicity in acidic paddy soils. The type of metal toxicity depends on the composition of the parent material from which the paddy soils are formed. For example, rice plants grown in paddy soils derived from Mn rich parental materials experience Mn toxicity during soil acidification. Generally non-acid forming cations such as $Ca^{2+}$ and $Mg^{2+}$ form soluble complexes with organic ligands and undergo leaching in acidic paddy soils [61]. Apart from this, organic matter contributes to soil acidity as it contains many functional groups that dissociate into $H^+$ ions. However, many of the acidic functional groups of organic matter dissociate at a higher soil pH. Oxidation of the sulfhydryl group (-SH) results in the formation of $H_2SO_4$ and accelerates the lowering of the paddy soil pH. Wet precipitation containing acids is also an agent of soil acidity. Thus rice plants grown in acid soil are confronted with a deficiency of macronutrients (Ca, Mg, K, P, N, and S) and micronutrients (Mo and B) due to leaching with runoff water as well as soil acidity. This situation leads to more availability of elements such as Al, Fe, Zn, Mn, Cu, Cd, Pb and Co in acidic paddy soils. Apart from all the above factors, intensive uptake of non-acidic mineral nutrients by rice plants as well as nitrogen fertilization also contributes to soil acidification in paddy soils that end up in TE toxicity. Hence corrective measures of soil pH play a critical role in the management of plant uptake of both essential and toxic TEs.

### 3.3. Alkalinity

The alkalinity of the soil often reflects the buffering capacity of the soil. Buffering activity is prominent in alkaline soils. Buffer reactions in paddy soils are mediated by carbonates, bicarbonates, carbonic acid and water [62,63]. Dissolution of carbonate minerals releases bicarbonate ions, which upon hydrolysis produce carbonic acid and hydroxyl ions as follows:

$$CaCO_3 + H_2O + H^+ <----> Ca^{2+} + H_3CO_3 + OH^- \tag{1}$$

When acid influx into bicarbonate buffered soil (usually alkaline soil) occurs, the reaction shifts to the right. On the other hand, the reaction will shift to left when the base content prevails in the soil. Both these reactions help in the maintenance of soil pH. Thus variations in soil pH seldom occur in alkaline paddy soil. Hence soil acidification due to nitrogen fertilization is often limited in alkaline paddy soils. In fact, applications of ammonium in alkaline soil help to mobilize TEs. On the other hand, buffering in acidic soil is usually maintained by reversible conversion of aluminum hydroxide ($AlOH_2^+$) to gibbsite (Al $(OH)_3$).Thus acidification of soil during fertilizer application is limited to a certain extent in acidic paddy soils due to the presence of hydroxyl groups containing Al compounds. Organic matter in the soil also assists in the maintenance of the buffering activity by protonation and deprotonation, which help to control TE chelation in the soil.

### 3.4. Cation Exchange Capacity

The cation exchange capacity playsa crucial role in TE mobility in soil solution [64,65]. Hydrogen ions will be attracted to cation exchange complexes where exchange of cations such as $Ca^{2+}$ occurs. Thus paddy soils having a higher cation exchange capacity tend to resist a decrease in soil pH (especially when pH > 6).The higher cation exchange capacity also indicates higher concentrations of cations, which can be adsorbed to soil. Thus paddy soils with a higher cation exchange capacity favor rice growth by providing more TE cations. Paddy soils with higher clay content were found to have more available TEs [66]. Clay hydroxides assist mineral nutrient retention in soil. However the adsorption of TEs in clay acts as a barrier for cation exchange. This accounts for the relatively lower mobility of TEs ions in alfisols that have more clay content. Hence alfisols retain more toxic TEs such as Cd in the soil solution because of the saturation of clay with TEs compared to vertisols that have lower clay content [67].

*3.5. Biological Processes*

Soil flora and fauna contribute to biological processes mediating TE accumulation in rice [55,68,69]. Biological activities bring changes in nutrient mobility, and most often these changes increase TE savailable to the plant. Many of the biological activities in paddy soil are found in the plow layer where root biomass is dense. Rhizosphere secretions such as organic acids act as energy sources for soil microbes. This situation increases soil microbial activity and stimulates redox reactions in the soil. These changes not only bring changes in chemical speciation of TEs but also affect solubility. Microbe mediated redox changes often influence solubility of multivalent elements such as As, Se and Fe. Arsenic, which is a toxic metalloid, occurs as As (III) and As (V) forms. Transformation of this metalloid in paddy soil due to oxidation, reduction and methylation is driven by microbes [70]. The addition of organic matter causes an increase in As methylation in paddy soils through microbial activity, and hence enhanced volatilization of As due to microbial activity, as may occur in the form of trimethylarsine, monomethylarsine, dimethylarsinic acid, and arsine. The form of occurrence depends on the concentration of As in the soil as well as the nature of microbes found in the soil.

Soil microbes also secrete compounds having metal binding ligands such as siderophores, which enhance plant uptake of TEs [71,72]. Plant growth promoting bacteria also promote both mobilization of TEs from paddy soils and root growth by hormonal activities [73–75]. Aerobic paddy fields are also benefitted by burrowers that enable air and water movement [76]. It is well known that the activity of earth worms incorporates organic matter such as plant residues into soil. Beetles also play a significant role in the dispersal and burial of organic matter that helps to prevent TE loss from green manures. Thus activities of soil fauna bring both temporal immobilizations of nutrients, especially TEs, and favor mineralization of organic matter. Soil microbes assist oxidation reactions that accelerate the formation of the ionic forms of nitrate, sulfates and phosphates in paddy soil which in turn release bound TEs into the soil solution [77]. Secretion of reducing agents such as NADPH, caffeic acids *etc.* by soil microbes also enhance solubility of TEs in the paddy soil. A characteristic feature of submerged paddy soil is the emission of greenhouse gases such as methane and nitrous oxide [78]. This phenomenon occurs due to anoxygenic degradation of organic matter. The periodic draining of paddy soils prevents anoxia, and substantially reduces greenhouse gases, for example methane. Apart from the microbial activities, cellulose degradation by termites also contributes to methane production in well-aerated paddy soils. The degradation of organic matter leads to a decrease of holding capacity of TEs especially that of K, Ca and Mg. Rice plants acclimatize to nutrient deficiency especially that of Fe, by secreting muginenic acid derivatives called phytosiderophores [79]. These compounds lead to dissolution of Fe in the soil. Phenolic secretions are also an adaptive strategy found in rice plants that mobilize Fe into the soil. Similarly, rice plants growing in acidic soil respond to aluminum toxicity by secretion of low molecular weight carboxylic acids such as citrate, malate and oxalate [80]. Thus regulations of biological activities in paddy soils help to control TE availability in rice plants.

## 3.6. Fertilization

Paddy soils are often spiked with fertilizer to replenish nutrients [81–83]. Many of the fertilizers are inorganic salts containing plant available mineral nutrients. Site-specific nutrient management is also carried out in paddy soils [84]. This kind of nutrient management practices incorporate nutrient elements that are deficient in a particular paddy soil. This also helps to avoid the application of wrong fertilizers which otherwise retard crop yield. For example, application of nitrogen in a field deficient in phosphorous delays plant growth more severely. The added nutrients enter complex nutrient cycles in the soil. Fertilizer application often increases cycling of TEs. This is due to the fact that plant uptake of TEs is from the TEs pool in the soil rather than from the fertilizer. Apart from this, the nutrients added in the form of fertilizer are not converted into biomass, and relatively little nutrients (10%–60%) are taken up by plants from fertilizer. This leads to the building up of nutrients in the soil, which stimulate cycling of nutrients including that of TEs.

Fertilizer application in the form of inorganic salts often possesses side effects. For example, nitrogen fertilization in the form of ammonium leads to accumulation of $H^+$ ions in the soil and promotes soil acidity [85]. Agricultural practices that increase crop yield through fertilization are often performed with the application of N, P and K in the field. Application of such a fertilizer mix not only enhances biomass but also depletes TEs. The reason for this effect is that more TEs are utilized to cope with macronutrient stimulated biomass production [86]. To avoid TE deficiency, organic farming or application of micronutrient fertilizers can be a practical solution [87,88]. However, the organic matter treatment must be monitored because of submergence mediated anoxia related changes in the paddy soil, which alter TE cycling. Application of vermicompost enhances soil porosity in aerobic paddy soils and hence supports root growth [89]. Green manures are a more reliable source of plant nutrients compared to composts. This is because about 70%–90% of nutrients taken up by animals appear in manure.

Organic matter as well as organic residue influence the availability of TEs in paddy soils [90,91]. Some organic matter form insoluble TE complexes and many of these complexes release TE cations slowly. Hence it can be predicted that addition of organic matter increases the chance of immobilization of TEs. Submerged paddy soils rich in organic matter often face deficiencies of Cu, Zn and Mn because of the immobilization of these elements by organic matter. TE deficiencies are commonly seen in aerated paddy soils due to high pH, high carbonates and low redox potential due to short-term submergence [92].

## 4. Nutrient Acquisition Directs TE Accumulation

Nutrient uptake in plants depends on nutrient availability, root architecture, presence of transition metal transporters in the roots and physiological processes such as transpiration [93–95]. The radial structure of rice root with aerenchyma in the cortex indicates the flexibility of the rice root to grow under both aerobic as well as anaerobic conditions [96]. It is well known that an extensive root system promotes the uptake of water and nutrients from the soil. Hence rice cultivars with a deeper and extensively branched root system pump more nutrients to aerial parts of the plant. Rice plants grown in upland conditions will have deeper roots than those grown in irrigated conditions. Thus the chance of accumulation of TEs is higher in upland rice compare with lowland rice. Rice plants pose a maximum root density at the time of heading, and the root density tends to decrease at the time of flowering. Hence TEs as fertilizers must be applied during the heading period. However, fertilizer application must be carried out under strict control to avoid limiting of plant growth due to excess mineral salts and resulting saline soils.

## 4.1. Root Phenology

The phenotypic characterization of rice germplasm indicates the existence of a variation in surface area, number of lateral roots, length, diameter, root hairs *etc.* among rice cultivars [97–100]. Rice cultivars with abundant root hairs require attention because root hairs not only help with water absorption but also with TE acquisition. Apart from this, root tip integrity is a critical factor for rice plant growth, as this is where the major quantity of ammonium is absorbed. Ammonium supports cell division in meristematic tissues in roots where carbohydrate is often limited. However, elements such as K, P and Fe are generally taken up by plants though any active locations in the root. Since root growth is an underground process, breeding approaches for utilization of difference in root architecture for better nutrient uptake are restricted, and the research on the development of better root phenotype are focused on quantitative trait loci (QTL) mapping [97]. QTLs clusters of rice root development have been observed in Chromosomes 1 (30–40 Mb), 2 (25–35 Mb), 3 (0–5Mb), 4(30–35 Mb), and 9 (15–20 Mb). Some of the major outcomes of QTLs studies on the control of root growth ctivity are presented in Table 3.Among the QTLs, over-expression of the *PSTOL1* (PHOSPHORUS-STARVATION TOLERANCE 1) gene was found to enhance root growth and resulted in a 60% increase of crop yield. Expression of this gene was also found to increase the content of P, N and K in rice plants. The increase of P uptake is well suited to inhibit toxic TE allocation to the rice grain because of the ability of P to form insoluble complexes with heavy metals. Expression of another candidate gene *DRO1* (DEEPER ROOTING 1) associated with the control of the root angle and deep rooting helps to increase the surface area for TE uptake. It is clear that genetic approaches that enhance root growth activities are highly promising to enhance nutrient fortification of rice grains.

**Table 3.** Rice root growth and development related genes.

| Gene | Function | Refs. |
|---|---|---|
| *OsGNOM1* | PIN protein expression and auxin response | [101] |
| *OsWOX3A* | Inhibition of PIN protein expression | [102] |
| *OsHO1* | Lateral root initiation | [103] |
| *PSTOL1* | Crown root initiation | [104] |
| *ARL1* | Adventitous root initiation | [105] |
| *OsABF2/OsTIR1/OsCYP2* | Inhibition of lateral roots emergence | [106–108] |
| *OsIAA11/OsIAA13* | Lateral root initiation | [109,110] |
| *OsLBD3-2* | Crown and latteral root initiation | [106,111] |
| *OsIAA23* | Quicent center inhibition | [112] |
| *OsSCR1* | Quicent center activation | [113] |
| *OsSHR1* | Endoderm diffewrentiation | [114] |
| *OsCAND1/OsPIN1* | Crown root emergence | [115] |
| *OsRPK1* | Inhibition of root growth | [116] |
| *DRO1* | Root growth | [117] |
| *OSGLU3/OSDGL1/OsEXPA8* | Cell elogation related root growth | [118–120] |
| *OsORC3* | Lateral root growth | [121] |
| *DES* | Root thickness | [122] |
| *OsRR1* | Crown root initiation | [123] |
| *RAL1,RAL2,RAL3* | Radicle initiation | [124,125] |

## 4.2. Molecular Physiological Process

Mineral nutrient uptake in plants depends on physiological processes [126–129]. Transpiration pull accounts for a mass flow of nutrients from the soil solution to the root. Transpiration is also believed to support upward movement of nutrient ions even though metabolic incorporation of ions occurs independently. Recent studies have shown that convective water transport in the xylem, brought about by root pressure, and the resultant guttation and Munch's phloem counter flow are in itself sufficient for long-distance mineral supply [130]. Negative charges on the root surface attracting cationic TEs also represent a critical factor that regulates TE uptake in plants [131,132]. The polarity of the root surface is maintained by $H^+$-ATPase. This proton pump takes part in pumping $H^+$ ions out of the root, which leads to changes in the rhizosphere pH, and an electrochemical potential difference that affects the movement of ions and solutes across the plasma membrane of the root. Activity of this pump requires ATP, and hence physiological and biochemical events in the root indirectly control plant uptake of TEs. Even though respiration brings ATP for $H^+$-ATPase activity, the process depends on photosynthesis, which produces substrates for respiration. Apart from this, sequestration of metals in the vacuole is mainly assisted with metal chelators generated through photosynthesis [133,134]. Hence maintenance of physiological processes such as photosynthesis is critical for limiting the transport of toxic TEs from the root to the grain. Regulation of action of transition metal transporters, which control the movement of TEs in the plants, is also critical for the enhancement of TE accumulation.

## 4.2.1. Photosynthesis

Among the physiological processes, photosynthesis plays an important role in the allocation and reallocation of TEs in the plant. The reason for this is that photosynthesis primarily accounts for the production of metabolites that are essential for the synthesis of metal chelators. Metal chelators in rice plants take part in extracellular complexation, rhizo-complexation and vacuolar complexation of metals. Secretion of tricarboxylic acids in response to metal toxicity is well studied among rice plants. Oxalate, citrate and malate are the tricarboxylic acids that are found to secrete into the rhizosphere and decrease aluminum toxicity in rice plants [135]. These acids tend to bind extracellular aluminum and prevent plant uptake of this metal. Organic acids involved in the above process are intermediates of primary metabolisms, especially respiration. Photosynthesis also influences root respiration by providing sugar skeletons, which act as substrates for respiration [136]. Thus rice cultivars that have higher photosynthetic efficiency and biomass productivity are able to avoid toxicity of metals because of their superior capacity of metal chelation. Extracellular chelation of metals is also noticed in rice plants with an excess of Cd, Cr, Fe, Pb and Mn in the rhizosphere [137–139]. Muginenic acids derivatives and phytosiderophores make up another class of extracellular chelators that is responsive to mineral nutrient deficiency and enhances plant uptake of TEs, especially Fe [140].

Apart from extracellular chelation of metals, plants have the ability to chelate metals within their roots that prevents the transport of the metals to the aerial parts. This is achieved by the transport and storage of metals in vacuoles of the root cells with help of metal chelators. Chelators involved in vacuolar sequestration of metals can be organic acids or phytochelatins [141,142]. Even though phytochelatins belong to the peptides, the dependency of the aminoacid metabolism on photosynthetic products points at the importance of photosynthesis in the maintenance of metal chelation activity in roots. Photosynthetic activity influenced Cd rhizo-complexation has been reported in rice plants [143]. Transport of TEs from root to leaf, and flag leaf to rice grain also occurs with help of metal chelators especially citrate and nicotianamine derivatives, respectively. Existence of citrate and nicotianamine transporters in plants is well characterized [144–146]. Hence the presence of metal chelators and corresponding transporters involved in cellular and vacuolar uptake of TEs complexes chiefly determines accumulation of TEs in rice grain. Site-specific over-expression of transition metal transporters helps to mobilize TEs into rice grains. It is also clear that any alteration in photosynthetic productivity affects production of metal chelators. Hence loss of photosynthetic activity restricts the trafficking of TEs in the plants, which ultimately disturbs the micronutrient composition of the grain.

## 4.2.2. Transition Metal Transporters

Plant uptake and transport of TEs is mediated by plasma membrane transporters. Transition metal transporter families such as ZIP, CDF, P1B-type ATPases, NRAMP, YSL, IREG1 and CCC1 are involved in the transition TE transport in rice [147] (Figure 3). Regulation of these transporters helps to control TE accumulation in rice plants. Most often deficiency of any of the TEs leads to strategic responses within plants. For example, rice plants secrete muginenic acids derivatives when they are subjected to Fe deficiency. But this will also promote the uptake of toxic TEs such as Cd because of the similarity of chemical properties that enables Cd to bind with these chelators. It is also reported that the OsNramp5 transporter assists with the uptake of both Mn and Cd in rice plants [148]. These reports suggest that the uptake and transport of both essential and toxic TEs are interrelated. This relationship can be used to manipulate plant uptake of specific TEs while limiting availability of toxic TEs. A transition metal ion is able to displace another metal ion from its specific binding site downstream in the Irving-Williams series in the following order: $Zn^{2+} < Cu^+ > Cu^{2+} > Ni^{2+} > Co^{2+} > Fe^{2+} > Mn^{2+} > Mg^{2+} > Ca^{2+}$. Hence, there is a chance of competitive binding of TEs at metal binding sites where excess of a TEs ion disrupt cellular functions due to displacement of cofactor often occurs.

Over-expression and tissue specific expression of transition metal transporters are potential solutions for both enhancement of TEs and restriction of toxic TEs content in rice plants [11]. A node-based switch in rice plants has been found that regulates Mn allocation in rice shoot [149]. The switch operates through expression and degradation of a Mn transporter, OsNramp3. When there is low Mn in the external medium, *OsNramp3* gene expression occurs in the nodes and allocates Mn to young leaves and the panicle. However, in the presence of excess Mn, the transporter degrades, which causes Mn to be allocated to older leaves. This study points to the existence of a feedback regulation of TE trafficking at the molecular level in rice plants. Up regulation of magnesium transporter OsMGT1 was found to enhance Mg accumulation and aluminum tolerance in rice plants [150]. This finding supports that up regulation of specific proteins helps to avoid metal toxicity and to increase the accumulation of beneficial TEs ions. Macronutrient uptake such as that of K also plays a critical role in TE uptake [151]. It has been reported that there is a synergetic relationship between K uptake and Fe and Mn uptake. On the other hand, K is reported to negatively influence plant uptake of Mo. Thus it is clear that uptake of macronutrients influences TE accumulation in rice. However, the advent of studies with transition metal transporters points out that the uptake and distribution of TEs can be regulated by manipulation of transition metal transporters.

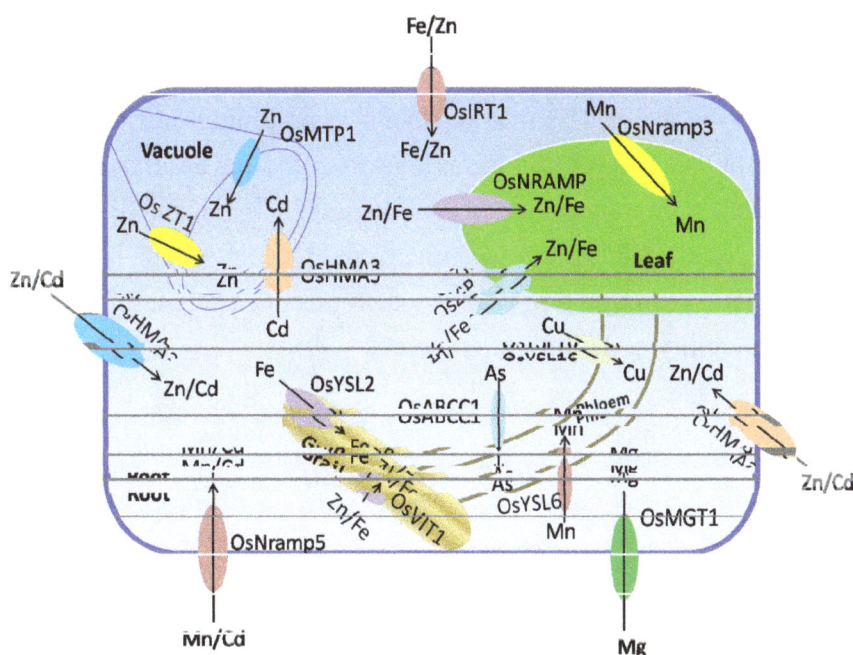

**Figure 3.** Transition metal transporters implicated for TEs management in rice.

Zinc is essential for rice plant growth. The influence of Zn on plant growth mainly relies on the functioning of redox-mediated processes. It has been reported that *OsHMA2* gene expression preferentially loads Zn to developing tissues of rice plants [152]. This transporter was also implicated in Cd transport. *OsIRT1* is another candidate gene reported to have a role in excess accumulation of both Zn and Fe in rice plants [153]. Over-expression of *OsHMA3* in rice was also found to have a dual role, i.e. enhancement of Zn uptake in the root and Cd tolerance [154]. This gene was found to enhance Zn uptake by over-expressing the ZIP family of metal transporters in the root. Decreased Cd accumulation in rice grain during *OsHMA3* over-expression was due to the vacuolar sequestration of Cd in the root. *OsYSL16* is a metal transporter involved in Cu transport in rice [155]. This transporter was found to enhance Cu allocation to younger leaves from older leaves and flag leaf to panicle with help of nicotianamine. The transport of Cu from flag leaf to panicle is important to form fertile flowers, and hence this transporter plays a crucial role in ensuring the crop yield. Arsenic accumulation in rice grain reduced during manipulation of the gene *OsABCC1* [156]. This gene was found to enhance vacuolar sequestration of As in companion cells of phloem of the upper nodes. This study also indicates the importance of site-specific localization of toxic TEs in rice plants which reduces the accumulation of toxic TEs in the grain.

Polished rice often contains limited concentrations of TEs. However, expression driven by insertion of promoter of *OsYSL2*, which causes over-expression of *OsYSL2*, has resulted in a 4.4-fold increase of Fe in rice grain. This transporter was also found to play a crucial role in long-distance transport of Fe and Mn in rice plants [157]. OsYSL6 is a transporter found in rice, which is involved in Mn detoxification [158]. This transporter mediates Mn transport with help of nicotianamine and is functional in the presence of excess Mn. Available reports with transition metal transporters indicate promising role of metal transporters in the restriction of toxic metals in rice grains while enhancing loading of TEs.

## 5. Trace Element Fortification

TEs such as Fe and Zn are the major TEs that are found to be limited in rice based diets [1,159]. Hence micronutrient fortifications of rice grain are mainly focused on Fe and Zn. Bio fortified rice grains with increased Zn and Fe amounts have been released in India and Bangladesh [160,161]. Bio fortification approaches are mainly focused on plant breeding and transgenic approaches (Figure 4). However, TEs nutrition through fertilization also requires a successful fortification program that provides the ambient availability of TEs for plant uptake.

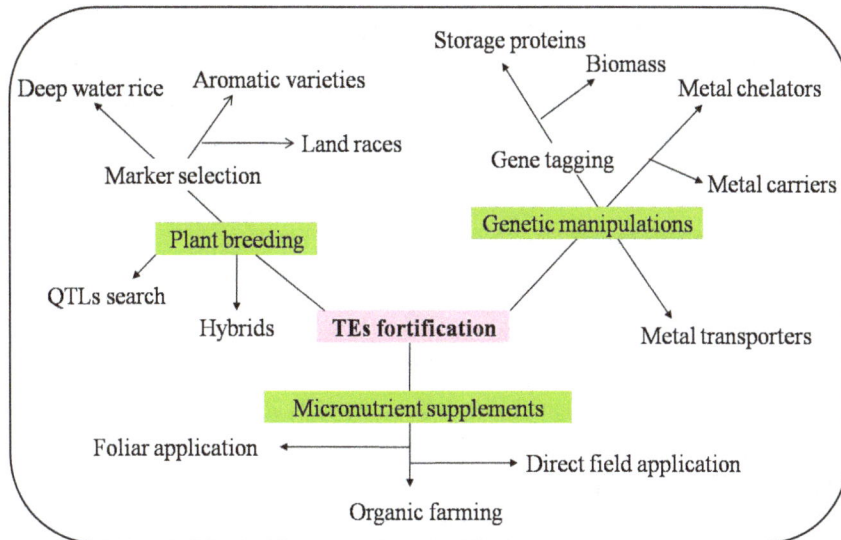

**Figure 4.** TEs fortification strategies for rice.

## 5.1. TEs Supplements

TE availability in the soil is essential for successful bio fortification programs. Some of the practical methods of application of TEs to rice plants are given in Table 4. It must be noted that toxic TEs also potentially enter into plants through transition metal transporters that are meant for TEs [162,163]. Hence more availability of TEs in the rhizosphere will have beneficial effects in the sense that these elements delay plant uptake of toxic TEs by competitive inhibition. Application of redox active TEs such as Mn and Fe in the soil requires attention because they are known to create metal toxicity. Rice plant grown in acidic soils are especially prone to toxicity of these elements [164,165]. In acidic soils, the process of submergence causes depletion of oxygen that results in the formation of reduced forms of Fe and Mn that are more toxic to plants compared to their oxidized form. It is recommended that TE application must be carried out with the help of metal chelators to prevent toxicity of metals. Foliar application of TEs is a widely accepted method. This method helps to avoid accumulation of TEs in the soil that can create metal toxicity. Thus along with prophylactic and a curative benefits, foliar application could be a practical solution for the enrichment of TEs in rice grain. Mineral salts of TEs in the form of sulfate, oxysulfate, oxide and chelate are the commonly applied forms of TEs. Metal chelates are well-suited for the application in soil because the presence of chelators slows down the reactivity of TEs with soil minerals. Application of organic manures is also a potential practical solution for ensuring ambient availability of TEs, especially, Zn and Cu. Silicon is an essential element of rice growth, and the application of silica based gels not only protects rice plants from abiotic stress but also helps to enhance crop productivity [166].

**Table 4.**TEs supplementpracticesapplicable to rice paddies.

| Nutrient | Mode of Application |
| --- | --- |
| Zn | Addition of compost, Zincated urea and Gypsum (20 Kg Zn/ha), Periodic draining, Foliar spraying or seed treatment of 0.2 to 0.5% $ZnSO_4$ |
| Fe | Addition of farm yard manure, FeSO4, Fe-EDTA, and Fe-EDDHA (25 Kg Fe/ha). Foliar spraying or seed treatment of 1%–2% $FeNH_4SO_4$ (pH5.2). |
| B | Application of Borax and$H_3BO_3$ (1Kg/ha), Foliar spray of 0.5% $H_3BO_3$ |
| Mn | Application of $MnSO_4$ (2–5 Kg/ha)or Farm yard manure (10 ton/hectare), Foliar Spray of 1% $MnSO_4$ at tillering |
| Mo | Liming of acid soils to a pH 6.2–6.5, Dustingof $(NH_4)_2MoO_4$ at 100–500 g/ha, Foliar spray of 0.1% $(NH_4)_2MoO_4$ |
| Cu | Addition of $CuSO_4$ (5 Kg/ha at an interval of 5 years), Root dipping in 1% $CuSO_4$ solution |
| Si | Irrigation with Si rich water, Application of Ca or K silicate (60 Kg/ha), Foliar spray of 0.1% $Na_2O_3Si$ |

## 5.2. Plant Breeding

Marker assisted selection in breeding programs with help of loci associated with nutrient accumulation in the grain allows the development of TE bio fortified rice as well as restriction of accumulation of toxic TEs. Among the TEs, increase in accumulation of Fe and Zn has gained attention because of the direct human health benefits. The existence of four fold differences in Fe and Zn concentrations among rice germplasm has been identified [167,168]. This indicates vast opportunities for plant breeding that can significantly increase the content of Fe and Zn in the grain. Studies with Indian rice varieties have shown the highest Fe contents to be found in the varieties Varsha (37.5 mg/kg) and Phou Dum (37.2 mg/kg) [160]. It was also found that basmati genotypes, deep-water rice and landraces havehigher iron and zinc contents in the grains compared with Swarna, Jaya, Uma *etc*. that are widely cultivated. Poornima, Ranbir Basmati, ADT 43 and Chittimutyalu were identified as the Indian rice varieties with the highest Zn content. The trait for high Fe content in grain was located on chromosomes 3, 4 and 8 in the Chittimutyalu variety. QTL mapping indicated four loci of high Fe content on chromosomes 3, 4, 6 and 12 in Ranbir Basmati.

Aromatic rice varieties such as Jalmagna, Zuchem, and Xua Bue Nuo are reported to be rich in both Fe and Zn [161]. The aromatic trait was not pleiotropic for grain-Fe or grain-Zn concentrations even though the linkage was broken at a low frequency. Thus the trait can be used to screen for high Fe and Zn levels in rice grain. Crossing of rice varieties such as BPT 5204 and Chittimutyalu was success to produce rice with higher Zn and Fe content [160]. Results of crossing between IR72 and Zawa Bonday resulted in a higher Fe content as well as an enhanced yield. The progeny of the cross possessed favorable characteristics such as tolerance of rice tungro virus, and P, Zn and Fe deficiency. Breeding of rice varieties is beneficial to prevent accumulation of toxic TEs such as Cd. Natural variation of a 13- to 23-fold difference in grain Cd concentration in diverse japonica rice germplasm points to plant breeding opportunities in the context of Cd minimization. Cd accumulation related QTLs in rice have been found on Chromosomes 2, 3, 4, 5, 6, 7, 8, 10 and 11 [169–171]. Cd minimization related alleles are also found to relate with crop yield. Thus approaches that are aimed to reduce Cd accumulation help to enhance crop yield too.

## 5.3. Genetic Manipulations

Gene based regulation of micronutrient transporters and carriers in the plant promise controlled accumulation of TEs in rice while restricting entry of toxic TEs. Nutrient content in the grain can be enhanced by expression of specific genes in the endosperm or expression of genes that take part in the transport of TEs to the grain. For example, rice endosperm expressing the soybean ferrtin gene, *SoyferH1*, was found to have a three-fold increase of Fe in seeds [172]. However, when the gene was expressed under endosperm specific multiple promoters such as*OsGlb1* and *OsGluB1*, there was no increase of Fe content in the seeds [173]. This study indicates that the enhancement of nutrient storage capacity alone cannot fulfill the requirement of more accumulation of TEs. Basmati rice (Pusasugandh II) plants over-expressing rice ferritin (*OsFer2*) under the control of the endosperm-specificGlutelinA2 (*OsGluA2*) promoter were found to accumulate more Fe and Zn [174]. Simultaneous over-expression of ferritin under the control of the endosperm-specific promotersglobulinb1(*OsGlb1*) and glutelin B1 (*OsGluB1*), NAS under the control of the *OsActin1*promoter, and *OsYSL2* under the control of *OsGlb1* promoter, and the *OsSUT1* transporter promoter significantly increased the Fe, Zn, Mn and Cu concentrations in polished rice [175]. These studies indicate that over-expression of ferritin and metallothionein in grain could enhance TE accumulation in grains because of the ability of these proteins to chelate metals. Over-expression of phytase, which degrades phytic acid, also helps to enhance availability of TE fortification in rice [176].

The nicotianamine synthesis related metabolic pathway plays a critical role in the regulation of metal homeostasis, especially Fe in plants [177]. Activation tagged line of *OsNAS3* showed a 3-fold increase of Fe content in polished rice grain [178]. Constitutive over-expression of the *OsNAS2* gene also resulted in Fe and Zn-bio fortification of rice endosperm [179]. It has been also demonstrated that over-expression of Fe homeostasis genes in rice such as *OsIRO2* increases Fe content in rice plants grown in calcareous soils [180]. Over-expression of deoxymugineic acid synthase gene *OsDMAS1* involved in synthesis of muginenic acid also could be a practical solution for enhancement of $Fe^{3+}$ uptake [181]. Thus it is clear that approaches with over-expression of carriers involved in metal translocation as well as metal homeostasis promotes accumulation of TEs in the grain. Metal transporters are also promising to enhance accumulation of TEs in plants. Hence focus should be placed on transgenic rice plants, which pose the capacity to express transition metal transporters and TEs carriers. *OsVIT1* and *OsNRAMP7* are potential candidate genes that can be targeted for Zn fortification in rice grain [182]. Over-expression of Fe transporters *OsIRT1* and *OsYSL15* were found to promote Fe content in grains of rice plants [183]. Expression of *OsNRAMP1*, *OsNRAMP7* and *OsNRAMP8* in flag leaves also was also shown to correlate with grain Fe and Zn content [181]. Up regulation of *OsZIP1*, *OsZIP4*, *OsZIP6* and *OsZIP8* was also reported in the above-mentioned study during Fe deficiency. Thus it can be assumed that over-expression of the Nramp and ZIP metal transporter family in the flag leaves of rice plants has the potential to enhance accumulation of Fe and Zn in the grain. However, studies with over expression of *OsZIP4*, *OsZIP5* and *OsZIP8* where a decrease of root-to-shoot translocation of Zn and of Zn concentrations in the seed were observed indicate that up regulation of proteins that are observed under nutrient deficiency does not always help to enhance plant uptake of TEs when plants grow with sufficient availability of ambient nutrients [184,185]. Non-specificity of transition metal transporters may lead to the accumulation of non-specified elements during a transgenic approach [186]. For example, over-expression of nicotianamine synthase and ferritin using CaMV 35S not only enhanced accumulation of Fe and Zn in rice but also enhanced Mn accumulation. Entry of toxic TEs can also be blocked in rice plants using transgenic approaches [11]. Cd excluder phenotypes screened in rice cultivars indicate potential Cd excluding traits in the rice genome that help to develop Cd excluder rice [187]. It has been found that transgenic tobacco carrying cDNALTC1, a nonspecific transporter, accumulated low concentrations of Cd [188].

## 6. Outlook

Rice-based diets often cause TE deficiency in humans. Approaches for enhancing TE content in rice grains can be either implemented on the field level or plant based. Nutrient cycling in paddy soils must be characterized with respect to TE content for ensuring TE availability for plant uptake. Optimal soil physiochemical characteristics for TE accumulation in grains need to be explored under varying environmental conditions especially during dry and wet periods in paddy soils. TE fertilization along with NPK fertilization is recommended. Fertilization often activates nutrient cycling in rice paddies, and hence the dynamics of TE availability to plants under fertilization must be monitored. Adequate management of organic matter in the soil and the soil's influence on TE accumulation in rice grains must be screened during biodynamic farming. Available data on rice root architecture must be screened for TE mining capacity. Genetic manipulation, especially with genes that control metal chelator production, and transition metal transporters are promising possibilities to enhance accumulation of TEs. Endosperm, representing the major part of the TE reserve in the grain, site-specific expression of metal chelators and transporters that load metal to endosperm are practical solutions for loading TEs into rice grains.

## Acknowledgements

The authors gratefully acknowledge the receipt of financial support ref. DST/INT/THAI/P-02/2012 dated 31-1-13.

## Conflict of Interest

The authors declare no conflict of interest.

## References

1.   Bashir, K.; Takahash, R.; Nakanishi, H.; Nishizawa, N.K. The road to micronutrient biofortification of rice: Progress and prospects. *Front. Plant Sci.* 2013, *4*, 15, doi: 10.3389/fpls.2013.00015.
2.   Lucca, P.; Hurrell, R.; Potrykus, I. Genetic engineering approaches to improve the bioavailability and the level of iron in rice grains. *Theor. Appl. Genet.* **2001**, *102*, 392–397.
3.   Witt, C.; Dobermann, A.; Abdulrachman, S.; Gines, H.C.; Wang, G.; Nagarajan, R.; Satawatananont, S.; Tran, T.S.; Pham, S.T.; Le, V.T.; *et al.* Internal nutrient efficiencies of irrigated lowland rice in tropical and subtropical Asia. *Field Crops Res.* **1999**, *63*, 113–138.

4. Andriesse, J.P.; Schelhaas, R.M. A monitoring study of nutrient cycles in soils used for shifting cultivation under various climatic conditions in tropical Asia. II. Nutrient stores in biomass and soil—Results of baseline studies. *Agric. Ecosyst. Environ.* **1987**, *19*, 285–310.

5. Chase, P.; Singh, O.P. Soil nutrients and fertility in three traditional land use systems of Khonoma, Nagaland, India. *Resour. Environ.* **2014**, *4*, 181–189.

6. Kögel-Knabner, I.; Amelung, W.; Cao, Z.; Fiedler, S.; Frenzel, P.; Jahn, R.; Kalbitz, K.;Kölbl, A.; Schloter, M. Biogeochemistry of paddy soils. *Geoderma* **2010**, *157*, 1–14.

7. Takahashi, Y.; Minamikawa, R.; Hattori, K.H.; Kurishima, K.; Kihou, N.;Yuita, K. Arsenic behavior in paddy fields during the cycle of flooded and non-flooded periods. *Environ. Sci. Technol.* **2004**, *38*, 1038–1044.

8. Carrillo-Gonzalez, R.; Simünek, J.; Sauvé, S.; Adriano, D. Mechanisms and pathways of trace element mobility in soils. *Adv. Agron.* **2006**, *91*, 113–180.

9. Wissuwa, M.; Ismail, A.M.; Yanagihara, S. Effects of zinc deficiency on rice growth and genetic factors contributing to tolerance. *Plant Physiol.* **2006**, *142*, 731–741.

10. Fieldler, S.; Vepraskas, M.J.; Richardson, J.L. Soil redox potential importance, field measurements and observations. *Adv. Agron.* **2007**, *94*, 41–57.

11. Sebastian, A.; Prasad, M.N.V. Cadmium minimization in rice: A review. *Agron. Sustain. Dev.* **2014**, *34*, 155–173.

12. Williams, C.H.; David, D.J. The accumulation in soil of cadmium residues from phosphate fertilizers and their effect on the cadmium content of plants. *Soil Sci.***1976**, *121*, 86–93.

13. Bolan, N.S.; Duraisamy, V.P. Role of inorganic and organic soil amendments on immobilization and phytoavailability of heavy metals: A review involving specific case studies. *Aust. J. Soil Res.* **2003**, *41*, 533–535.

14. Czarnecki, S.; Düring, R.A. Influence of long-term mineral fertilization on metal contents and properties of soil samples taken from different locations in Hesse, Germany. *Soil* **2015**, *1*, 23–33.

15. Pan, J.; Plant, J.A.; Voulvoulis, N.; Oates, C.J.; Ihlenfeld, C. Cadmium levels in Europe: Implications for human health. *Environ. Geochem. Health* **2010**, *32*, 1–12.

16. Vitousek, P. Nutrient cycling and nutrient use efficiency. *Am. Nat.***1987**, *119*, 553–572.

17. Salt, D.E.; Blaylock, M.; Kumar, N.P.B.A.; Dushenkov, V.; Ensley, D.; Chet, I.; Raskin, I. Phytoremediation: A novel strategy for the removal of toxic metals from the environment using plants. *Biotechnology* **1995**, *13*, 468–474.

18. Sessitsch, A.; Kuffner, M.; Kidd, P.; Vangronsveld, J.; Wenzel, W.W.; Fallmann, K.; Puschenreiter, M. The role of plant-associated bacteria in the mobilization and phytoextraction of trace elements in contaminated soils. *Soil Biol. Biochem.* **2013**, *60*, 182–194.

19. Tsheboeng, G.; Bonyongo, M.; Murray-Hudson, M. Flood variation and soil nutrient content in flood plain vegetation communities in the Okavango Delta. *S. Afr. J. Sci.* **2014**, *110*, 1–5.

20. Chen, Y.; Stevenson, F.J. Soil organic matter interactions with trace elements. *Dev. Plant Soil Sci.* **1986**, *25*, 73–116.

21. Chaoui, H.I.; Zibilske, L.M.; Ohno, T. Effects of earthworm casts and compost on soil microbial activity and plant nutrient availability. *Soil Biol. Biochem.* **2003**, *35*, 295–302.

22. Kabarta-Pendias, A. Behavioural properties of trace metals in soils. *Appl. Geochem.* **1993**, *2*, 3–9.

23. Lu, Y.H.; Watanabe, A.; Kimura, M. Contribution of plant-derived carbon to soil microbial biomass dynamics in a paddy rice microcosm. *Biol. Fertil. Soils* **2002**, *36*, 136–142.

24. Gao, Y.; Kan, A.T.; Tomson, M.B. Critical evaluation of desorpion phenomena of heavy metals from natural sediments. *Environ. Sci. Technol.* **2003**, *37*, 5566–5573.

25. Meunier, N.; Drogui, P.; Montané, C.; Hausler, R.; Mercier, G.; Blais, J.F. Comparison between electrocoagulation and chemical precipitation for metals removal from acidic soil leachate. *J. Hazard. Mater.* **2006**, *137*, 581–590.

26. Rajmohan, N.; Elango, L. Mobility of major ions and nutrients in the unsaturated zone during paddycultivation: A field study and solute transport modelling approach. *Hydrol. Process.* **2007**, *21*, 2698–2712.

27. Blair, N.; Faulkner, R.D.; Till, A.R.; Poulton, P.R. Long-term management impacts on soil C, N and physical fertility. Part I. Broadbalk experiment. *Soil Tillage Res.* **2006**, *91*, 30–38.

28. Kothawala, D.N.; Moore, T.R. Adsorption of dissolved nitrogen by forest mineral soils. *Can. J. For. Res.* **2009**, *39*, 2381–2390.

29. Kanda, K.; Tsuruta, H.; Minami, K. Emission of dimethyl sulfide, carbonyl sulfide and carbon disulfide from paddy fields. *Soil Sci. Plant Nutr.* **1992**, *38*, 709–716.

30. Whiticar, M.J. Carbon and hydrogen isotope systematics of bacterial formation and oxidation of methane. *Chem. Geol.* **1999**, *161*, 291–314.

31. Kuo, S. Application of modified Langmuir isotherm to phosphate sorption by some acid soils. *Soil Sci. Soc. Am. J.* **1988**, *52*, 97–102.

32. Brennan, R.F.; Bolland, M.D.A.; Bowden, J.W. Potassium deficiency, and molybdenum deficiency and aluminium toxicity due to soil acidification have become problems for cropping sandy soils in south-western Australia. *Aust. J. Exp. Agric.* **2004**, *44*, 1031–1039.

33. Inal, A.; Gunes, A.; Sahin, O.; Taskin, M.B.; Kaya, E.C. Impacts of biochar and processed poultry manure, applied to a calcareous soil, on the growth of bean and maize. *Soil Use Manag.* **2015**, *31*, 106–113.

34. Green, C.H.; Heil, D.M.; Cardon, G.E.; Butters, G.L.; Kelly, E.F. Solubilization of manganese and trace metals in soils affected by acid mine runoff. *J. Environ. Qual.* **2003**, *32*, 1323–1334.

35. Greipsson, S. Effect of iron plaque on roots of rice on growth of plants in excess zinc and accumulation of phosphorus in plants in excess copper or nickel. *J. Plant Nutr.* **1995**, *18*, 1659–1665.

36. Adriano, D.C.; Wenzel, W.W.; Vangronsveld, J.; Bolan, N.S. Role of assisted natural remediation in environmental cleanup. *Geoderma* **2004**, *122*, 121–142.

37. Coleman, D.C.; Reid, C.P.P.; Cole, C.V. Biological strategies of nutrient cycling in soil systems. *Adv. Ecol. Res.* **1983**, *13*, 1–55.

38. Rauthan, B.S.; Schnitzer, M. Effects of a soil fulvic acid on the growth and nutrient content of cucumber (cucumis sativus) plants. *Plant Soil* **1981**, *63*, 491–495.

39. Chiang, K.; Wang, Y.; Wang, M.; Chiang, P. Low-molecular-weight organic acids and metal speciation in rhizosphere and bulk soils of a temperate rain forest in Chitou, Taiwan. *Taiwan J. For. Sci.* **2006**, *21*, 327–337.

40. Vinolas, L.C.; Healey, J.R.; Jones, D.L. Kinetics of soil microbial uptake of free amino acids. *Biol. Fertil. Soils* **2001**, *33*, 67–74.

41. Sivapullaiah, P.V.; Prakash, B.S.N.; Suma, S. Electrokinetic removal of heavy metals from soil. *J. Electrochem. Sci. Eng.* **2015**, *5*, 47–65.

42. Chapin, F.S.; Bloom, A.J.; Field, C.B.; Waring, R.H. Plant responses to multiple environmental factors. *J. Biosci.* **1987**, *37*, 49–57.

43. Ehrenfeld, J.G. Effects of exotic plant invasions on soil nutrient cycling processes. *Ecosystems* **2003**, *6*, 503–523.

44. Hattenschwiler, S.; Vitousek, P. The role of polyphenols in terrestrial ecosystem nutrient cycling. *Trends Ecol. Evol.* **2000**, *15*, 238–243.

45. Katoh, M.; Murase, J.; Hayashi, M.; Matsuya, K.; Kimura, M. Nutrient leaching from the plow layer by water percolation and accumulation in the subsoil in an irrigated paddy field. *Soil Sci. Plant Nutr.* **2004**, *50*,721–729.

46. Zhou, W.; Lv, T.; Chen, Y.; Westby, A.P.; Ren, W. Soil physicochemical and biological properties of paddy-upland rotation: A review. *Sci. World J.* **2014**, *2014*, doi:10.1155/2014/856352.

47. Shaheen, S.M.; Tsadilas, C.D.; Rinklebe, J. Immobilization of soil copper using organic and inorganic amendments. *Plant Nutr. Soil Sci.* **2015**, *178*,112–117.

48. Sharma, A.; Weindorf, D.C.; Wang, D.; Chakraborty, S. Characterizing soils via portable X-ray fluorescence spectrometer: 4. Cation exchange capacity (CEC). *Geoderma* **2015**, *239–240*, 130–134.

49. Charlet, L.; Markelova, E.; Parsons, C.; Couture, R.; Madé, B. Redox oscillation impact on natural and engineered biogeochemical systems: Chemical resilience and implications for contaminant mobility. *Proced. Earth Planet. Sci.* **2013**, *7*,135–138.

50. Hoffman, M.R.; Yost, E.C.; Eisenrich, S.J.; Maier, W.J. Characterization of soluble and colloidal phase metal complexes in river water by ultra-filtration: A mass balance approach. *Environ. Sci. Technol.* **1981**, *15*, 655–661.

51. Yuan, C.; Wu, H.B.; Xi, Y.; Lo, X.W. Mixed transition-metal oxides: Design, synthesis, and energy-related applications. *Angew. Chem.* **2014**, *53*, 1488–1504.

52. Scharpenseel, H.W.; Pfeiffer, E.M.; Becker-Heidmann, P. Organic carbon storage in tropical hydromorphic soils. In *Structure Andorganic Matter Storage in Agricultural Soils*; Carter, M.R., Stewart, B.A., Eds.; Lewis Publishers: Boca Raton, FL, USA, 1996; pp. 361–392.

53. Diaz, R.J. Overview of hypoxia around world. *J. Environ. Qual.* **2001**, *30*, 275–281.

54. Qureshi, S.; Richards, B.K.; Steenhuis, T.S.; McBride, M.B.; Baveye, P.; Dousset, S. Microbial acidification and pH effects on trace element release from sewage sludge. *Environ. Pollut.* **2004**, *132*, 61–71.

55. Olaniran, A.O.; Balgobind, A.; Pillay, B. Bioavailability of heavy metals in soil: Impact on microbial biodegradation of organic compounds and possible improvement strategies. *Int. J. Mol. Sci.* **2013**, *14*, 10197–10228.

56. Brinkman, R. *Ferrolysis: Chemical and Mineralogical Aspects of Soil Formation in Seasonally Wet Acid Soils, and Some Practical Implications*; International Rice Research Institute: Los Banos, Philippines, 1978; pp. 295–303.

57. Van Ranst, E.; Dumon, M.; Tolossa, A.R.; Cornelis, J.T.; Stoops, G.; Vandenberghe, R.E.; Deckers, J. Revisiting ferrolysis processes in the formation of Planosols for rationalizing the soils with stagnic properties in WRB. *Geoderma* **2011**, *163*, 265–274.

58. Fageria, N.K.; Carvalho, G.D.; Santos, A.B.; Ferreira, E.P.B.; Knupp, A.M. Chemistry of lowland rice soils and nutrient availability. *Commun. Soil Sci. Plant Anal.* **2011**, *42*, 1913–1933.

59. Sahrawat, K.L.; Sika, M. Comparative tolerance of *Oryza sativa*and *Oryza glaberrima*rice cultivars for iron toxicity in West Africa. *Int. Rice Res. Notes* **2002**, *27*, 30–31.

60. Barak, P.; Jobe, B.O.; Krueger, A.R.; Peterson, L.A.; Laird, D.A. Effects of long term soil acidification due to nitrogen inputs in Wisconsin. *Plant Soil* **1997**, *197*, 61–69.

61. Behera, S.K.; Shukla, A.K. Spatial distribution of surface soil acidity, electrical conductivity, soil organic carbon content and exchangeable potassium, calcium and magnesium in some cropped acid soils of India. *Land Degrad. Dev.* **2015**, *26*, 71–79.

62. Magdoff, F.R.; Bartlett, R.J. Soil pH buffering revisited. S*oil Sci. Soc. Am. J.* **1985**, *49*, 145–148.

63. Pan, Y.; Koopmans, G.F.; Bonten, L.T.; Song, J.; Luo, Y.; Temminghoff, E.J.; Comans, R.N. Influence of pH on the redox chemistry of metal (hydr) oxides and organic matter in paddy soils. *J. Soils Sediments***2014**, *14*, 1713–1726.

64. Evangelou, V.P.; Philipps, R.E. Cation exchange in soils. In *Chemical Processes in Soil*; Tabatabai, M.A., Sparks, D.L., Eds.; Soil Science Society of America: Madison, WI, USA, 2005; pp. 343–410.

65. Deumlich, D.; Thiere, J.; Altermann, M. Characterization of cation exchange capacity (CEC) for agricultural land-use areas Arch. *Agron. Soil Sci.* **2015**, *6*, 767–784.

66. Anda, M.; Suryani, E.; Husnain.; Subardja, D. Strategy to reduce fertilizer application in volcanic paddy soils: Nutrient reserves approach from parent materials. *Soil Till. Res.* **2015**, *150*, 10–20.

67. Sebastian, A.; Prasad, M.N.V. Vertisol prevents cadmium accumulation in rice: Analysis by ecophysiological toxicity markers. *Chemosphere* **2014**, *108*, 85–92.

68. Zheng, R.; Sun, G.; Zhu, Y.G. Effects of microbial processes on the fate of arsenic in paddy soil. *Chin. Sci. Bull.***2013**, *58*,186–193.

69. Nayak, A.K.; Raja, R.; Rao, K.S.; Shukla, A.K.; Mohanty, S.; Shahid, M.; Tripathi, R.; Panda, B.B.; Bhattacharyya, P.; Kumar, A.; *et al.* Effect of fly ash application on soil microbial response and heavy metal accumulation in soil and rice plant. *Ecotoxicol. Environ. Saf.* **2014**, doi:10.1016/j.ecoenv.2014.03.033i.

70. Lomax, C.; Liu, W.J.; Wu, L.; Xue, K.; Xiong, J.; Zhou, J.; McGrath, S.P.; Meharg, A.A.; Miller, A.J.; Zhao, F.J. Methylated arsenic species in plants originate from soil microorganisms. *New Phytol.* **2012**, *193*, 665–672.

71. Rajkumar, M.; Ae, N.; Prasad, M.N.V.; Freitas, H. Potential of siderophore-producing bacteria for improving heavy metal phytoextraction. *Trends Biotechnol.* **2010**, *28*, 142–149.

72. Loaces, I.; Ferrando, L.; Scavino, A.F. Dynamics, diversity and function of endophytic siderophore-producing bacteria in rice. *Microb. Ecol.* **2010**, *61*, 606–618.

73. Vessey, J.K. Plant growth promoting rhizobacteria as biofertilizers. *Plant Soil* **2003**, *255*, 571–586.

74. Noori, M.S.S.; Saud, H.M. Potential plant growth-promoting activity of *Pseudomonas*sp. isolated from paddy soil in Malaysia as biocontrol agent. *J. Plant Pathol. Microb.* **2012**, *3*, 120, doi: 10.4172/2157-7471.1000120.

75. Deshwal, V.K.; Kumar, P. Plant growth promoting activity of Pseudomonads in Rice crop. *Int. J. Curr. Microbiol. App. Sci.* **2013**, *2*, 152–157.

76. Doi, R.; Pitiwut, S. From maximization to optimization: A paradigm shift in rice production in thailand to improve overall quality of life of stakeholder. *Sci. World J.* **2014**, doi:10.1155/2014/604291.

77. Bodelier, P.L.E.; Sorrell, B.; Drake, H.L.; Küsel, K.; Hurek, T.; Reinhold-Hurek, B.; Lovell, C.; Megonigal, P.; Frenzel, P. Ecological aspects of microbes and microbial communities inhabiting the rhizosphere of wetland plants. In *Wetlands as a Natural Resource*; Bobbink, R., Beltman, B., Verhoeven, J.T.A., Whigham, D.F., Eds.; Springer-Verlag: New York, NY, USA, 2006; pp. 205–238.

78. Redeker, K.; Wang, N.; Low, J.; McMillan, A.; Tyler, S.; Cicerone, R. Emissions of methyl halides and methane from rice paddies. *Science* **2000**, *290*, 966–969.

79. Kobayashi, T.; Nishizawa, N.K. Iron uptake, translocation, and regulation in higher plants. *Annu. Rev. Plant Biol.* **2012**, *63*,131–152.

80. Kochian, L.V.; Hoekenga, A.O.; Pineros, A.M. How do crop plants tolerate acid soils? Mechanism of Aluminium tolerance and Phosphorous efficiency. *Annu. Rev. Plant Biol.* **2004**, *55*, 459–493.

81. Ding, L.-J.; Su, J.-Q.; Xu, H.-J.; Jia, Z.-J.; Zhu, Y.-G. Long-term nitrogen fertilization of paddy soil shifts iron-reducing microbial community revealed by RNA-$^{13}$C-acetate probing coupled with pyrosequencing. *ISME J.* **2014**, doi:10.1038/ismej.2014.159.

82. Zhang, A.F.; Bian, R.J.; Pan, G.X.; Cui, L.Q.; Hussain, Q.; Li, L.Q.; Zheng, J.W.; Zheng, J.F.; Zhang, X.H.; Han, X.J.; *et al.* Effect of biochar amendment on soil quality, crop yield and greenhouse gas emission in a Chinese rice paddy: A field study of 2 consecutive rice growing cycles. *Field Crop Res.* **2012**, *127*, 153–160.

83. Yang, C.; Yang, L.; Yang, Y.; Ouyang, Z. Rice root growth and nutrient uptake as influenced by organic manure in continuously and alternately flooded paddy soils. *Agric. Water Manag.* **2004**, *70*, 67–81.

84. Dobermann, A.; Witt, C.; Dawe, D. Performance of site-specific nutrient management in intensive rice cropping systems of Asia. *Better Crops Int.* **2002**, *16*, 25–30.

85. Eriksson, J.E. Effects of nitrogen-containing fertilizers on solubility and plant uptake of cadmium. *Water Air Soil Pollut.* **1990**, *49*, 355–368.

86. Das, S.K. Role of micronutrient in rice cultivation and management strategy in organic agriculture—A reappraisal. *Agric. Sci.* **2014**, *5*,765–769.

87. Wong, M.C.; Ma, K.K.; Fang, K.M.; Cheung, C. Utilization of a manure compost for organic farming in Hong Kong. *Biores. Technol.* **1999**, *67*, 43–46.

88. Tang, J.C.; Inoue, Y.; Yasuta, T.; Yoshida, S.; Katayama, A. Chemical and microbial properties of various compost products. *Soil Sci. Plant Nutr.* **2003**, *49*, 273–280.

89. Sebastian, A.; Prasad, M.N.V. Cadmium accumulation retard activity of functional components of photo assimilation and growth of rice cultivars amended with vermicompost. *Int. J. Phytoremediation* **2013**, *15*, 965–978.

90. Shibaharaa, F.; Inubushiab, K. Effects of organic matter application on microbial biomass and available nutrients in various types of paddy soils. *Soil Sci. Plant Nutr.* **1997**, *43*, 191–203.

91. Kumpiene, J.; Lagerkvist, A.; Maurice, C. Stabilization of As, Cr, Cu, Pb and Zn in soil using amendments—A review. *Waste Manag.* **2008**, *28*, 215–225.

92. Li, Z.H.; Horikawa, Y. Stability behavior of soil colloidal suspensions in relation to sequential reduction of soils. II Turbidity changes by submergence of paddy soils at different temperatures. *Soil Sci. Plant Nutr.* **1997**, *43*, 911–919.

93. Baligar, V.C.; Fageria, N.K.; He, Z. Nutrient use efficiency in plants. *Commun. Soil Sci. Plant Anal.* **2001**, *31*, 921–950.

94. Fageria, N.K.; Baligar, V.C. Improving nutrient use efficiency of annual crops in Brazilian acid soils for sustainable crop production. *Commun. Soil Sci. Plant Anal.* **2001**, *32*, 1303–1319.

95. Raun, W.R.; Johnson, G.V. Improving nitrogen use efficiency for cereal production. *Agron. J.* **1999**, *91*, 357–363.

96. Yoo, Y.; Choi, H.; Jung, K. Genome-wide identification and analysis of genes associated with lysigenous aerenchyma formation in rice roots. *J. Plant Biol.* **2015**, *58*, 117–127.

97. Mai, C.D.; Phung, N.T.P.; To, H.T.M.; Gonin, M.; Hoang, G.T.; Nguyen, K.L.; Do, V.N.; Courtois, B.; Gantet, P. Genes controlling root development in rice. *Rice* **2014**, *7*, 30.

98. Wu, W.; Cheng, S. Root genetic research, an opportunity and challenge to rice improvement. *Field Crops Res.* **2014**, *165*, 111–124.

99. Gowdaa,V.R.P.; Henrya, A.; Yamauchi, A.; Shashidhar, H.E.; Serraj, R. Root biology and genetic improvement for drought avoidance in rice. *Field Crops Res.* **2011**, *122*, 1–13.

100. Thanh, N.D.; Zheng, H.G.; Dong, N.V.; Trinh, L.N.; Ali, M.L.; Nguyen, H.T. Genetic variation in root morphology and microsatellite DNA loci in upland rice (*Oryza sativa*L.) from Vietnam. *Euphytica* **1999**, *105*, 43–55.

101. Liu, S.; Wang, J.; Wang, L.; Wang, X.; Xue, Y.; Wu, P.; Shou, H. Adventitious root formation in rice requires *OsGNOM1* and is mediated by the *OsPINs* family. *Cell Res.* **2009**, *19*, 1110–1119.

102. Yoo, S.C.; Cho, S.H.; Paek, N.C. Rice *WUSCHEL*-related homeobox 3A (*OsWOX3A*) modulates auxin-transport gene expression in lateral root and root hair development. *Plant Signal Behav.* **2013**, *8*, e25929.

103. Hsu, Y.Y.; Chao, Y.-Y.; Kao, C.H. Biliverdin-promoted lateral root formation is mediated through heme oxygenase in rice. *Plant Signal Behav.* **2012**, *7*, 885–887.

104. Gamuyao, R.; Chin, J.H.; Pariasca-Tanaka, J.; Pesaresi P.; Catausan, S.; Dalid, C.; Slamet-Loedin, I.; Tecson-Mendoza, E.M.; Wissuwa, M.; Heuer, S. The protein kinase Pstol1 from traditional rice confers tolerance of phosphorus deficiency. *Nature* **2012**, *488*, 535–539.

105. Liu, H.; Wang, S.; Yu, X.; Yu, J.; He, X.; Zhang, S.; Shou, H.; Wu, P. ARL1, a LOB-domain protein required for adventitious root formation in rice. *Plant J.* **2005**, *43*, 47–56, doi:10.1111/j.1365-313X.2005.02434.x.

106. Hossain, M.A.; Cho, J.I.; Han, M.; Ahn, C.H.; Jeon, J.S.; An, G.; Park, P.B. The ABRE-binding transcription factor OsABF2 is a positive regulator of abiotic stress and ABA signaling in rice. *J Plant Physiol.* **2010**, *167*, 1512–1520.

107. Xia, K.; Wang, R.; Ou, X.; Fang, Z.; Tian, C.; Duan, J.; Wang, Y.; Zhang, M. OsTIR1and OsAFB2downregulation via *OsmiR393* overexpression leads to more tillers, early flowering and less tolerance to salt and drought in rice. *PLoS ONE* **2012**, *7*, e30039.

108. Kang, B.; Zhang, Z.; Wang, L.; Zheng, L.; Mao, W.; Li, M.; Wu, Y.; Wu, P.; Mo, X. OsCYP2, a chaperone involved in degradation of auxin-responsive proteins, plays crucial roles in rice lateral root initiation. *Plant J.* **2013**, *74*, 86–97.

109. Zhu, Z.X.; Liu, Y.; Liu, S.J.; Mao, C.Z.; Wu, Y.R.; Wu, P. A gain-of-function mutation in OsIAA11 affects lateral root development in rice. *Mol. Plant* **2012**, *5*, 154–161.

110. Kitomi, Y.; Inahashi, H.; Takehisa, H.; Sato, Y.; Inukai, Y. OsIAA13-mediated auxin signaling is involved in lateral root initiation in rice. *Plant Sci.* **2012**, *190*, 116–122.

111. Wu, J.; Liu, S.; Guan, X.; Chen, L.; He, Y.; Wang, J.; Lu, G. Genome-wide identification and transcriptional profiling analysis of auxin response-related gene families in cucumber. *BMC Res. Notes* **2014**, *7*, 218, doi: 10.1186/1756-0500-7-218.

112. Jun, N.; Gaohang, W.; Zhenxing, Z.; Huanhuan, Z.; Yunrong, W.; Ping, W. *OsIAA23*-mediated auxin signaling defines postembryonic maintenance of QC in rice. *Plant J.* **2011**, *68*, 433–442.

113. Ni, J.; Shen, Y.; Zhang, Y.; Wu, P. Definition and stabilization of the quiescent center in rice roots. *Plant Biol.* **2014**, *16*, 1014–1019.

114. Cui, H.; Levesque, M.P.; Vernoux, T.; Jung, J.W.; Paquette, A.J.; Gallagher, K.L.; Wang, J.Y.; Blilou, I.; Scheres, B.; Benfey, P.N. An evolutionarily conserved mechanism delimiting SHR movement defines a single layer of endodermis in plants. *Science* **2007**, *316*, 421–425.

115. Wang, X.F.; He, F.F.; Ma, X.X.; Mao, C.Z.; Hodgman, C.; Lu, C.G.; Wu, P. *OsCAND1* is required for crown root emergence in rice. *Mol. Plant* **2010**, *4*, 289–299.

116. Xu, M.; Zhu, L.; Shou, H.; Wu, P. A PIN1 family gene, *OsPIN1*, involved in auxin-dependent adventitious root emergence and tillering in rice. *Plant Cell Physiol.* **2005**, *46*, 1674–1681.

117. Uga, Y.; Sugimoto, K.; Ogawa, S.; Rane, J.; Ishitani, M.; Hara, N.; Kitomi, Y.; Inukai, Y.; Ono, K.; Uga, Y.; *et al.* Dro1, a major QTL involved in deep rooting of rice under upland field conditions. *J. Exp. Bot.* **2011**, *62*, 2485–2494.

118. Zhang, J.W.; Xu, L.; Wu, Y.R.; Chen, X.A.; Liu, Y.; Zhu, S.H.; Ding, W.N.; Wu, P.; Yi, K.K. OsGLU3, a putative membrane-bound endo-1,4-beta-glucanase, is required for root cell elongation and division in rice (Oryza sativa L.). *Mol. Plant* **2012**, *5*, 176–186.

119. Qin, C.; Li, Y.; Gan, J.; Wang, W.; Zhang, H.; Liu, Y.; Wu, P. OsDGL1, a homolog of an oligosaccharyltransferase complex subunit, is involved in N-glycosylation and root development in rice. *Plant Cell Physiol.* **2013**, *54*,129–137.

120. Ma, N.; Wang, Y.; Qiu, S.; Kang, Z.; Che, S.; Wang, G.; Huang, J. Overexpression of *OsEXPA8*, a root-specific gene, improves rice growth and root system architecture by facilitating cell extension. *PLoS ONE* **2013**, *8*, e75997.

121. Chen, X.; Shi, J.; Hao, X.; Liu, H.; Wu, Y.; Wu, Z.; Chen, M.; Wu, P.; Mao, C. OsORC3 is required for lateral root development in rice. *Plant J.* **2013**, *74*, 339–350.

122. Wan, J.; Nakazaki, T.; Ikehashi, H. Analyses of genetic loci for diameter of seminal root with marker genes in rice. *Breed. Sci.* **1996**, *46*, 75–77.

123. Kitomi, Y.; Ito, H.; Hobo, T.; Aya, K.; Kitano, H.; Inukai, Y. The auxin responsive *AP2/ERF* transcription factor *CROWN ROOTLESS5* is involved in crown root initiation in rice through the induction of *OsRR1*, a type-A response regulator of cytokinin signaling. *Plant J.* **2011**, *67*, 472–484.

124. Scarpella, E.; Rueb, E.; Meijer, A.H. The *RADICLELESS1* gene is required for vascular pattern formation in rice. *Development* **2003**, *130*, 645–665.

125. Hong, S.K.; Aoki, T.; Kitano, H.; Satoh, H.; Nagato, Y. Phenotypic diversity of 188 rice embryo mutants. *Dev. Genet.* **1995**, *16*, 298–310.

126. Siebers, N.; Godlinski, F.; Leinweber, P. Bone char as phosphorus fertilizer involved in cadmium immobilization in lettuce, wheat, and potato cropping. *J. Plant Nutr. Soil Sci.* **2014**, *177*, 75–83.

127. Clark, R.B. Physiology of cereals for mineral nutrient uptake, use and efficiency. In *Crops Enhancers of Nutrient Use*; Baligar, V.C., Duncan, R.R., Eds.; Academic Press: San Diego, CA, USA, 1990; pp. 131–209.

128. Gahoonia, T.S.; Nielsen, N.E. Root traits as tools for creating phosphorus efficient crop varieties. *Plant Cell Environ.* **2004**, *260*, 47–57.

129. Gourley, C.J.P.; Allan, D.L.; Russlle, M.P. Plant nutrient effi-ciency: A comparison of definitions and suggested improvement. *Plant Soil* **1994**, *158*, 29–37.

130. Tanner, W.; Beevers, H. Transpiration, a prerequisite for long-distance transport of minerals in plants? *Proc. Natl. Acad. Sci. USA* **2001**, *98*, 9443–9447.

131. Palmgren, M.G. Plant plasma membrane H+-ATPases: Powerhouses for nutrient uptake. *Annu. Rev. Plant Physiol. Plant Mol. Biol.* **2001**, *52*, 817–845.

132. Sondergaard, T.E.; Schulz, A.; Palmgren, M.G. Energization of transport processes in plants. Roles of the plasma membrane $H^+$-ATPase. *Plant Physiol.* **2004**, *136*, 2475–2482.

133. Pinto, E.; Siguad-Kutner, T.; Leitão, M.; Okamoto, O.K.; Morse, D.; Colepicolo, P. Heavy metal-induced oxidative stress in algae. *J. Phycol.* **2003**, *39*, 1008–1018.

134. Rauser, W.E. Structure and function of metal chelators produced by plants: The case for organic acids, amino acids, phytin and metallothioneins. *Cell Biochem. Biophys.* **1999**, *31*, 19–48.

135. Azura, A.E.; Shamshuddin, J.; Fauziah, C.I. Root elongation, root surface area and organic acid by rice seedling under $Al^{3+}$ and/or $H^+$ Stress. *Am. J. Agric. Biol. Sci.* **2011**, *6*, 324–331.

136. Wertin, T.M.; Teskey, R.O. Close coupling of whole-plant respiration to net photosynthesis and carbohydrates. *Tree Physiol.* **2008**, *28*, 1831–1840.

137. Wang, Y.; Wu, Y.; Liu, P.; Zheng, G.; Zhang, J.; Xu, G. Effects of potassium on organic acid metabolism of Fe-sensitive and Fe-resistant rices (*Oryza sativa* L.). *Aust. J. Crop Sci.* **2013**, *7*, 843–848.

138. Dong, J.; Mao, W.H.; Zhang, G.P.; Wu, F.B.; Cai, Y. Root excretion and plant tolerance to cadmium toxicity—A review. *Plant Soil Environ.* **2007**, *53*,193–200.

139. Zhu, X.F.; Zheng, C.; Hu, Y.T.; Jiang, T.; Liu, Y.; Dong, N.Y.; Yang, J.L.; Zheng, S.J. Cadmium-induced oxalate secretion from root apex is associated with cadmium exclusion and resistance in Lycopersicon esulentum. *Plant Cell Environ.* **2011**, *34*, 1055–1064.

*140.* Negishi, T.; Nakanishi, H.; Yazaki, J.; Kishimoto, N.; Fujii, F.; Shimbo, K.; Yamamoto, K.; Sakata, K.; Sasaki, T.; Kikuchi, S.;*et al.* cDNA microarray analysis of gene expression during Fe-deficiency stress in barley suggests that polar transport of vesicles is implicated in phytosiderophore secretion in Fe-deficient barley roots. *Plant J.* **2002**, *30*, 83–94.

141. Raskin, I.; Kumar, P.B.A.N.; Dushenkov, S.; Salt, D.E. Bioconcentration of heavy metals by plants. *Curr. Opin. Biotechnol.* **1994**, *5*, 285–290.

142. Krämer, U. Metal hyperaccumulation in plants. *Ann. Rev. Plant Biol.* **2010**, *61*, 517–534.

143. Sebastian, A.; Prasad, M.N.V. Red and blue lights induced oxidative stress tolerance promote cadmium rhizocomplexation in *Oryza sativa*. *J. Photochem. Photobiol. B* **2014**, *137*, 135–143.

144. Stephan, U.W.; Schmidke, I.; Stephan, V.W.; Scholz, G. The nicotianamine molecule is made-to-measure for complexation of metal micronutrients in plants. *Biometals* **1996**, *9*, 84–90.

145. Von Wirén, N.; Klair, S.; Bansal, S.; Briat, J.F.; Khodr, H.; Shioiri, T.; Leigh, R.A.; Hider, R.C. Nicotianamine chelates both $Fe^{III}$ and $Fe^{II}$. Implications for metal transport in plants. *Plant Physiol.* **1999**, *119*, 1107–1114.

146. Errécalde, O.; Seidl, M.; Campbell, P.G.C. Influence of a low molecular weight metabolite (citrate) on the toxicity of cadmium and zinc to the unicellular green alga *Selenastrum capricornutum*: An exception to the free ion activity model. *Water Res.* **1998**, *32*, 419–429.

147. Kramer U.; Talke I.N.; Hanikenne, M. Transition metal transport. *FEBS Lett.* **2007**, *581*, 2263–2272.

148. Sasaki, A.; Yamaji, N.; Yokosho, K.; Ma, J.F. Nramp5 is a major transporter responsible for manganese and cadmium uptake in rice. *Plant Cell* **2012**, *24*, 2155–2167.

149. Yamaji, N.; Sasaki, A.; Xia, J.X.; Yokosho, K.; Ma, J.F. Anode-based switch for preferential distribution of manganese in rice. *Nat. Commun.* **2013**, *4*, 2442, doi:10.1038/ncomms3442.

150. Chen, Z.C.; Yamaji, N.; Motoyama, R.; Nagamura, Y.; Ma, J.F. Up-regulation of a magnesium transporter gene *OsMGT1* is required for conferring aluminum tolerance in rice. *Plant Physiol.* **2012**, *159*, 1624–1633.

151. Ranade-malvi, U. Interaction of micronutrients with major nutrients with special reference to potassium. *KarnatakaJ. Agric. Sci.* **2011**, *24*, 106–109.

152. Yamaji, N.; Xia, J.X.; Mitani-Ueno, N.; Yokosho, K.; Ma, J.F. Preferential delivery of Zn to developing tissues in rice is mediated by a P-type ATPases, OsHMA2. *Plant Physiol.* **2013**, *162*, 927–939.

153. Lee, S.; An, G. Over-expression of OsIRT1 leads to increased iron and zinc accumulations in rice. *Plant Cell Environ.* **2009**, *32*, 408–416.

154. Sasaki, A.; Yamaji, N.; Ma, J.F. Overexpression of *OsHMA3* enhances Cd tolerance and expression of Zn transporter genes in rice. *J. Exp. Bot.* **2014**, *65*, 6013–6021.

155. Zheng, L.; Yamaji, N.; Yokosho, K.; Ma, J.F. YSL16 is a phloem-localized transporter of the copper-nicotianamine complex that is responsible for copper distribution in rice. *Plant Cell* **2012**, *37*, 3767–3782.

156. Song, W.; Yamaki, T.; Yamaji, N.; Ko, D.; Jung, K.; Fujii-Kashino, M.; An, G.; Martinoia, E.; Lee, Y.; Ma, J.F. A rice ABC transporter, OsABCC1, reduces arsenic accumulation in the grain. *Proc. Natl. Acad. Sci. USA* **2014**, *111*, 15699–15704.

157. Ishimaru, Y.; Masuda, H.; Bashir, K.; Inoue, H.; Tsukamoto, T.; Takahashi, M.; Nakanishi, H.; Aoki, N.; Hirose, T.; Ohsugi, R.; *et al*. Rice metal-nicotianamine transporter, OsYSL2, is required for the long-distance transport of iron and manganese. *Plant J.* **2010**, *62*, 379–390.

158. Sasaki, A.; Yamaji, N.; Xia, J.; Ma, J.F. OsYSL6 is involved in the detoxification of excess manganese in rice. *Plant Physiol.* **2011**, *157*, 1832–1840.

159. Beyer, P. Golden Rice and "Golden" crops for human nutrition. *New Biotechnol.* **2010**, *27*, 478–481.

160. Ravindrababu, V. Importance and advantages of rice biofortification with iron and zinc. *J. SAT Agric. Res.* **2013**, *11*, 1–6.

161. Piccoli, N.B.; Grede, N.; de Pee, S.; Singhkumarwong, A.; Roks, E.; Moench-Pfanner, R.; Bloem, M.W. Rice fortification: Its potential for improving micronutrient intake and steps required for implementation at scale. *Food Nutr.Bull.* **2012**, *33*, S360–S372.

162. Clemens, S. Molecular mechanisms of plant metal tolerance and homeostasis. *Planta* **2001**, *212*, 475–486.

163. Hall, J.L.; Williams, L.E. Transition metal transporters in plants. *J. Exp. Bot.* **2003**, *54*, 2601–2613.

164. Sahrawat, K.L. Soil fertility in flooded and non-flooded irrigated rice systems. *Arch. Agron. Soil Sci.* **2012**, *58*, 423–436.

165. Shahid, M.; Nayak, A.K.; Shukla, A.K.; Tripathi, R.; Kumar, A.; Raja, R.; Panda, B.B.; Meher, J.; Bhattacharyya, P.; Dash, D. Mitigation of iron toxicity and iron, zinc, and manganese nutrition of wetland rice cultivars (*Oryza sativa* L.) grown in iron-toxic soil. *Soil Air Water* **2014**, *42*, 1604–1609.

166. Chen, W.; Yao, X.; Cai, K.; Chen, J. Silicon alleviates drought stress of rice plants by improving plant water status, photosynthesis and mineral nutrient absorption. *Biol. Trace Elem. Res.* **2011**, *142*, 67–76.

167. Gregorio, G.B. Progress in breeding for traceminerals in staple crops. *J. Nutr.* **2002**, *132*, 500–502.

168. Patne, N.; Ravindrababu, V.; Usharani, G.; Reddy, T.D. Grain iron and zinc association studies in rice (Oryza sativa L.) F1 progenies. *Arch. Appl. Sci. Res.* **2012**, *4*, 696–702.

169. Ishikawa, S.; Abe, T.; Kuramata, M.; Yamaguchi, M.O.T.; Yamamoto, T.; Yano, M. Major quantitative trait locus for increasing cadmium specific concentration in rice grain is located on the short arm of chromosome 7. *J. Exp. Bot.* **2010**, *61*, 923–934.

170. Ueno, D.; Koyama, E.; Kono, I.O.T.; Yano, M.; Ma, J.F. Identification of a novel major quantitative trait locus controlling distribution of Cd between roots and shoots in rice. *Plant CellPhysiol.* **2009**, *50*, 2223–2233.

171. Xue, D.; Chen, M.; Zhang, G. Mapping of QTLs associated with cadmium tolerance and accumulation during seedling stage in rice (*Oryza sativaL.*). *Euphytica* **2009**, *165*, 587–596.

172. Goto, F.; Yoshihara, T.; Shigemoto, N.; Toki, S.; Takaiwa, F. Iron fortification of rice seed by the soybean ferritin gene. *Nat. Biotechnol.* **1999**, *17*, 282–286.

173. Qu, L.Q.; Yoshihara, T.; Ooyama, A.; Goto, F.; Takaiwa, F. Iron accumulation does not parallel the high expression level of ferritin in transgenic rice seeds. *Planta* **2005**, *222*, 225–233.

174. Paul, S.; Ali, N.; Gayen, D.; Datta, S.K.; Datta, K. Molecular breeding of *Osfer2* gene to increase iron nutrition in rice grain. *GM Crops Food* **2012**, *3*, 310–316.

175. Masuda, H.; Kobayashi, T.; Ishimaru, Y.; Takahashi, M.; Aung, M.S.; Nakanishi, H.; Mori, S.; Nishizawa, N.K. Iron-biofortification in rice by the introduction of three barley genes participated in mugineic acid biosynthesis with soybean ferritin gene.*Front. Plant Sci.* **2013**, doi:10.3389/fpls.2013.00132.

176. Wirth, J.; Poletti, S.; Aeschlimann, B.; Yakandawala, N.; Drosse, B.; Osorio,S.; Tohge, T.; Fernie, A.R.; Günther, D.; Gruissem, W.; *et al.* Rice endosperm iron biofortification by targeted and synergistic action of nicotianamine synthase and ferritin. *Plant Biotechnol. J.* **2009**, *7*, 631–644.

177. Inoue, H.; Higuchi, K.; Takahashi, M.; Nakanishi, H.; Mori, S.; Nishizawa, N.K. Three rice nicotianamine synthase genes, *OsNAS1*, *OsNAS2*and *OsNAS3* are expressed in cells involved in long-distance transport of iron and differentially regulated by iron. *Plant J.* **2003**, *36*, 366–381.

178. Lee, S.; Jeon, U.S.; Lee, S.J.; Kim, Y.K.; Persson, D.P.; Husted, S.; Schjørring, J.K.; Kakei, Y.; Masuda, H.; Nishizawa, N.K.; *et al.* Iron fortification of rice seeds through activation of the nicotianamine synthase gene. *Proc. Natl. Acad. Sci. USA* **2009**, *106*, 22014–22019.

179. Johnson, A.A.T.; Kyriacou, B.; Callahan, D.L.; Carruthers, L.; Stangoulis, J.; Lombi, E.; Tester, M. Constitutive overexpression of the *OsNAS* gene family reveals single-gene strategies for effective iron- and zinc-biofortification of rice endosperm. *PLoS ONE* **2011**, *6*, e24476.

180. Ogo, Y.; Itai, R.N.; Kobayashi, T.; Aung, M.S.; Nakanishi, H.; Nishizawa, N.K. *OsIRO2* is responsible for iron utilization in rice and improves growth and yield in calcareous soil. *Plant Mol. Biol.* **2011**, *75*, 593–605.

181. Agarwal, S.; Venkata, T.V.G.N.; Kotla, A.; Mangrauthia, S.K.; Neelamraju, S. Expression patterns of QTL based and other candidate genes in Madhukar × Swarna RILs with contrasting levels of iron and zinc in unpolished rice grains. *Gene* **2014**, *546*, 430–436.

182. Chandel, G.P.; Samuel, M.; Dubey, R.; Meena, R.K. In silico expression analysis of QTL specific candidate genes for grain micronutrient (Fe/Zn) content using ESTs and MPSS signature analysis in rice (*Oryza sativa L.*). *Plant Genet. Transgenics* **2011**, *2*, 11–22.

183. Lee, S.; Chiecko, J.C.; Kim, S.A.; Walker, E.L.; Lee, Y.; Guerinot, M.L.; An, G. Disruption of OsYSL15 leads to iron inefficiency in rice plants. *Plant Physiol.* **2009**, *150*, 786–800.

184. Ishimaru, Y.; Masuda, H.; Suzuki, M.; Bashir, K.; Takahashi, M.; Nakanishi, H.; Mori, S.; Nishizawa, N.K. Overexpression of the OsZIP4 zinc transporter confers disarrangement of zinc distribution in rice plants. *J. Exp. Bot.* **2007**, *58*, 2909–2915.

185. Lee, S.; Jeong, H.; Kim, S.; Lee, J.; Guerinot, M.; An, G. OsZIP5 is a plasma membrane zinc transporter in rice. *Plant Mol. Biol.* **2010**, *73*, 507–517.

186. Wang, H.; Huang, J.; Ye, Q.; Wu, D.; Chen, Z. Modified accumulation of selected heavy metals in Bt transgenic rice. *J. Environ. Sci.* **2009**, *21*, 1607–1612.

187. Zhan, J.; Wei, S.; Niu, R.; Li, Y.; Wang, S.; Zhu, J. Identification of rice cultivar with exclusive characteristic to Cd using a field-polluted soil and its foreground application. *Environ. Sci. Pollut. Res. Int.* **2013**, *20*, 2645–2650.

188. Antosiewicz, D.M.; Henning, J. Overexpression of LTC1in tobacco enhances the protective action of calcium against cadmium toxicity. *Environ. Pollut.* **2004**, *129*, 237–245.

**2**

# Differences in Aluminium Accumulation and Resistance between Genotypes of the Genus *Fagopyrum*

**Benjamin Klug [1], Thomas W. Kirchner [2] and Walter J. Horst [2,\*]**

[1]  Aglukon Spezialdünger GmbH & Co. KG, Düsseldorf; Germany;
     E-Mail: Benjaminklug@aglukon.com
[2]  Institute of Plant Nutrition, Leibniz Universität Hannover, Germany;
     E-Mail: kirchner@pflern.uni-hannover.de

\*  Author to whom correspondence should be addressed; E-Mail: horst@pflern.uni-hannover.de

Academic Editor: Gareth Norton

---

**Abstract:** Aluminium (Al) toxicity is a major factor reducing crop productivity worldwide. There is a broad variation in intra- and inter-specific Al resistance. Whereas the Al resistance mechanisms have generally been well explored in Al-excluding plant species, Al resistance through Al accumulation and Al tolerance is not yet well understood. Therefore, a set of 94 genotypes from three *Fagopyrum* species with special emphasis on *F. esculentum* Moench were screened, with the objective of identifying genotypes with greatly differing Al accumulation capacity. The genotypes were grown in Al-enriched peat-based substrate for 21 days. Based on the Al concentration of the xylem sap, which varied by a factor of five, only quantitative but not qualitative genotypic differences in Al accumulation could be identified. Aluminium and citrate and Al and Fe concentrations in the xylem sap were positively correlated suggesting that Fe and Al are loaded into and transported in the xylem through related mechanisms. In a nutrient solution experiment using six selected *F. esculentum* genotypes differing in Al and citrate concentrations in the xylem sap the significant correlation between Al and iron transport in the xylem could be confirmed. Inhibition of root elongation by Al was highly significantly correlated with root oxalate-exudation and leaf Al accumulation. This suggests that Al-activated oxalate exudation and rapid transport of Al to the shoot are prerequisites for the protection of the root apoplast from Al injury and thus overall Al resistance and Al accumulation in buckwheat.

**Keywords:** aluminum; buckwheat; citrate translocation; genotypic differences; oxalate exudation; tolerance

## 1. Introduction

Acid soils represent a significant percentage of the world's arable lands [1]. Aluminium toxicity is one of the major constrains for crop production on acid soils [2]. Great inter- and intra-species differences exist in Al resistance, which have successfully been exploited for the breeding of better yielding Al-resistant cultivars of crop species on acid, Al-toxic soils [3–5]. The understanding of the physiological and molecular mechanisms underlying Al resistance could contribute to further intensify the breeding of Al-resistant crop cultivars [6,7]. Generally, plant Al resistance may be achieved either through Al exclusion from uptake and binding in the apoplast of the most Al-sensitive root apex or Al uptake and detoxification within the root symplast, thus Al tolerance [2,8]. For most crop species, Al exclusion through sequestration of Al in the root apoplast with organic acid anions released from the root symplast is the most common mechanism of Al resistance [9]. Less binding of Al in the cell walls owing to a lower content of unmethylated pectins [10,11] or hemicellulose [12] of the cell wall may also contribute to Al exclusion. There is recent evidence that both Al exclusion and Al tolerance contribute to Al resistance in a coordinated way in Arabidopsis [13] and particularly in the Al-resistant cereal crop rice [14]. In some plant species Al tolerance is combined with the capacity to translocate Al to the shoots where Al is accumulated in large concentrations in the leaves [15,16]. Among these plant species are tea (*Camellia sinensis* var. *sinensis*), buckwheat (*Fagopyrum esculentum*) and hortensia (*Hydrangea macrophylla*).

Whereas the physiological and molecular understanding of resistance mechanisms in Al-excluding plant species has made considerable progress in recent years [7,14,17], the understanding of Al accumulation in relation to Al tolerance still widely lags behind. The Al accumulator buckwheat is characterized by both, Al exclusion from uptake through Al-induced release by root tips of oxalate [18] and symplastic sequestration of Al by oxalate [19]. Although the role of oxalate release from the roots in Al exclusion is well established, references [20,21] could not relate differences in oxalate exudation to genotypic differences in Al resistance between buckwheat genotypes. Comparing genotypes of *Fagopyrum tataricum* Yang *et al.* [21] also concluded that root exudation of oxalate could not explain genotypic differences in Al resistance. They suggested that the differences in Al accumulation of root tips were related to differences in cell-wall negativity conferred by lower pectin contents and a higher degree of pectin methylation. Al accumulation in shoots requires radial transfer of Al from the root surface to the xylem where Al is transported primarily as Al-citrate to the shoot [22,23]. Based on sophisticated physiological approaches references [24–26] proposed a hypothesis reconciling Al exclusion and Al accumulation in which an Al oxalate $(Ox)^+$ plasma-membrane transporter in the root cortex and a xylem-loading Al citrate $(Cit)^{n-}$ transporter in the xylem parenchyma cells are key elements of Al tolerance and accumulation in buckwheat.

The aim of the present study was to enhance the physiological understanding of Al exclusion on the one hand and Al accumulation on the other hand by exploiting differences in Al resistance and

Al accumulation within a large set of genotypes of the species *F. esculentum* (buckwheat). To even widen the genetic background we included into the genotype selection also genotypes from further species of the *Fagopyrum* genus namely *F. tataricum* and *F. acutatum*.

## 2. Results

Addition of 2–8 g $Al_2(SO_4)_3 \cdot 18H_2O$ decreased the pH of the substrate from 5.6 to 4.0 approximately linearly (Figure 1). In contrast, the soluble and mononuclear Al concentrations increased exponentially from an Al application of 6 g on. Neither the pH reduction, nor the related increase of the soluble Al in the substrate lead to reduced growth of the plants of the buckwheat cultivar "Lifago" (Figure 2).

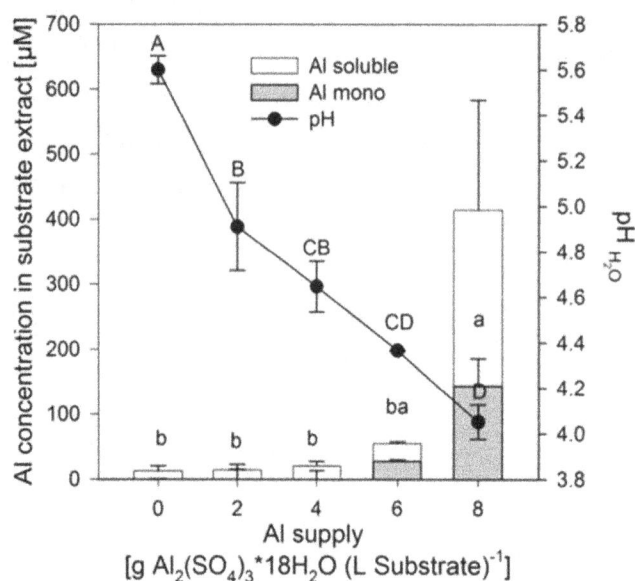

**Figure 1.** Total soluble, mononuclear Al (Al mono) concentrations and substrate pH in the aqueous 1:3 (*v/v*) substrate extract as affected by increasing supply of $Al_2(SO_4)_3 \cdot 18H_2O$. Bars represent means ± SE, $n = 3$. Different capital and lower case letters denote differences between treatment durations for pH values and for Al mono concentrations, respectively, at $p < 0.05$ (Tukey test).

Increasing Al supply did not induce mineral element deficiency-symptoms in shoots of the buckwheat cultivar "Lifago". Also growth, height, and fresh matter of shoots were not significantly affected (Figure 2). There was a tendency of optimum shoot growth at 4 g Al sulfate supply which was also especially true for root growth (Figure S1). This might indicate a growth-promoting effect of small amounts of Al in buckwheat.

Although the soluble Al concentration in the substrate only significantly increased at 8 g Al sulfate supply (see above, Figure 1), the Al concentrations in the primary leaves steadily increased from the lowest Al application rate to reach about 500 μg (g dry weight)$^{-1}$ at the highest Al supply (Figure 3A).

Aluminium is readily transported from the roots to the shoots via the xylem typical for Al accumulators. The Al concentration in the xylem sap increased with increasing Al supply to a level of about 400 μM at an Al sulfate application of 6 g (Figure 3A). The xylem-sap citrate-concentration which was more than 10 times higher than the Al concentration increased with the Al supply

(Figure 3B). In spite of great variability between individual plants, there was a significant positive correlation between Al and citrate xylem-sap concentrations.

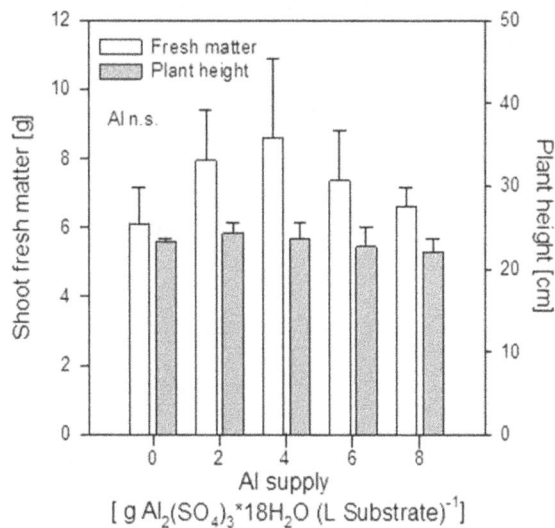

**Figure 2.** Shoot fresh matter production and height of buckwheat *cv.* "Lifago" plants as affected by increasing $Al_2(SO_4)_3 \cdot 18H_2O$ supply to the substrate, three weeks after germination. Bars represent means ± SE, $n = 3$.

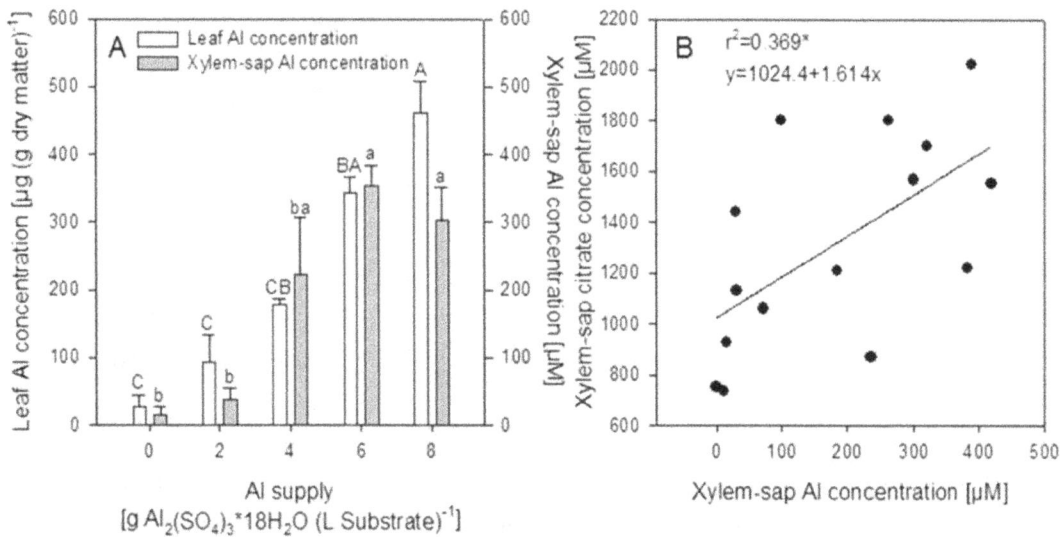

**Figure 3.** Primary leaf and xylem-sap Al concentrations of buckwheat *cv.* "Lifago" plants as affected by increasing $Al_2(SO_4)_3 \cdot 18H_2O$ supply to the substrate, three weeks after germination (A), and relationship between xylem-sap Al and citrate concentrations (B). Bars represent means ± SE, $n = 3$. In (A), different capital and lower-case letters denote significant ($p < 0.05$) differences between leaf and xylem-sap Al concentrations, respectively; In (B), *** denote significance of the regression coefficient at $p < 0.001$.

Based on this experiment the whole set of genotypes of the genus *Fagopyrum* was cultivated at an Al supply of 8 g $Al_2(SO_4)_3 \cdot 18H_2O$. The fresh matter production of the 94 genotypes varied between 5 g and 10 g per plant after three weeks of cultivation. The number of leaves per plant was particularly variable. Some genotypes showed about five times more leaves than others. Since this biased the comparison of the Al concentrations of the leaves owing to dilution or concentration effects in the analyzed leaf tissue, the xylem-sap Al concentration appeared to be the more suitable parameter for the characterization of the Al accumulation capacity. The comparison of the xylem-sap Al concentrations reveals a broad genotypic variation (Figure 4). The highest Al concentrations reached about 350 μM. None of the genotypes showed Al concentrations lower than 100 μM which could have been indicative of Al excluders. Thus all genotypes can be classified as Al accumulators. However the Al accumulation capacity differed greatly between the genotypes across and within the *Fagopyrum* species.

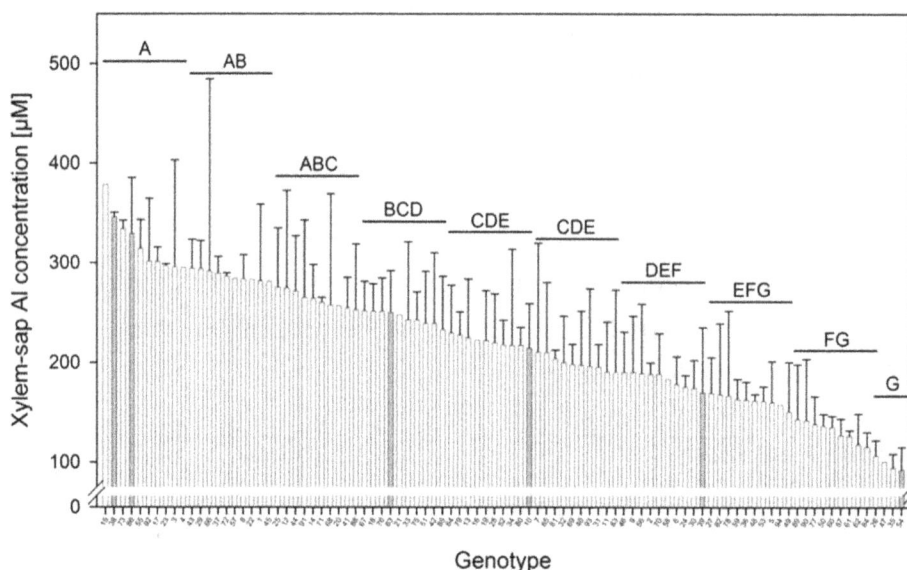

**Figure 4.** Xylem-sap Al concentration of 94 genotypes of the genus *Fagopyrum* after three weeks of substrate culture with an Al supply of 8 g $Al_2(SO_4)_3 \cdot 18H_2O$. Xylem exudates of three plants per 1 L pot were combined to one composite sample. Bars represent means ± SE, $n = 3$. Different letters denote significant differences between Al xylem concentrations at $p < 0.05$ (Tukey test).

In the xylem sap not only Al but also citrate and other mineral elements were determined. In spite of high variability of the xylem-sap Al and citrate concentrations, a highly significant positive correlation between the Al and citrate concentrations across the genotypes existed (Figure 5A). The mean xylem-sap citrate-concentration was 10 times higher than the Al concentration. Among all mineral elements determined in the xylem sap, only Fe was highly significantly positively correlated with the Al concentration (Figures 5B and S2). The mean Fe xylem-sap concentration was 10 times lower than the Al concentration. Also citrate and Fe xylem-sap concentrations were highly significantly positively correlated ($p < 0.001$, Figure S3). However, the slope of the regression suggests that on average across genotypes, 93 times more citrate than Fe was loaded into the xylem.

For a further in-depth study in hydroponics, six *F. esculentum* genotypes (grey columns in Figure 4) were selected on the basis of significantly different xylem-sap Al concentrations. In the hydroponic system the ranking of the genotypes for Al accumulation based on the screening in substrate could only partially be confirmed: the genotypes 38 and 86 proved to be more efficient Al accumulators than the genotypes 63 and 10 based on high xylem (Figure 6A) and leaf (Figure 6B) Al concentrations. However, the genotypes 54 and 39 with the lowest xylem Al concentrations in the substrate experiment had equally high Al accumulation capacity compared to the genotypes 38 and 86. This indicates that the conditions in the substrate and the hydroponic experiment differed greatly with regard to form and concentration of plant-available Al.

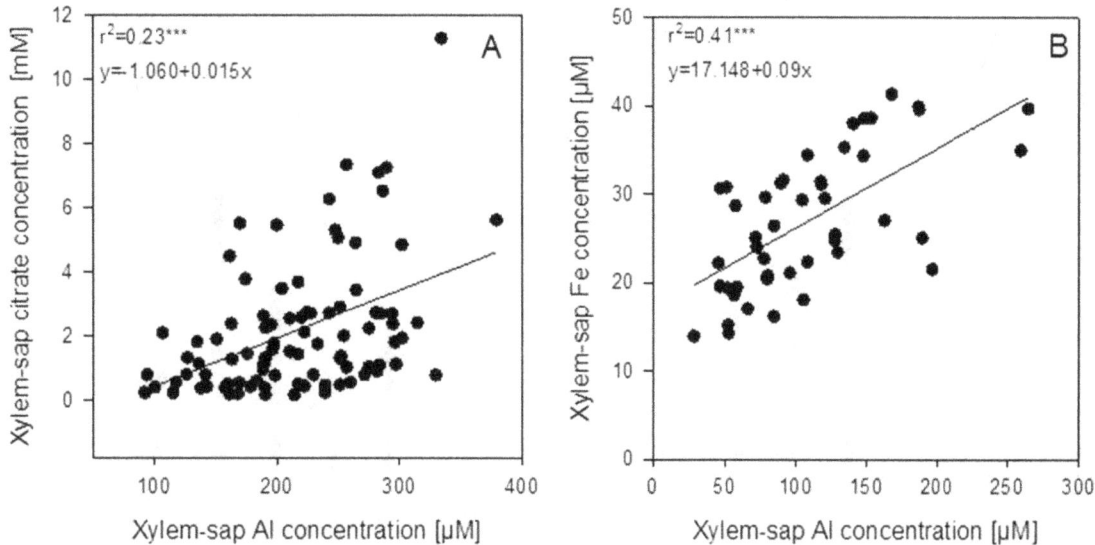

**Figure 5.** Correlations between xylem-sap Al concentrations and the xylem-sap citrate (**A**) and iron (**B**) concentrations of 94 (**A**) or 46 (**B**) genotypes of the genus *Fagopyrum* after 3 weeks of substrate culture with an Al supply of 8 g $Al_2(SO_4)_3 \cdot 18H_2O$. Xylem sap was collected for 30 min after cutting off the shoots. Points represent means $\pm$ SE, $n = 3$. *** denote significance of the regression coefficient at $p < 0.001$.

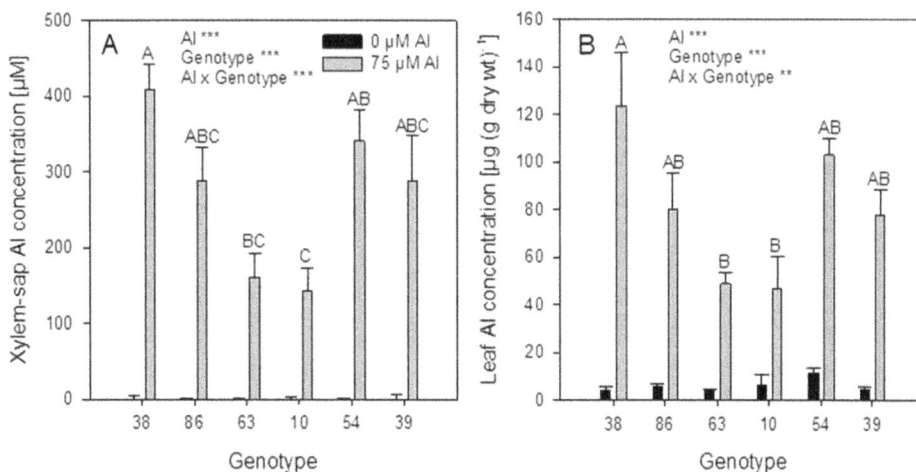

**Figure 6.** Xylem-sap (**A**) and leaf (**B**) Al concentrations of six selected *Fagopyrum esculentum* genotypes. Plants were pre-cultured in complete nutrient solution and then transferred for 24 h to simplified nutrient solution without or with 75 µM Al. Bars represent means $\pm$ SE, $n = 3$. For the ANOVA, ** and *** denote significant effects at $p < 0.01$ and $p < 0.001$, respectively.

Al treatment significantly reduced the Fe concentrations in the roots (Figure 7A) and even more in the xylem-sap (Figure 7B). The genotypes differed significantly in both, root and xylem-sap Fe concentrations, however, they responded comparably to Al treatment (no Al/genotype interaction). In spite of the clear negative effect of Al supply on the Fe root and xylem-sap concentrations the Al and Fe concentrations were highly significantly positively correlated across genotypes and Al treatments (Figure 8) as in the substrate experiment (see above Figure 5B).

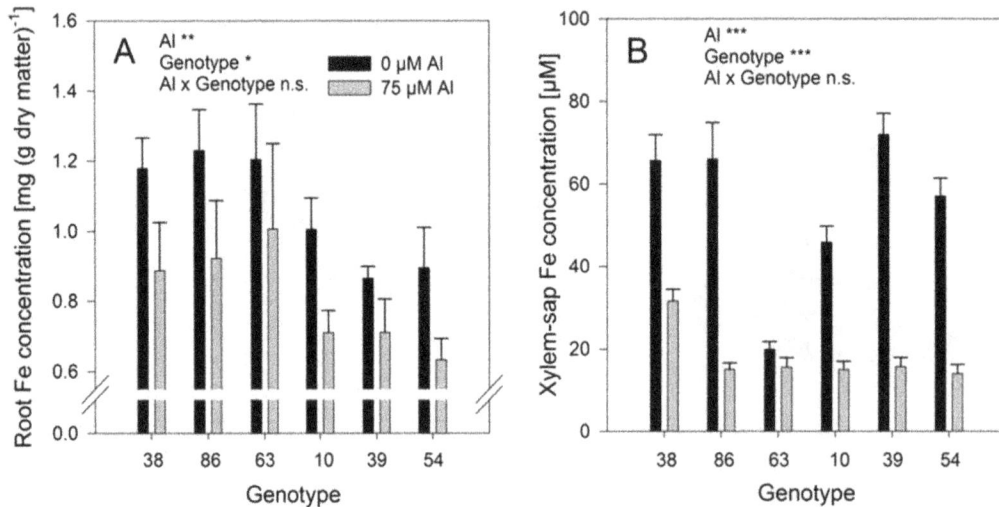

**Figure 7.** Iron concentrations of the bulk-root dry matter (**A**) and xylem-sap Fe concentrations (**B**) of six selected *Fagopyrum esculentum* genotypes as affected by Al supply. Plants were pre-cultured in complete nutrient solution containing at 60 μM Fe-EDDHA and then transferred for 24 h to simplified nutrient solution without or with 75 μM Al. Bars represent means ± SE, $n = 3$. For the ANOVA, *, **, and *** denote significant effects at $p < 0.05$, 0.01 and 0.001, respectively, n.s. non-significant.

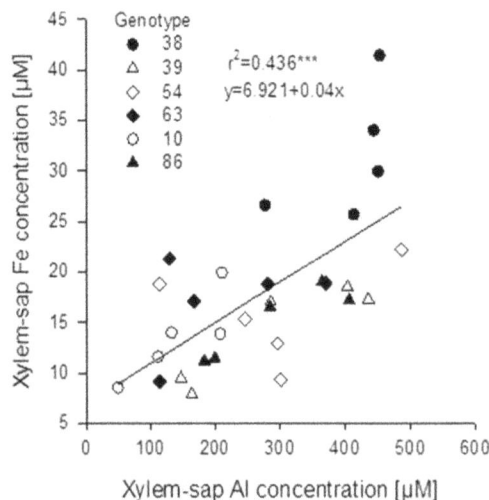

**Figure 8.** Relationship between xylem-sap Al and Fe concentrations of six selected *Fagopyrum esculentum* genotypes ($n = 5$). Plants were pre-cultured in complete nutrient solution containing 60 μM Fe-EDDHA and then transferred for 24 h to simplified nutrient solution without or with 75 μM Al. *** denote significance of the regression coefficient at $p < 0.001$.

The root-growth rates of the controls not treated with Al were in the range of $0.7–1$ mm·h$^{-1}$ (Figure 9A). Aluminium supply (75 µM) decreased the root growth in some genotypes. The genotypes 38, 86, and 63 showed no root-growth inhibition (thus can be classified as Al-resistant), while the genotypes 10, 39, and 54 were inhibited by 30%–50% (thus can be classified as Al-sensitive).

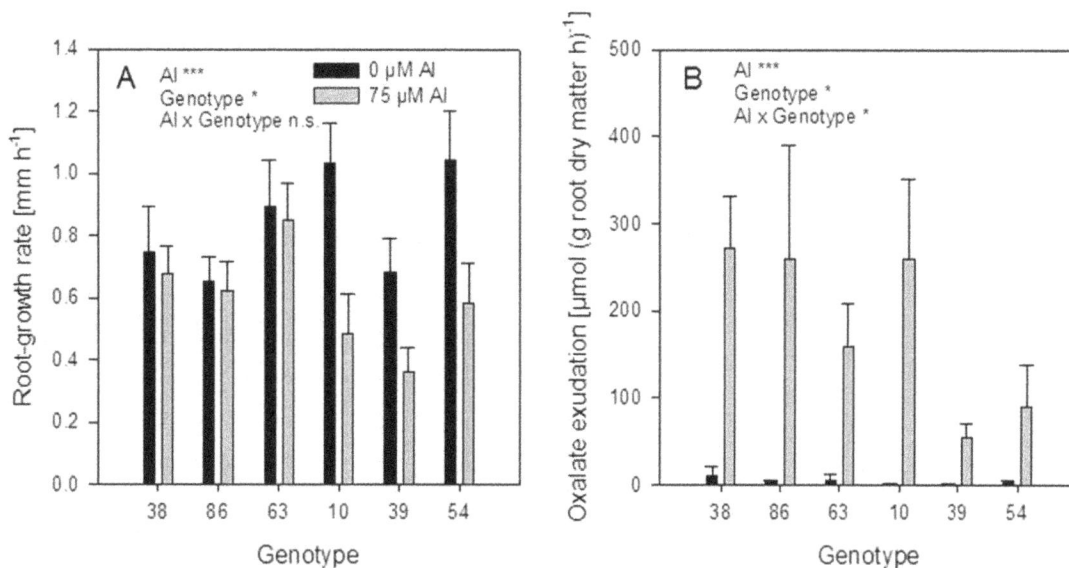

**Figure 9.** Root-growth rate (**A**) and oxalate exudation-rate (**B**) of six selected buckwheat cultivars. Plants were pre-cultured in complete nutrient solution and then transferred for 24 h to simplified nutrient solution without or with 75 µM Al. Root growth-rate was determined by marking the root 15 mm behind the tip before the treatment and measuring the distance again after 24 h. Root oxalate exudation rate was determined during the first 4 h of Al treatment. Bars represent means $\pm$ SE, $n = 3$. For the ANOVA, *, ** and *** denote significant effects at $p < 0.05$, 0.01 and 0.001, respectively.

Root oxalate exudation, was significantly activated by Al application in all genotypes (Figure 9B). Citrate was exuded in minor amounts and showed no Al-activated exudation pattern). The Al-resistant genotypes 38 and 86 showed significantly higher oxalate exudation rates than the Al-sensitive genotypes 39 and 54. However, the genotypes 63 and 10 showed lower or higher oxalate exudation-rates, respectively, than could be expected from their Al resistance classification.

## 3. Discussion

The main focus of this study was to determine differences in Al accumulation in the shoots among a large number of genotypes across three *Fagopyrum* species. The comparison revealed that all tested species and genotypes accumulated Al in the shoot, thus no qualitative differences were found. In a recent study comparing Al tolerance and Al accumulation of three *Fagopyrum species* (*F. esculentum, F. tataricum, F. homotropicum*) reference [27] also concluded that Al accumulation is a conserved trait in *Fagopyrum*. However, there were significant quantitative differences in Al accumulation capability. Genotypes differed in xylem Al concentration by a factor of about four (Figure 4). The xylem-sap Al concentration was chosen as the primary parameter for the comparison of the genotypes for Al accumulation capacity. The xylem sap was sampled after only 0.5 h, because various studies showed that only shortly after cutting of the shoot representative data on the *in vivo* xylem-sap composition can be assessed [27]. Moreover, the same technique has been used for the characterization of heavy metal hyper-accumulation. Xylem-sap Cd and Zn concentrations were used to show a considerable scope for the selection of advanced-hyper-accumulating ecotypes with the objective of increasing phyto-extraction efficiency and remediation of metal-contaminated soils [28]. The Al-accumulation trait could not be related to the geographic origin of the genotypes. The genotypes with the highest Al concentration in the xylem sap originate from different regions as for example Belarus (genotype 15), Iran (genotype 38), North Korea (genotype 73), and Italy (genotype 86).

A similar range of quantitative differences in Al accumulation has also been found within the family of *Melastomataceae* [29]. Species from this family accumulated Al in great amounts in the shoot. However, the variation within the family was shown to be in the range of 6–66 mg Al·g$^{-1}$ dry matter. In contrast, a comparison of members of the taxa *Polygonaceae*, of which the genus *Fagopyrum* is a member, showed that some species differing in Al resistance did not accumulate Al in their shoots [30]. This might suggest that the trait of Al accumulation is not spread over the whole family of the *Polygonaceae* and is rather typical for the genus *Fagopyrum*. Here, this genus showed only quantitative but not qualitative differences in Al accumulation. Based on semi-quantitative tests reference [29], mapped a recent angiosperm phylogeny for the trait of Al accumulation. As indicated by references [15,31] this classification might suggest that Al hyper-accumulation is a simple, primitive trait that has arisen independently several times during evolution but was lost independently in many derived taxa. These authors suggest that the trait of Al accumulation shows low incidence in evolutionary advanced groups which appeared to be correlated with the herbaceous habit. Therefore, the herbaceous genus *Fagopyrum* might represent an exceptional case.

An unexpected result of the study was the confirmation of an interrelationship between Al and Fe transport in buckwheat. Across all *Fagopyrum* genotypes studied xylem-sap Al and Fe concentrations were highly significantly positively correlated (Figure 5B). A similar relationship could also be found in the hydroponic experiment with six selected *F. esculentum* genotypes (Figure 8). This was unexpected, because Al supply strongly reduced root Fe concentrations (Figure 7A) and xylem-sap Fe concentrations (Figure 7B). Since during the 24h Al treatment period in simplified nutrient solution no Fe was applied, this cannot be explained by inhibition of Fe uptake by Al. It rather appears that the Al-induced release of oxalate from the root ([18], Figure 9B) mobilizes apoplastically immobilized Fe. It has been reported that a large amount of Fe may be bound in the root apoplast which may be mobilized under conditions of Fe deficiency through Fe deficiency-induced release of Fe complexors [32]. However, the mobilization of apoplastic Fe during the Al treatment period did not lead to higher Fe uptake and translocation in the xylem. It thus appears that Fe-oxalate is not readily taken up and loaded into the xylem, where, in fact, Fe is transported as citrate complex [33,34].

Using NMR, reference [35] showed that in buckwheat also Al is complexed and detoxified by citrate anions in the xylem sap, and reference [23] confirmed the dominance of Al-citrate as the main Al transport form not only in *F. esculentum* but also in *F. tataricum* and the wild buckwheat *F. homotropicum*. In the study by reference [35] the xylem citrate-concentration did not respond to externally applied or internally transported Al. These results are in agreement with the analysis of the effect of Al on the global transcriptome in *F. esculentum* [36] and *F. tataricum* [37] root tips which did not show major effects on genes involved in organic acid metabolism suggesting that citrate was not a rate-limiting step for Al transport into and in the xylem. However, in these studies a major difference in citrate concentration and synthesis between the root cortex and the central cylinder has not been taken into consideration. This assumption is supported by own data on the oxalate and citrate contents in surgically partitioned cortical and stele tissues [25]. In contrast, in our study a significant positive correlation between xylem-sap citrate and Al concentrations existed when the Al supply to substrate grown buckwheat cultivar Lifago was increased (Figure 3B) and across the *Fagopyrum* genotypes (Figure 5A) confirming previous results by references [24,25]. Also in the Al accumulator *H. macrophylla* Al and citrate xylem-sap concentrations were positively correlated [38]. However, only a low percentage of the variation of the citrate concentration could be explained by the Al concentration ($r^2 = 0.23$ ***), and citrate was more abundant than Al by a factor of ten (Figure 5A). Iron is known to be transported in the xylem coupled to citrate which is loaded into the xylem through MATE/FRD-proteins coded by *FRD3* in Arabidopsis [39], *OsFRDL1* in rice [40] or *HvAACT1* in barley [41]. The above mentioned recent global transcriptome analysis of Al-induced genes in root tips clearly revealed that among the strongly Al up-regulated genes were the homologues of these genes, *FtFRDL1* and *FtFRDL2* in *F. tataricum* [37] and *FeMATE1* and *FeMATE2* in *F. esculentum* [36]. Their role in citrate and Al loading into the xylem remains to be investigated in the future.

It is particularly important to clarify whether Al (and Fe) is loaded into the xylem in a co-transport or as the negatively charged Al-citrate complex as has been postulated by reference [25]. It is intriguing that the genotypic differences in xylem-sap Al concentrations in the presence of Al are closely related to the Fe concentrations in the absence of Al (compare Figures 6A and 7B). Together with the strong reduction of Fe xylem-sap concentrations by Al (Figure 7B) this may indicate that both Al and Fe are transported through the same transporter, and Fe transport is competitively inhibited by the more abundant Al (factor ten). The positive correlation between xylem-sap Al and Fe concentrations may then reflect the Al-enhanced transporter abundance in agreement with the higher expression of the genes coding for these transporters [36,37].

The six *F. esculentum* genotypes selected for differences in Al xylem-sap concentrations also showed great variability in Al-induced root-growth inhibition and thus Al resistance (Figure 9A). The Al-induced root oxalate-exudation is a well-established Al resistance and Al exclusion mechanism (see Introduction). Among the six genotypes Al resistance was highly significantly positively correlated with Al-induced root oxalate-exudation (Table 1). However, oxalate exudation explained only about 50% of the variation in Al resistance. Also shoot Al accumulation was highly significantly correlated with Al resistance indicating that Al accumulation and Al tolerance also contribute to overall Al resistance of buckwheat as has been suggested by reference [25]. Evaluating the role of both oxalate exudation and Al accumulation in Al resistance using multiple regression analysis (Table 1) revealed that more than 70% ($r^2 = 0.725$ ***) of the variation in Al-induced root-growth inhibition can be explained by these two variables. It thus appears that both Al exclusion by oxalate root-exudation and rapid oxalate-facilitated Al uptake and citrate-facilitated Al translocation to the shoot are involved in the protection of the root-tip apoplast, the main target of Al rhizotoxicity as elaborated by reference [42]. The genes and proteins (particularly transporters involved in Al exclusion and Al tolerance/accumulation) have still not yet been identified. It is expected that recent transcriptomic studies will pave the way for a better understanding of the molecular basis of Al resistance of buckwheat [36,37,43]. In the model plants Arabidopsis and rice, an interplay between Al exclusion by root exudation of organic acid anions and the rapid uptake and accumulation in the root vacuole has been widely established [17,44]. However, these plant species do not translocate Al to the shoots, a typical feature of Al accumulators such as *Fagopyrum*.

**Table 1.** Regression analysis of the relationships between relative root-growth rate (75 μM Al supply for 24 h in simplified nutrient solution compared to the 0 μM Al control) and root oxalate exudation or/and leaf Al concentration of six *F. esculentum* genotypes. *** denote the level of significance.

| Variable parameters | Function of the regression | $r^2$ | $p$ value |
|---|---|---|---|
| Oxalate exudation *vs.* relative root-growth rate | $y = 37.09 + 0.19x$ | 0.538 | <0.001 *** |
| Leaf Al accumulation *vs.* relative root-growth rate | $y = -23.14 + 0.7x$ | 0.485 | <0.001 *** |
| Multiple regression: Oxalate exudation and leaf Al accumulation *vs.* relative root-growth rate | $y = -18.534 + (0.139 \cdot \text{oxalate exudation}) + (0.480 \text{ leaf Al accumulation})$ | 0.725 | <0.001 *** |

## 4. Experimental Section

### 4.1. Plant Material

The screening experiments were conducted using a reference cultivar of buckwheat (*Fagopyrum esculentum* Moench *cv.* "Lifago") known from former experiments (Klug and Horst, 2010a, b and c) which was provided by Deutsche Saatveredelung AG (Lipstadt, Germany). Additionally, a set of 94 *Fagopyrum* accessions was kindly provided by the gene bank of the Leibniz Institute of Plant Genetics and Crop Plant Research (IPK, Gatersleben, Germany). These *Fagopyrum* accessions included primarily genotypes of *F. esculentum* Moench var. *esculentum* (76); *F. esculentum* Moench var. *Emarginatum* (Roth) Alef. (5); *F. tataricum* (L.) Gaertn (13), and *F. acutatum* (Lehm.) Mansf. ex Hammer (1). The overall genetic origin of these genotypes is spread over Europe and Asia (Table S1). For the subsequent in-depth analysis of Al resistance mechanisms in hydroponics, genotypes with differing Al translocation patterns according to the screening experiment were chosen from the species *F. esculentum* Moench *esculentum*.

### 4.2. Plant Cultivation

#### 4.2.1. Pot Experiments

For the main pot experiment, plants were sown into peat-based substrate containing 30% clay without the addition of Al avoiding Al toxicity during the process of germination. In the following 3 weeks of cultivation Al was added in 4 steps each consisting of supplies of 2 g $Al_2(SO_4)_3 \cdot 18H_2O$ (162 mg Al) resulting in a total supply of 8 g of $Al_2(SO_4)_3 \cdot 18H_2O$ (648 mg Al) per L substrate. This Al amount was based on a substrate specific Al-buffer curve with a target pH of the aqueous substrate extract of 4.3, as recommended for the production of blue-colored *Hydrangea macrophylla* [38].

The reference *F. esculentum* variety "Lifago" was additionally exposed to different Al concentrations (0; 2; 4; 6, and 8 g of $Al_2(SO_4)_3 \cdot 18H_2O$ corresponding to 162, 324, 486, and 648 mg Al, respectively, per L substrate) to evaluate the response of the plants to increasing Al supply. Three plants were grown in each pot, containing 1 L of substrate for 21 days. Plant tissue and xylem exudate samples of these three plants were combined to one composite sample. The experimental design was a randomized block design with three replications. Plants were grown in a greenhouse at natural light intensity and day length and 20/18 °C day and night temperatures, respectively. Plants were fertilized once a week with 100 mL of a 30 $g \cdot L^{-1}$ solution of a commercial phosphorus-free fertilizer containing 18% N, 18.3% K., 2% Mg, 0.02% B, 0.04% Cu, 0.1% Fe, 0.05% Mn, 0.01% Mo, and 0.01% Zn.

## 4.2.2. Nutrient Solution Experiments

*F. esculentum* genotypes differing in Al accumulation were chosen for further investigations in a nutrient solution experiment. Selected *F. esculentum* genotypes were germinated for 7 days in a foam filter-paper sandwich-system. After germination the seedlings were transferred to low ionic-strength nutrient solution with the following composition (μM): 500 $KNO_3$, 162 $MgSO_4$, 30 $KH_2PO_4$, 250 $Ca(NO_3)_2$, 8 $H_3BO_3$, 0.2 $CuSO_4$, 0.2 $ZnSO_4$, 5 $MnSO_4$, 0.2 $(NH_4)_6Mo_7O_{24}$, 50 NaCl, and 30 Fe-EDDHA. The seedlings were grown for 2 weeks in nutrient solution. Afterwards, the pH of the nutrient solution was reduced in three steps to 4.3 resulting in at least 12 h for adaptation to the low pH before the beginning of the Al treatment (± 75 μM $AlCl_3$ at pH 4.3) in minimal nutrient solution, containing 500 μM $CaCl_2$, 100 μM $K_2SO_4$ and 8 μM $H_3BO_3$ to avoid mineral interactions during the short-term Al treatment.

## 4.3. Substrate Analysis

The pH, and the concentration of soluble total and monomeric Al and mineral nutrients in the substrate were determined in a 1:3 water extract after 1 day incubation of the substrate at room temperature. The extracts were passed through a filter with a pore size of 0.45 μm. Al and mineral nutrients were determined by optical inductively coupled plasma-emission spectroscopy (ICP-OES) (Spectro Analytical Instruments GmbH, Kleve, Germany). Monomeric Al concentration was determined following the method of [45] using aluminon. The extinction of the Al-aluminon complex was spectrophotometrically measured at 532 nm.

## 4.4. Sampling of Xylem Sap

Sampling of xylem sap was performed following the method described by Ma and Hiradate (2000) with some modifications. The stem was severed 2 cm above the root and the xylem sap was collected for not more than 0.5 h to yield reliable data on *in vivo* concentrations of solutes in the xylem sap. The cut surface was rinsed with double distilled water (ddH2O) and blotted off. The xylem sap was collected in 1 mL micro-pipette tips which were trimmed to fit the cut stem. The volumes of the exudates were determined by 1000 μL micropipettes. Exudates were immediately frozen in liquid $N_2$.

## 4.5. Measurement of Root Elongation Rate

Plants were pre-cultured in complete nutrient solution and then transferred for 24 h to simplified nutrient solution with or without 75 μM Al. Root growth-rate was determined by marking the root tips with Plaka color (Pelikan, Feusisberg, Switzerland) using a fine paint brush 15 mm behind the tip before the treatment and measuring the distance again after 24 h.

## 4.6. Collection of Root Exudates

For collection of root exudates adventitiously rooted cuttings were produced as follows: Plants were grown for 4 weeks in a green house at 25/20 °C day/night temperature. After this period of growth the shoots were cut 1 cm below the first node with adventitious root initials and additionally above the primary leaf to reduce transpiration. These shoot cuttings were transferred to low ionic strength nutrient solution with the following composition [μM]: 500 $KNO_3$, 162 $MgSO_4$, 30 $KH_2PO_4$, 250 $Ca(NO_3)_2$, 8 $H_3BO_3$, 0.2 $CuSO_4$, 0.2 $ZnSO_4$, 5 $MnSO_4$, 0.2 $(NH_4)_6Mo_7O_{24}$, 50 NaCl, and 30 Fe-EDDHA for 4 days keeping the shoots at 100% relative humidity (rH) until adventitious roots had emerged. The following day the plants were adapted to lower rH by reducing air humidification. Another day later the pH of the nutrient solution was reduced in three steps resulting in at least 12 h for adaptation to pH 4.3 before the beginning of the Al treatment. Afterwards, the plants were transferred to 10 mL of a simplified continuously aerated nutrient solution (500 μM $CaCl_2$, 8 μM $H_3BO_3$; 100 μM $K_2SO_4$, pH 4.3) supplemented either with 0 μM or 75 μM $AlCl_3$ inhibiting root growth by 50%–60% (Klug and Horst, 2010b) and activating Al exclusion and tolerance mechanisms [24]. The pH was controlled frequently and, when necessary, re-adjusted to 4.3 using 0.1 M HCl or 0.1 M KOH.

## 4.7. Mineral Element Analysis

Mineral element contents in the bulk-leaf tissue were determined in the primary leaf after dry ashing at 480 °C for 8 h, dissolving the ash in concentrated 1:3 diluted $HNO_3$, and then diluting (1:10 $v/v$) with $ddH_2O$. Measurements were carried out by inductively coupled plasma-emission mass spectrometry (ICP-OES, 7500 CX, Agilent Technologies, Santa Clara, CA, USA). The Al concentration in the xylem sap was determined after appropriate dilution with $ddH_2O$ by graphite furnace atomic absorption spectrometry (GF-AAS) (Unicam 939 QZ, Analytical Technologies, Cambridge, UK). The composition of other mineral elements in the xylem sap was determined after dilution by ICP-OES.

## 4.8. Determination of Organic Acids

The organic acid (OA) concentrations of the root exudates as well as the xylem sap were measured by isocratic High Pressure Liquid Chromatography (HPLC, Kroma System 3000, Kontron Instruments, Munich, Germany). The OAs were injected through a 20 μL loop-injector (Auto-sampler 360) of the HPLC, separating different OAs on an Animex HPX-87H (300 × 7.8 mm) column (BioRad, Laboratories, Richmond, CA, USA), supplemented with a cation $H^+$ micro-guard cartridge, using 10 mM perchloric acid as eluent at a flow rate of 0.5 mL per minute, at a constant temperature of 35 °C (Oven 480), and 74 hPa of atmospheric pressure. Measurements were performed at a wavelength

$\lambda = 214$ nm (UV Detector 320). Prior to the analysis of exuded OA the nutrient solution samples run through a cation exchange-column (hydrochloric form) (AG® 50W-X8; BioRad; Life Science Group; Hercules, CA, USA) followed by concentration to dryness using a centrifugal evaporator (RCT 10–22T, Jouan,Saint-Herblain, France).

## 5. Conclusions

The screening in Al-enriched peat substrate of 94 accessions within the genus *Fagopyrum* suggested that Al accumulation is a conserved trait in *Fagopyrum* and no qualitative but quantitative differences existed between the accessions. A correlation between Al and citrate and Fe concentrations in the xylem sap which could be confirmed in a nutrient solution experiment with selected *F. esculentum* genotypes suggest that Fe and Al are loaded into and transported in the xylem through related mechanisms involving citrate. This is in agreement with recent global transcriptomic analysis showing major up-regulation by Al of citrate transporters. The mechanism of citrate-enhanced Al loading into the xylem remains to be elucidated. The significant relationships between oxalate root-exudation and Al accumulation in the shoot suggest that Al exclusion from the root and rapid Al uptake and loading into the xylem both contribute to the protection of the root apoplast from Al injury and thus Al resistance in *Fagopyrum*. The recent progress through global transcriptome analysis of Al-induced genes will allow further characterization of Al resistance and Al accumulation in buckwheat in the near future. However many of the related genes are constitutively expressed in buckwheat requiring the availability of isolines/mutants lacking the Al exclusion/Al accumulation phenotypes.

## Acknowledgments

We wish to thank the gene bank of the Leibniz Institute of Plant Genetics and Crop Plant Research and the Deutsche Saatveredelung AG for providing the seeds.

## Author Contributions

Benjamin Klug and Thomas Kirchner conducted the experiments, evaluated and presented the results. Walter Horst initiated the research and provided the necessary research support and guidance. Benjamin Klug and Walter Horst wrote the manuscript with the support of Thomas Kirchner.

## Conflict of Interest

The authors declare no conflict of interest.

## References

1.  Von Uexküll, H.R.; Mutert, E. Global extent, development and economic impact of acid soils. *Plant Soil* **1995**, *171*, 1–15.
2.  Kochian, L.V.; Hoekenga, O.A.; Piñeros, M.A. How do crop plants tolerate acid soils? Mechanisms of aluminum tolerance and phosphorus efficiency. *Annu. Rev. Plant Biol.* **2004**, *55*, 459–493.

3.  Eticha, D.; Thé, C.; Welcker, C.; Narro, L.; Staß, A.; Horst, W.J. Aluminium-induced callose formation in root apices: Inheritance and selection trait for adaptation of tropical maize to acid soils. *Field Crops Res.* **2005**, *93*, 252–263.

4.  Welcker, C.; Thé, C.; Andréaub, B.; De Leon, C.; Parentoni, S.N.; Bernal, J.; Félicité, J.; Zonkeng, C.; Salazar, F.; Narro, L.; *et al*. Heterosis and combining ability for maize adaptation to tropical acid soils. *Crop Sci.* **2005**, *45*, 2405–2413.

5.  Kochian, L.V.; Pineros, M.A.; Liu, J.; Magalhaes, J.V. Plant adaptation to acid soils: The molecular basis for crop aluminium resistance. *Annu. Rev. Plant Biol.* **2015**, *66*, 571–598.

6.  Magalhaes, J.V. How a microbial drug transporter became essential for crop cultivation on acid soils: Aluminium tolerance conferred by the multidrug and toxic compound extrusion (MATE) family. *Ann. Bot.* **2010**, *106*, 199–203.

7.  Ryan, P.R.; Tyerman, S.D.; Sasaki, T.; Furuichi, T.; Yamamoto, Y.; Zhang, W.H.; Delhaize, E. The identification of aluminium resistance genes provides opportunities for enhancing crop production on acid soils. *J. Exp. Bot.* **2011**, *62*, 9–20.

8.  Ma, J.F. Role of organic acids in detoxification of aluminum in higher plants. *Plant Cell Physiol.* **2000**, *41*, 383–390.

9.  Delhaize, E.; Gruber, B.D.; Ryan, P.R. The roles of organic anion permeases in aluminium resistance and mineral nutrition. *FEBS Lett.* **2007**, *581*, 2255–2262.

10.  Schmohl, N.; Horst, W.J. Pectin methylesterase modulates aluminium sensitivity in *Zea mays* and *Solanum tuberosum*. *Physiol. Plant.* **2000**, *109*, 419–427.

11.  Eticha, D.; Staß, A.; Horst, W.J. Cell-wall pectin and its degree of methylation in the maize root-apex: Significance for genotypic differences in aluminium resistance. *Plant Cell Environ.* **2005**, *28*, 1410–1429.

12.  Yang, J.L.; Zhu, X.F.; Peng, Y.X.; Zheng, C.; Li, G.X.; Liu, Y.; Shi, Y.Z.; Zheng, S.J. Cell wall hemicellulose contributes significantly to aluminum adsorption and root growth in Arabidopsis. *Plant Physiol.* **2011**, *155*, 1885–1892.

13.  Zhu, X.F.; Lei, G.J.; Wang, Z.W.; Shi, Y.Z.; Braam, J.; Li, G.X.; Zheng, S.J. Coordination between apoplastic and symplastic detoxification confers plant aluminum resistance. *Plant Physiol.* **2013**, *162*, 1947–1955.

14.  Ma, J.F.; Chen, Z.C.; Shen, R.F. Molecular mechanisms of Al tolerance in gramineous plants. *Plant Soil* **2014**, *381*, 1–12.

15.  Jansen, S.; Broadley, M.R.; Robbrecht, E.; Smets, E. Aluminum hyperaccumulation in angiosperms: A review of its phylogenetic significance. *Bot. Rev.* **2002**, *68*, 235–269.

16.  Metali, F.; Salim, K.A.; Burslem, D.F.R.P. Evidence of foliar aluminium accumulation in local, regional and global datasets of wild plants. *New Phytol.* **2012**, *193*, 637–649.

17.  Delhaize, E.; Ma, J.M.; Ryan, P.R. Transcriptional regulation of aluminium tolerance genes. *Trends Plant Sci.* **2012**, *17*, 341–348.

18.  Zheng, S.J.; Ma, J.F.; Matsumoto, H. High aluminum resistance in buckwheat: I. Al-induced specific secretion of oxalic acid from root tips. *Plant Physiol.* **1998**, *117*, 745–751.

19.  Ma, J.F.; Hiradate, S.; Matsumoto, H. High aluminum resistance in buckwheat: II. Oxalic acid detoxifies aluminum internally. *Plant Physiol.* **1998**, *117*, 753–759.

20. Peng, X.; Yu, L.; Yang, C.; Liu, Y. Genotypic difference in aluminum resistance and oxalate exudation of buckwheat. *J. Plant Nutr.* **2003**, *26*, 1767–1777.

21. Yang, J.L.; Zhu, X.F.; Zheng, C.; Zhang, Y.J.; Zheng, S.J. Genotypic differences in Al resistance and the role of cell-wall pectin in Al exclusion from the root apex in *Fagopyrum tataricum*. *Ann. Bot.* **2011**, *107*, 371–378.

22. Shen, R.; Ma, J.F. Distribution and mobility of aluminium in an Al-accumulator, *Fagopyrum esculentum* Moench. *J. Exp. Bot.* **2001**, *52*, 1683–1687.

23. Wang, H.; Chen, R.F.; Iwashita, T.; Shen, R.F.; Ma, J.F. Physiological characterization of aluminum tolerance and accumulation in tartary and wild buckwheat. *New Phytol.* **2015**, *205*, 273–279.

24. Klug, B.; Horst, W.J. Spatial characteristics of aluminium uptake and translocation in roots of buckwheat (*Fagopyrum esculentum* Moench). *Pysiol. Plant.* **2010**, *139*, 181–191.

25. Klug, B.; Horst, W.J. Oxalate exudation into the root-tip water free space confers protection from Al toxicity and allows Al accumulation in the symplast in buckwheat (*Fagopyrum esculentum*). *New Phytol.* **2010**, *187*, 380–391.

26. Klug, B.; Specht, A.; Horst, W.J. Aluminium localization in root tips of the aluminium-accumulating plant species buckwheat (*Fagopyrum esculentum* Moench). *J. Exp. Bot.* **2011**, *62*, 5453–5462.

27. Siebrecht, S.; Tischner, R. Changes in the xylem exudate composition of poplar (*Populus tremula* x *P. alba*)- dependent on the nitrogen and potassium supply. *J. Exp. Bot.* **1999**, *50*, 1797–1806.

28. Lombi, E.; Zhao, F.J.; McGrath, S.P.; Young, S.D.; Sacchi, G.A. Physiological evidence for a high affinity cadmium transporter highly expressed in a *Thlaspi caerulescens* ecotype. *New Phytol.* **2001**, *149*, 53–60.

29. Jansen, S.; Watanabe, T.; Smets, E. Aluminium accumulation in leaves of 127 species in *Melastomataceae*, with comments on the order *Myrtales*. *Ann. Bot.* **2002**, *90*, 53–64.

30. You, J.F.; He, Y.F.; Yang, J.L.; Zheng, S.J. A comparison of aluminum resistance among *Polygonum* species originating on strongly acidic and neutral soils. *Plant Soil* **2005**, *276*, 143–151.

31. White, P.J. Al's families: The phylogeny of aluminium accumulation in angiosperms. *Trends Plant Sci.* **2002**, *7*, 526.

32. Pavlovic, J.; Samardzic, J.; Maksimović, V.; Timotijevic, G.; Stevic, N.; Laursen, K.H.; Hansen, T.H.; Husted, S.; Schjoerring, J.K.; Liang, Y.; *et al.* Silicon alleviates iron deficiency in cucumber by promoting mobilization of iron in the root apoplast. *New Phytol.* **2013**, *198*, 1096–1107.

33. White, M.C.; Baker, F.D.; Dédaldéchamp, F.; Gaymard, F.; Guerinot, M.; Briat, J.-F.; Curie, C. Metal complexes in xylem fluid: II. Theoretical equilibrium model and computational computer program. *Plant Physiol.* **1981**, *67*, 301–310.

34. Tiffin, L.O. Iron translocation I: Plant culture, exudate sampling, iron-citrate analysis. *Plant Physiol.* **1966**, *45*, 280–283.

35. Ma, J.F.; Hiradate, S. Form of aluminium for uptake and translocation in buckwheat (*Fagopyrum esculentum* Moench) *Planta* **2000**, *211*, 355–360.

36. Yokosho, K.; Yamaji, N.; Ma, J.F. Global transcriptome analysis of Al-induced genes in an al-accumulating species, common buckwheat (*Fagopyrum esculentum* Moench). *Plant Cell Physiol.* **2014**, *55*, 2077–2091.

37. Zhu, H.; Wang, H.; Zhu, Y.; Zou, J.; Zhao, F.-J.; Huang, C.-F. Genome-wide transcriptomic and phylogenetic analyses reveal distinct aluminum-tolerance mechanisms in the aluminum-accumulating species buckwheat (*Fagopyrum tataricum*). *BMC Plant Biol.* **2015**, *15*, 16.

38. Naumann, A.; Horst, W.J. Effect of aluminium supply on aluminium uptake, translocation and blueing of *Hydrangea macrophylla* (Thunb.) Ser. cultivars in a peat-clay substrate. *J. Hort. Sci. Biotech.* **2003**, *78*, 463.

39. Durrett, T.P.; Gassmann, W.; Rogers, E.E. The FRD3-mediated efflux of citrate into the root vasculature is necessary for efficient iron translocation. *Plant Physiol.* **2007**, *144*, 197–205.

40. Yokosho, K.; Yamaji, N.; Ueno, G.; Mitani, N.; Ma, J.F. OsFRDL1 is a citrate transporter required for efficient translocation of iron in rice. *Plant Physiol.* **2009**, *149*, 297–305.

41. Fujii, M.; Yokosho, K.; Yamaji, N.; Saisho, D.; Yamane, M.; Takahashi, H.; Sato, K.; Nakazono, M., Ma, J.F. Acquisition of aluminium tolerance by modification of a single gene in barley. *Nat. Commun.* **2012**, *3*, 713.

42. Horst, W.J.; Wang, Y.; Eticha, D. The role of the root apoplast in aluminium-induced inhibition of root elongation and in aluminium resistance of plants: A review. *Ann. Bot.* **2010**, *106*, 185–197.

43. Reyna-Llorens, I.; Corrales, I.; Poschenrieder, C.; Barcelo, J.; Cruz-Ortega, R. Both aluminum and ABA induce the expression of an ABC-like transporter gene (*FeALS3*) in the Al-tolerant species *Fagopyrum esculentum*. *Environ. Exp. Bot.* **2015**, *111*, 74–82.

44. Li, J.-Y.; Liu, J., Dong, D.; Jia, X., McCouch, S.R.; Kochian, L.V. Natural variation underlies alterations in Nramp aluminum transporter (NRAT1) expression and function that play a key role in rice aluminum tolerance. *PNAS* **2014**, *111*, 6503–6508.

45. Kerven, G.L.; Edwards, D.G.; Asher, C.J.; Hallman, P.S.; Kokot, S. Aluminium determination in soil solution. II. Short-term colorimetric procedures for the measurement of inorganic monomeric aluminium in the presence of organic acid ligands. *Aust. J. Soil Res.* **1989**, *27*, 91–102.

**3**

# Heavy Metals in Crop Plants: Transport and Redistribution Processes on the Whole Plant Level

**Valérie Page and Urs Feller ***

Institute of Plant Sciences, University of Bern, Bern 3012, Switzerland;
E-Mail: valerie_page1@yahoo.fr

*  Author to whom correspondence should be addressed; E-Mail: urs.feller@ips.unibe.ch

Academic Editor: Gareth Norton

**Abstract:** Copper, zinc, manganese, iron, nickel and molybdenum are essential micronutrients for plants. However, when present in excess they may damage the plant or decrease the quality of harvested plant products. Some other heavy metals such as cadmium, lead or mercury are not needed by plants and represent pollutants. The uptake into the roots, the loading into the xylem, the acropetal transport to the shoot with the transpiration stream and the further redistribution in the phloem are crucial for the distribution in aerial plant parts. This review is focused on long-distance transport of heavy metals via xylem and phloem and on interactions between the two transport systems. Phloem transport is the basis for the redistribution within the shoot and for the accumulation in fruits and seeds. Solutes may be transferred from the xylem to the phloem (e.g., in the small bundles in stems of cereals, in minor leaf veins). Nickel is highly phloem-mobile and directed to expanding plant parts. Zinc and to a lesser degree also cadmium are also mobile in the phloem and accumulate in meristems (root tips, shoot apex, axillary buds). Iron and manganese are characterized by poor phloem mobility and are retained in older leaves.

**Keywords:** heavy metals; micronutrients; pollutants; transport; xylem; phloem; redistribution

# 1. Heavy Metals: Micronutrients or Pollutants?

Several heavy metals (e.g., Fe, Mn, Cu, Zn, Ni, and Mo) are required in traces as micronutrients by plants [1–3], while others (e.g., Cd, Cr, Hg, and Pb) are known as pollutants [4–12]. Major functions of heavy metals in plant metabolism are based on their involvement in oxidation/reduction processes [1,2,13]. Fe is essential as central ion in heme proteins (e.g., in cytochromes, nitrate reductase, catalase, and peroxidase), in siroheme proteins (e.g., nitrite reductase and sulfite reductase), in iron-sulfur proteins (e.g., ferredoxin) and in other iron-containing proteins (e.g., lipoxygenases) [1,2,13–16]. Ferritin is located in plastids and represents important intracellular storage form for Fe [17–19]. Mn is essential for the oxygen evolution in photosystem II and for a series of enzymatic reactions (e.g., phosphoenolpyruvate carboxykinase, and superoxide dismutase) [1,2,13,20–22]. Cu is present in the plastidial plastocyanin, in the mitochondrial cytochrome c oxidase, in Cu-Zn superoxide dismutase as well as in a series of other proteins [1,2,13,23–26]. Zn is essential for several enzymes (e.g., metalloproteinase, carbonic anhydrase, and Cu-Zn superoxide dismutase) [1,2,13,26–30]. Ni is known for its involvement in urease activity [1,2,13,14,31–34]. Mo (present in the soil solution as molybdate anion) is part of the molybdenum cofactor, which is required for nitrate reductase or xanthine dehydrogenase activity [1,2,13,14,35–37]. Co is an interesting element in this context, since it is not required by the higher plant itself, but is essential for the microorganisms involved in symbiotic nitrogen fixation [2,38,39]. Even those heavy metals required as micronutrients by all plants may become pollutants at elevated levels [2,40,41]. This review is focused on long-distance transport of heavy metals in crop plants, since the redistribution via xylem and phloem is important for the micronutrient supply of various plant parts and also for the accumulation of undesirably large quantities damaging the plant or negatively influencing the quality of harvested plant products.

The importance of heavy metal homeostasis in plants is illustrated in Figure 1. Heavy metals present at elevated levels may cause an increased production of reactive oxygen species (ROS) in plant cells on one hand [42–51] and may be involved in the enzymatic detoxification of ROS on the other hand [2,3,42,52–54]. Stresses (e.g., drought, heat and high light intensity) may also cause an accumulation of ROS and of ROS-damaged cell constituents [54,55]. These findings underline the importance of heavy metal homeostasis for plant cells. Several forms of superoxide dismutases are present in plants: Cu/Zn superoxide dismutases (usually present under several forms) were reported to be present in various subcellular compartments, while Mn superoxide dismutase isoforms were localized in the mitochondria and Fe superoxide dismutase was reported to be a plastidial enzyme [1,2,54]. Especially Mn superoxide dismutase activity was found to be increased under drought stress [54].

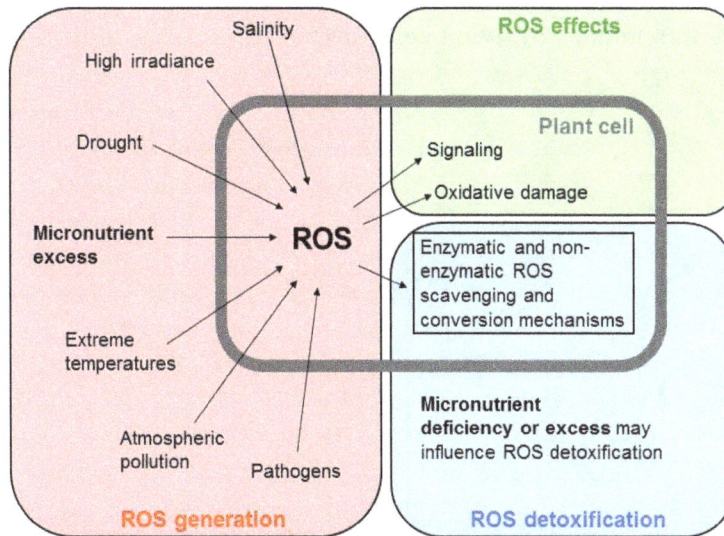

**Figure 1.** Involvement of micronutrients (e.g., Fe, Mn, Zn, Cu and Ni) in the generation and detoxification of ROS in plant cells under abiotic and biotic stress. Micronutrient deficiency and excess influence ROS levels and as a consequence damages caused by them.

## 2. Transport with the Transpiration Stream in the Xylem

Numerous processes in the soil, which are not subject of this review, may influence heavy metal solubility and as a consequence the availability for plants [2]. In some cases, heavy metals (e.g., Co, Cr, and Fe) are retained in the roots and only a minor portion reaches the shoot [56–61]. This retention can be due to insolubilization (e.g., at the root surface and in the root apoplast) [56,57] or to a compartmentation in cells avoiding the release to the xylem [58,59]. Oxygen released from the roots may cause oxidation and insolubilization of iron in the root apoplast of rice grown in waterlogged soil [57].

A strong retention in the roots was detected in wheat and lupin for Co [60,61] and in lupin also for Cd [60]. These elements were present in soluble form, but only a minor portion reached the vascular cylinder suggesting that compartmentation played a major role [58–60]. The abundance of ligands and the formation of heavy metal complexes with organic acids [62,63], with phytochelatin [64] or with nicotianamine [65] were found to be important for the retention in the roots. Good evidence was presented for the involvement of Ni-histidine complexes in the vacuolar compartmentation of Ni and as a consequence in the retention of this heavy metal in roots [59]. A heavy metal ATPase was suggested to be involved in Cd accumulation in vacuoles of root cells causing Cd retention in roots and decreasing the transport to the shoot [66].

Heavy metals are transported with the transpiration stream in the xylem from the roots to transpiring shoot parts (e.g., photosynthesizing leaves) [60,61,67]. After the release into the root xylem, free or chelated ions flow with the xylem sap upwards (Figure 2). Important for the concentration of heavy metals in the transpiration stream are xylem loading in the roots, interactions with cell walls during acropetal transport and selective removal from the xylem sap [65,66,68,69]. If there would be no further redistribution, the heavy metals would accumulate primarily in photosynthetically active (transpiring) leaves. Such an accumulation with no or only a minor redistribution was observed for the micronutrient Mn as well as for the macronutrient Ca and to a minor extent for Fe (Figures 2 and 3) [2,67–70]. Although the xylem vessels are dead, membranes of living cells around the vessels allow a selective removal of ions from the xylem sap [68,69].

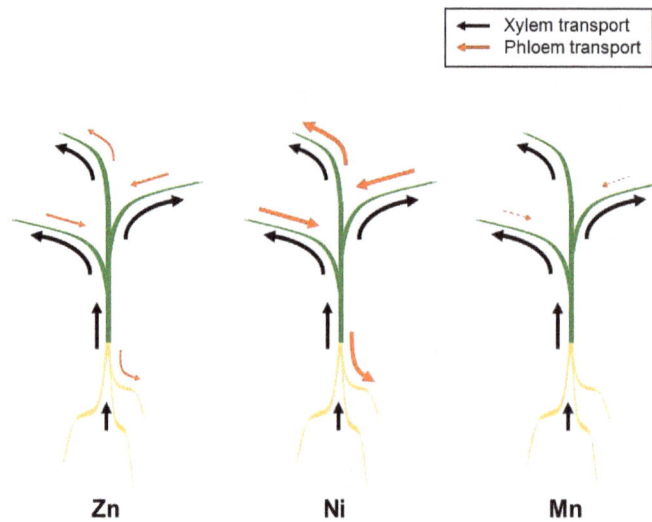

**Figure 2.** Xylem and phloem transport in the intact plant. Zn, Ni and Mn are readily transported via xylem to the shoot. Mn is essentially immobile in phloem, while Ni is rapidly redistributed to the youngest (expanding) plant parts and Zn is more slowly redistributed via the phloem (accumulation in meristems, but also well mobile in the phloem).

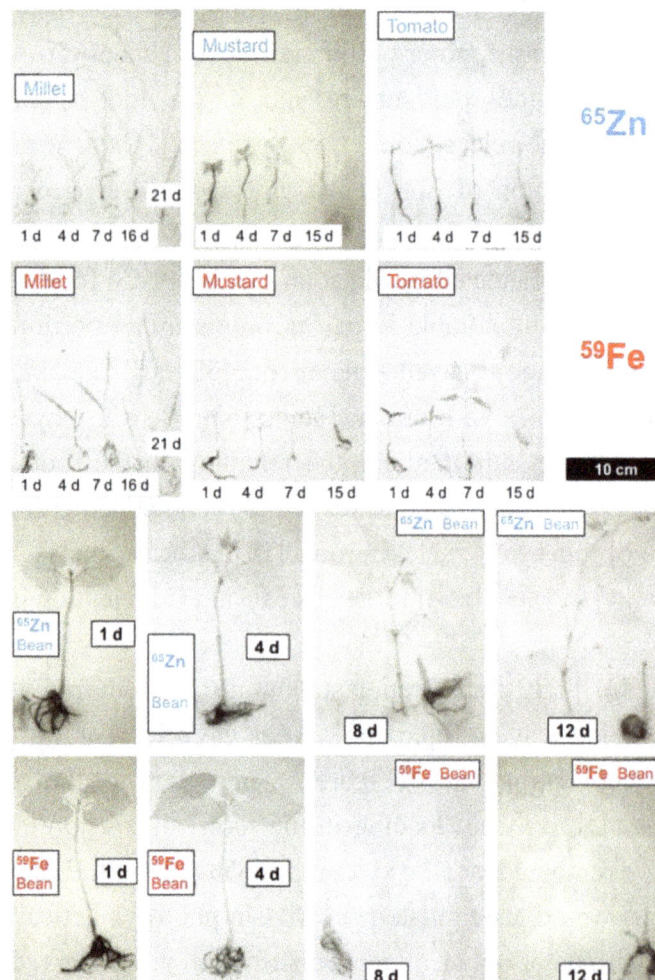

**Figure 3.** Autoradiographs documenting $^{65}Zn$ and $^{59}Fe$ distribution and redistribution in millet, mustard, tomato and dwarf bean. Young plants were labeled for one day with nutrient solution containing $^{65}Zn$ or $^{59}Fe$. The roots of the labeled plants were rinsed several times with unlabeled nutrient solution and then incubated for up to 12 days (bean), 15 days (mustard, tomato) or 21 days (millet). Plants collected throughout the incubation period were dried, fixed on cardboard and then exposed in darkness to an X-ray film. The relative distribution of the label in a plant must be considered and not the absolute label content in a particular plant part, since initial labeling was not identical.

## 3. Redistribution via the Phloem

The symplastic transport via the phloem allows a redistribution of nutrients, assimilates and pollutants within the plants and depends on the actual source/sink network (Figure 2) [71–73]. Phloem mobility of heavy metals varies in a wide range. Ni is highly phloem-mobile and can be repeatedly redistributed in crop plants throughout vegetative growth and the reproductive phase (Figure 2) [60,61,69]. Zn is also mobile in plants and can be transported via the phloem to growing plant parts (Figure 2) [60,61,68,69]. Cd (a pollutant and not a nutrient for plants) is less phloem-mobile than Zn, but it is also to some extent redistributed via the phloem [60,61,69]. An extremely poor redistribution via the phloem was found for Mn and Fe [60,61,69]. Although these two elements are micronutrients, they are not easily transported from mature organs to major sinks (e.g., developing roots, emerging leaves, maturing fruits). Distribution patterns for several heavy metals are summarized in Figure 4. Approaches to identify redistribution processes via the phloem include balance sheets for contents in various organs [60,61,70,74], introduction of radiolabeled heavy metals into a defined leaf via a flap [69] and phloem interruption (e.g., by steam-girdling) [68,69,74].

Source strength and sink strength are important for the direction and the velocity of phloem sap flow, while phloem loading is essential for the transported compounds [71–73]. The delivery through plasmodesmata in symplastic phloem loaders and the uptake through the membrane of the sieve tube-companion cell complex are crucial for the composition of the phloem sap [71–73]. Within the shoot, heavy metals can be redistributed from senescing leaves via the phloem to sinks (e.g., growing vegetative parts and maturing fruits). Another possibility is the transfer to the phloem before the xylem sap reaches mesophyll cells [2,68,69]. Several types of metal-binding compounds including nicotianamine and phytochelatins were reported to be relevant for the transport of heavy metals in the phloem [75–77]. A transporter for the copper–nicotianamine complex was found to be located in the phloem of rice leaves and was proposed to be important for the translocation of Cu from leaves to developing organs and maturing seeds [78].

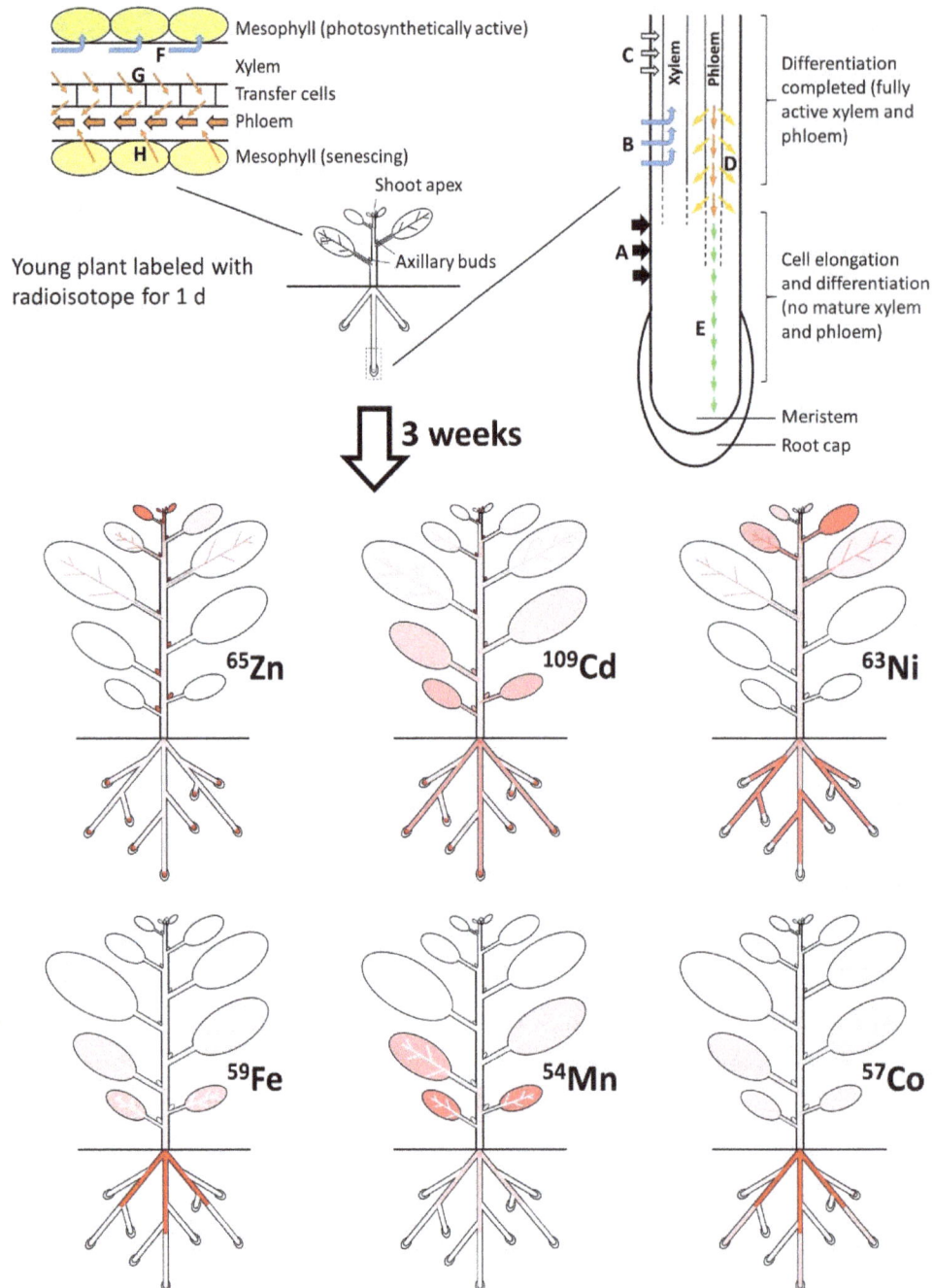

**Figure 4.** Scheme summarizing the processes involved in the distribution and redistribution of heavy metals in plants. The relative distribution patterns (red color) of $^{65}$Zn, $^{109}$Cd, $^{63}$Ni, $^{59}$Fe, $^{54}$Mn and $^{57}$Co are based on autoradiographs with various plant species emphasizing important findings. The following processes are involved in generating these radionuclide-specific distribution patterns: insolubilization at the roots surface and in the root apoplast (**A**); uptake and loading into the root xylem (**B**); uptake and retention in root cells (**C**); basipetal transport to the roots via the phloem (**D**); cell-to-cell transport in meristems and regions without functional phloem (**E**); transport with the transpiration stream in the xylem to mesophyll cells (**F**); transfer from xylem to phloem via transfer cells in small bundles (**G**); and remobilization from senescing leaves and transport to sinks via the phloem (**H**).

The redistribution from expanded leaves to sinks represents besides the flux from the roots to the shoot an additional control mechanism for the heavy metal contents in emerging organs and maturing fruits and seeds [68,70,74,79]. Fe and Mn are reduced in waterlogged soil to the well water-soluble $Fe^{++}$ and $Mn^{++}$ ions increasing for these two elements the concentration in the soil solution [70]. As a consequence, higher Fe and Mn contents can be detected in wheat shoots under such conditions [70]. The contents are drastically increased in leaves and glumes, while the grains are far less affected [70]. These findings indicate that the control of redistribution processes via the phloem is important for the composition of harvested grains [70]. Elevated concentrations of Zn, Ni, Co or Cd cause somewhat increased contents in wheat grains, but leaves and glumes are far more affected suggesting again a control of heavy metal delivery to the grains via the phloem [68,80,81].

## 4. Xylem/Phloem Interactions

Xylem and phloem are two different long-distance transport systems with different properties. Functional xylem and phloem are often separated by a cambium and not yet fully differentiated elements (e.g., in stems of dicotyledonous plants or in gymnosperms) [82]. Rays may allow a xylem–phloem exchange of solutes [84]. Functional xylem and phloem elements are close together and may interact more directly in the small stem bundles of cereals or in minor leaf veins without cambium and without formation of secondary vascular tissues [82,83]. This arrangement allows the selective transfer of solutes from the xylem to the phloem [2,84]. Transfer cells with a proliferated cell surface are presumably key players for the selective xylem-to-phloem transfer of heavy metals and other solutes (Figure 4) [2,84,85]. From steam-girdling experiments and heavy metal balance sheets it became evident that in the stem of wheat some heavy metals (e.g., the micronutrients Zn and Ni and the pollutant Cd) are transferred from the xylem to the phloem, while other heavy metals (e.g., the micronutrient Mn or Fe) are not or only much less efficiently loaded into the phloem [68,69,74,80]. Since the transfer of $^{65}$Zn from the xylem to the phloem was also observed in the wheat peduncle without node, it can be concluded that this type of phloem loading occurs in the small bundles in internodes and that the node is not a prerequisite [68]. Such a selective xylem-to-phloem transfer (e.g., in the peduncle of wheat or in minor veins of lupin) allows a channeling of solutes to the maturing grains/seeds. In leaves of legumes labeled via the roots for a short period (e.g., for one day) and then incubated for several days in unlabeled medium clear differences between the labeling patterns with $^{65}$Zn and $^{59}$Fe can be observed (Figures 3 and 4). The veins in bean leaves are strongly labeled after $^{65}$Zn introduction, while other leaf cells contain far less $^{65}$Zn. This pattern can be explained by the transfer from the rapidly flowing xylem to the more slowly flowing phloem (Figure 4). In contrast, $^{59}$Fe is mainly present in interveinal regions, while the veins are far less labeled (Figures 3 and 4). Such a labeling pattern can be explained by the flux with the transpiration stream to minor leaf veins followed by an accumulation in non-vascular tissues (Figure 4), as was proposed previously for some amino acids [84].

Some heavy metals (e.g., Ni, Zn, Co, and Cd) may be delivered to roots by basipetal transport in the phloem (Figure 4) [69]. Afterwards they may be partially loaded again into the xylem and flow upwards with the transpiration stream. $^{63}$Ni, $^{65}$Zn and $^{109}$Cd are redistributed via the phloem and can be translocated to previously unlabeled roots and to newly formed leaves [60,61]. However, the distribution of $^{63}$Ni differs from that of $^{65}$Zn and $^{109}$Cd [60,61,86]. While $^{63}$Ni accumulates in expanding leaves and in root parts behind the meristem, $^{65}$Zn and $^{109}$Cd strongly accumulate in root tips, apical shoot meristems and axillary buds [60,61,86]. This difference might be caused by a transport of $^{63}$Ni, $^{65}$Zn and $^{109}$Cd via the phloem followed by an efficient cell-to-cell transport of $^{65}$Zn and $^{109}$Cd (but not of $^{63}$Ni) to the meristem through the not yet fully differentiated regions without a functional phloem [60,61,86].

## 5. Selective Accumulation of Heavy Metals in Harvested Plant Parts

The contents of heavy metals in plant products for human nutrition or in fodder plants are important for the quality of the harvest [87–98]. It must be distinguished between seed/grain crops such as cereals or soybean and plants of which the whole shoot is harvested such as plants used for silage or grasslands. Good phloem mobility is a prerequisite for the redistribution of heavy metals from leaves to maturing seeds. When whole shoots are used for human or animal nutrition, the total content in the shoot is very important, while redistribution via the phloem is far less relevant. In contrast, phloem transport is highly relevant when only cereal grains or seeds of dicotyledenous plants are harvested.

In certain regions a sufficient supply of heavy metals for humans (e.g., Fe and Zn) is an issue [87–91]. The uptake of heavy metals into the plants, the transport to the shoot and redistribution processes within the shoot are involved in controlling the contents in harvested products. Breeding is a key aspect in the context of biofortification of plant products with heavy metals to overcome deficiencies in humans and animals [88,90]. Furthermore, negative effects of antinutrients interfering with heavy metal availability (e.g., phytate, oxalate) must be considered when evaluating the bioavailability of essential heavy metals [90,91].

Elevated contents of heavy metals—especially of pollutants such as Cd, Cr or Pb—can also decrease the quality of harvested plant products [92–98]. Undesirably high or even toxic levels are a rather local, but nevertheless an important problem. Low uptake rates for the pollutant(s), a slow root-to-shoot transfer and a minimized redistribution to harvested shoot parts is desirable for crops grown on polluted soil. Contaminated irrigation water, the use of sewage sludge for fertilization, the release of pollutants from industrial processes and an improved solubilization of some heavy metals under extreme climatic conditions (e.g., flooding) may cause undesirably high concentrations in plant products [93–98]. Even when the input of heavy metals in contaminated soil is drastically lowered, fields may still contain high heavy metal contents for decades. A cautious selection of crop species and varieties, optimized nutrient supply, and phytoremediation of the field may contribute to improve the situation.

## 6. Relevance of Long-Distance Transport for Phytoremediation

Hyperaccumulators of heavy metals are important for phytoremediation of polluted soils [99–101]. Such plants should have properties clearly differing from desirable properties of crop plants. Hyperaccumulators should efficiently take up heavy metals from the soil, should transport them to the shoot and not retain them in the belowground plant parts. Furthermore, they should tolerate high contents without showing symptoms of toxicity [99–104]. Cd, Co and Zn are efficiently transported from the roots to the shoot of the hyperaccumulator *Solanum nigrum* and are further redistributed within the shoot via the phloem [103]. Rapid uptake and release into the xylem are key properties of this species, while the redistribution via the phloem would be less important when whole shoots would be collected for phytoremediation [103].

Ligands were found to be important in hyperaccumulators for vacuolar sequestration and for the transport of heavy metals in the xylem [59,62,65]. Genetic engineering of the metabolism may lead to increased concentrations of ligands important for xylem loading and to decreased levels of ligands leading to vacuolar sequestration causing a retention of heavy metals in the roots [59,62,65,105,106]. Transporters for heavy metals or heavy metal complexes represent other targets for classical breeding and genetic engineering to improve properties of plants envisaged for phytoremediation [107,108].

## 7. Conclusions

Some translocators for heavy metals or heavy metal complexes are known, but we are far away from knowing the network of translocator proteins involved in the uptake into the roots, the subcellular distribution in roots and shoots, the release from the root symplast to the xylem, the uptake from the apoplast into leaf cells and the loading into the phloem for further redistribution [78,109,110]. A more complete picture of transporters for heavy metals is highly desirable and may serve as a basis for genotype selection and breeding. Genetic tools available nowadays will presumably allow a rapid progress in this field.

The chemical forms (e.g., chelating agents) in various subcellular compartments as well as during long-distance transport in xylem and phloem should be known in more detail. The availability of complexing agents in a plant cell may be relevant for the retention in the cell (e.g., transport across the tonoplast and storage in the vacuole), the transfer to neighbor cells through plasmodesmata and the release into the apoplast for xylem or phloem loading. Sensitive localization techniques for heavy metals and metabolic studies for ligand availability may serve as a basis for more detailed studies [59,62–66,71,78,111].

Often interactions between various biotic and/or abiotic stresses affect plant growth and the composition of the biomass [112,113]. High salt concentrations or climatic factors such as drought or heat should be considered in more detail in the context of global change.

It will be a challenge for agronomists and for plant breeders to select appropriate genotypes for a given environment and to breed new genotypes with desired properties [114,115]. The harvested plant parts as well as the utilization of the plant biomass and not only heavy metal availability in the soil or other environmental factors are relevant in this context. Biofortification depends on the accumulation of desired micronutrients and on avoiding high levels of undesired heavy metals in plant parts harvested for human or animal nutrition [88,91]. Selective redistribution and accumulation processes for various heavy metals are the basis for well-balanced contents and must be considered for agronomic strategies, including plant breeding, genotype selection and agronomic practices [88,91]. Criteria for seed crops, grassland plants and heavy metal hyperaccumulators differ considerably. This must be adequately considered in basic research as well as in agronomic applications and breeding programs.

## Acknowledgments

The authors thank Franzsika von Lerber for providing the previously unpublished $^{59}$Fe and $^{65}$Zn autoradiographs with various plant species. Experimental work was supported by the Institute of Plant Sciences, University of Bern.

**Author Contributions**

Valérie Page analyzed and reviewed heavy metal transport processes in cereals and legumes and contributed to writing the manuscript. Urs Feller analyzed and reviewed xylem–phloem interactions and heavy metal transport in hyperaccumulators and contributed to writing the manuscript.

**Conflicts of Interest**

The authors declare no conflict of interest.

**References**

1.   Hänsch, R.; Mendel, R.R. Physiological functions of mineral micronutrients (Cu, Zn, Mn, Fe, Ni, Mo, B, Cl). *Curr. Opin. Plant Biol.* **2009**, *12*, 259–266, doi:10.1016/j.pbi.2009.05.006.

2.   Marschner, H. *Mineral Nutrition of Higher Plants*, 2nd ed.; Academic Press: London, UK, 1995, p. 38.

3.   Grotz, N.; Guerinot, M.L. Molecular aspects of Cu, Fe and Zn homeostasis in plants. *Biochim. Biophys. Acta-Mol. Cell Res.* **2006**, *1763*, 595–608, doi:10.1016/j.bbamcr.2006.05.014.

4.   Prasad, M.N.V. Cadmium toxicity and tolerance in vascular plants. *Environ. Exp. Bot.* **1995**, *35*, 525–545, doi:10.1016/0098-8472(95)00024-0.

5.   Pal, M.; Horvath, E.; Janda, T.; Paldi, E.; Szalai, G. Physiological changes and defense mechanisms induced by cadmium stress in maize. *J. Plant Nutr. Soil Sci.* **2006**, *169*, 239–246, doi:10.1002/jpln.200520573.

6.   Liu, W.; Yang, Y.S. Li, P.J.; Zhou, Q.X.; Xie, L.J.; Han, Y.P. Risk assessment of cadmium-contaminated soil on plant DNA damage using RAPD and physiological indices. *J. Hazard. Mater.* **2009**, *161*, 878–883, doi:10.1016/j.jhazmat.2008.04.038.

7.   Rascio, N.; Vecchia, F.D.; La Rocca, N.; Barbato, R.; Pagliano, C.; Raviolo, M.; Gonnelli, C.; Gabbrielli, R. Metal accumulation and damage in rice (*cv.* Vialone nano) seedlings exposed to cadmium. *Environ. Exp. Bot.* **2008**, *62*, 267–278, doi:10.1016/j.envexpbot.2007.09.002.

8.   Singh, S.; Sinha, S.; Saxena, R.; Pandey, K.; Bhatt, K. Translocation of metals and its effects in the tomato plants grown on various amendments of tannery waste: Evidence for involvement of antioxidants. *Chemosphere* **2004**, *57*, 91–99, doi:10.1016/j.chemosphere.2004.04.041.

9.   Zhou, Z.S.; Zhao, S.; Wang, S.J.; Yang, Z.M. Biological detection and analysis of mercury toxicity to alfalfa (*Medicago sativa*) plants. *Chemosphere* **2008**, *70*, 1500–1509, doi:10.1016/j.chemosphere.2007.08.028.

10.  Cenkci, S.; Cigerci, I.H.; Yildiz, M.; Ozay, C.; Bozdag, A.; Terzi, H. Lead contamination reduces chlorophyll biosynthesis and genomic template stability in *Brassica rapa* L. *Environ. Exp. Bot.* **2010**, *67*, 467–473, doi:10.1016/j.envexpbot.2009.10.001.

11.  Pinho, S.; Ladeiro, B. Phytotoxicity by lead as heavy metal focus on oxidative stress. *J. Bot.* **2012**, *2012*, doi:10.1155/2012/369572.

12.  Sharma, D.C.; Sharma, C.; Tripathi, R.D. Phytotoxic lesions of chromium in maize. *Chemosphere* **2003**, *51*, 63–68, doi:10.1016/S0045-6535(01)00325-3.

13. Welch, R.M. Micronutrient nutrition of plants. *Crit. Rev. Plant Sci.* **1995**, *14*, 49–82, doi:10.1080/713608066.

14. Campbell, W.H. Nitrate reductase structure, function and regulation: Bridging the gap between biochemistry and physiology. *Annu. Rev. Plant Physiol. Plant Mol. Biol.* **1999**, *50*, 277–303, doi:10.1146/annurev.arplant.50.1.277.

15. Prescott, A.G.; John, P. Dioxygenases: Molecular structure and role in plant metabolism. *Annu. Rev. Plant Physiol. Plant Mol. Biol.* 1996, 47, 245–271, doi:10.1146/annurev.arplant.47.1.245.

16. Siedow, J.N. Plant lipoxygenase—Structure and function. *Annu. Rev. Plant Physiol. Plant Mol. Biol.* **1991**, *42*, 145–188, doi:10.1146/annurev.arplant.42.1.145.

17. Briat, J.F.; Lobreaux, S. Iron transport and storage in plants. *Trends Plant Sci.* **1997**, *2*, 187–193, doi:10.1016/S1360-1385(97)85225-9.

18. Briat, J.F.; Lobreaux, S.; Grignon, N.; Vansuyt, G. Regulation of plant ferritin synthesis: How and why. *Cell. Mol. Life Sci.* **1999**, *56*, 155–166, doi:10.1007/s000180050014.

19. Ravet, K.; Touraine, B.; Boucherez, J., Briat, J.F.; Gaymard, F.; Cellier, F. Ferritins control interaction between iron homeostasis and oxidative stress in *Arabidopsis. Plant J.* **2009**, *57*, 400–412, doi:10.1111/j.1365-313X.2008.03698.x.

20. Lidon, F.C.; Barreiro, M.G.; Ramalho, J.C. Manganese accumulation in rice: Implications for photosynthetic functioning. *J. Plant Physiol.* **2004**, *161*, 1235–1244, doi:10.1016/j.jplph.2004.02.003.

21. Lanquar, V.; Ramos, M.S.; Lelievre, S.; Barbier-Brygoo, H.; Krieger-Liszkay, A.; Kramer, U.; Thomine, S. Export of vacuolar manganese by AtNRAMP3 and AtNRAMP4 is required for optimal photosynthesis and growth under manganese deficiency. *Plant Physiol.* **2010**, *152*, 1986–1999, doi:10.1104/pp.109.150946.

22. Filiz, E.; Tombuloglu, H. Genome-wide distribution of superoxide dismutase (SOD) gene families in *Sorghum bicolor. Turk. J. Biol.* **2015**, *39*, 49–59, doi:10.3906/biy-1403-9.

23. Yruela, I. Copper in plants: Acquisition, transport and interactions. *Funct. Plant Biol.* **2009**, *36*, 409–430, doi:10.1071/FP08288.

24. Redinbo, M.R.; Yeates, T.O.; Merchant, S. Plastocyanin—Structural and functional analysis. *J. Bioenerg. Biomembr.* **1994**, *26*, 49–66, doi:10.1007/BF00763219.

25. Bueno, P.; Varela, J.; Gimenezgallego, G.; Delrio, L.A. Peroxisomal copper,zinc-superoxide dismutase—Characterization of the isoenzyme from watermelon cotyledons. *Plant Physiol.* **1995**, *108*, 1151–1160, doi:10.1104/pp.108.3.1151.

26. Yruela, I. Transition metals in plant photosynthesis. *Metallomics* **2013**, *5*, 1090–1109, doi:10.1039/c3mt00086a.

27. Mishra, P.; Dixit, A.; Ray, M.; Sabat, S.C. Mechanistic study of CuZn-SOD from Ipomoea carnea mutated at dimer interface: Enhancement of peroxidase activity upon monomerization. *Biochimie* **2014**, *97*, 181–193, doi:10.1016/j.biochi.2013.10.014.

28. Hacisalihoglu, G.; Hart, J.J., Wang, Y.H.; Cakmak, I.; Kochian, L.V. Zinc efficiency is correlated with enhanced expression and activity of zinc-requiring enzymes in wheat. *Plant Physiol.* **2003**, *131*, 595–602, doi:10.1104/pp.011825.

29. Delorme, V.G.R.; McCabe, P.F.; Kim, D.J.; Leaver, C.J. A matrix metalloproteinase gene is expressed at the boundary of senescence and programmed cell death in cucumber. *Plant Physiol.* **2000**, *123*, 917–927, doi:10.1104/pp.123.3.917.

30. Takatsuji, H. Zinc-finger transcription factors in plants. *Cell. Mol. Life Sci.* **1998**, *54*, 582–596, doi:10.1007/s000180050186.

31. Witte, C.P. Urea metabolism in plants. *Plant Sci.* **2011**, *180*, 431–438, doi:10.1016/j.plantsci.2010.11.010.

32. Sirko, A.; Brodzik, R. Plant ureases: Roles and regulation. *Acta Biochim. Pol.* **2000**, *47*, 1189–1195.

33. Polacco, J.C.; Freyermuth, S.K.; Gerendas, J.; Cianzio, S.R. Soybean genes involved in nickel insertion into urease. *J. Exp. Bot.* **1999**, *50*, 1149–1156, doi:10.1093/jexbot/50.336.1149.

34. Psaras, G.K.; Constantinidis, T.; Cotsopoulos, B.; Manetas, Y. Relative abundance of nickel in the leaf epidermis of eight hyperaccumulators: Evidence that the metal is excluded from both guard cells and trichomes. *Ann. Bot.* **2000**, *86*, 73–78, doi:10.1006/anbo.2000.1161.

35. Mendel, R.R.; Schwarz, G. Molybdoenzymes and molybdenum cofactor in plants. *Crit. Rev. Plant Sci.* **1999**, *18*, 33–69.

36. Mendel, R.R. Biology of the molybdenum cofactor. *J. Exp. Bot.* **2007**, *58*, 2289–2296, doi:10.1093/jxb/erm024.

37. Schwarz, G.; Boxer, D.H.; Mendel, R.R. Molybdenum cofactor biosynthesis—The plant protein Cnx1 binds molybdopterin with high affinity. *J. Biol. Chem.* **1997**, *272*, 26811–26814, doi:10.1074/jbc.272.43.26811.

38. O'Hara, G.W. Nutritional constraints on root nodule bacteria affecting symbiotic nitrogen fixation: A review. *Aust. J. Exp. Agric.* **2001**, *41*, 417–433, doi:10.1071/EA00087.

39. Jayakumar, K.; Vijayarengan, P.; Changxing, Z.; Gomathinayagam, M.; Jaleel, C.A. Soil applied cobalt alters the nodulation, leg-haemoglobin content and antioxidant status of *Glycine max* (L.) Merr. *Colloid Surf. B-Biointerfaces* **2008**, *67*, 272–275, doi:10.1016/j.colsurfb.2008.08.012.

40. Long, X.X.; Yang, X.E.; Ni, W.Z.; Ye, Z.Q.; He, Z.L.; Calvert, D.V.; Stoffella, J.P. Assessing zinc thresholds for phytotoxicity and potential dietary toxicity in selected vegetable crops. *Commun. Soil Sci. Plant Anal.* **2003**, *34*, 1421–1434, doi:10.1081/CSS-120020454.

41. Gupta, U.C.; Gupta, S.C. Trace element toxicity relationships to crop production and livestock and human health: Implications for management. *Commun. Soil Sci. Plant Anal.* **1998**, *29*, 1491–1522, doi:10.1080/00103629809370045.

42. Foyer, C.H.; Lelandais, M.; Kunert, K.J. Photooxidative stress in plants. *Physiol. Plant.* **1994**, *92*, 696–717, doi:10.1111/j.1399-3054.1994.tb03042.x.

43. Dat, J.; Vandenabeele, S.; Vranova, E.; Van Montagu, M.; Inze, D.; Van Breusegem, F. Dual action of the active oxygen species during plant stress responses. *Cell. Mol. Life Sci.* **2000**, *57*, 779–795, doi:10.1007/s000180050041.

44. Xu, X.Y.; Shi, G.X.; Wang, J.; Zhang, L.L.; Kang, Y.N. Copper-induced oxidative stress in Alternanthera philoxeroides callus. *Plant Cell Tissue Organ Cult.* **2011**, *106*, 243–251, doi:10.1007/s11240-010-9914-2.

45. Rodriguez-Serrano, M.; Romero-Puertas, M.C.; Zabalza, A.; Corpas, F.J.; Bomet, M.; del Rio, L.A.; Sandalio, L.M. Cadmium effect on oxidative metabolism of pea (*Pisum sativum* L.) roots. Imaging of reactive oxygen species and nitric oxide accumulation *in vivo*. *Plant Cell Environ.* **2006**, *29*, 1532–1544, doi:10.1111/j.1365-3040.2006.01531.x.

46. Tewari, P.K.; Kumar, P; Sharma, P.N. Antioxidant responses to enhanced generation of superoxide anion radical and hydrogen peroxide in the copper-stressed mulberry plants. *Planta* **2006**, *223*, 1145–1153, doi:10.1007/s00425-005-0160-5.

47. Lei, Y.B.; Korpelainen, H.; Li, C.Y. Physiological and biochemical responses to high Mn concentrations in two contrasting *Populus cathayana* populations. *Chemosphere* **2007**, *68*, 686–694, doi:10.1016/j.chemosphere.2007.01.066.

48. Kumar, P.; Tewari, P.K.; Sharma, P.N. Modulation of copper toxicity-induced oxidative damage by excess supply of iron in maize plants. *Plant Cell Rep.* **2008**, *27*, 399–409, doi:10.1007/s00299-007-0453-1.

49. Gajewska, E.; Sklodowska, M. Effect of nickel on ROS content and antioxidative enzyme activities in wheat leaves. *Biometals* **2007**, *20*, 27–36, doi:10.1007/s10534-006-9011-5.

50. Shi, Q.H.; Zhu, Z.J.; Xu, M.; Qian, Q.Q.; Yu, J.Q. Effect of excess manganese on the antioxidant system in *Cucumis sativus* L. under two light intensities. *Environ. Exp. Bot.* **2006**, *58*, 197–205, doi:10.1016/j.envexpbot.2005.08.005.

51. Demirevska-Kepova, K.; Simova-Stoilova, L.; Stoyanova, Z.; Holzer, R.; Feller, U. Biochemical changes in barley plants after excessive supply of copper and manganese. *Environ. Exp. Bot.* **2004**, *52*, 253–266, doi:10.1016/j.envexpbot.2004.02.004.

52. Alscher, R.G.; Erturk, N.; Heath, L.S. Role of superoxide dismutases (SODs) in controlling oxidative stress in plants. *J. Exp. Bot.* **2002**, *53*, 1331–1341, doi:10.1093/jexbot/53.372.1331

53. Gill, S.S.; Tuteja, N. Reactive oxygen species and antioxidant machinery in abiotic stress tolerance in crop plants. *Plant Physiol. Biochem.* **2010**, *48*, 909–930, doi:10.1016/j.plaphy.2010.08.016.

54. Huseinova, I.M.; Aliyeva, D.R.; Aliyev, J.A. Subcellular localization and responses of superoxide dismutase isoforms in local wheat varieties subjected to continuous soil drought. *Plant Physiol. Biochem.* **2014**, *81*, 54–60, doi:10.1016/j.plaphy.2014.01.018.

55. Jiang, M.Y.; Zhang, J.H. Water stress-induced abscisic acid accumulation triggers the increased generation of reactive oxygen species and up-regulates the activities of antioxidant enzymes in maize leaves. *J. Exp. Bot.* **2002**, *53*, 2401–2410, doi:10.1093/jxb/erf090.

56. Kosegarten, H.; Koyro, H.W. Apoplastic accumulation of iron in the epidermis of maize (*Zea mays*) roots grown in calcareous soil. *Physiol. Plant.* **2001**, *113*, 515–522, doi:10.1034/j.1399-3054.2001.1130410.x.

57. Bravin, M.N.; Travassac, F.; Le Floch, M.; Hinsinger, P.; Garnier, J.M. Oxygen input controls the spatial and temporal dynamics of arsenic at the surface of a flooded paddy soil and in the rhizosphere of lowland rice (*Oryza sativa* L.): A microcosm study. *Plant Soil* **2008**, *312*, 207–218, doi:10.1007/s11104-007-9532-x.

58. Yang, X.; Li, T.Q.; Yang, J.C.; He, Z.L.; Lu, L.L.; Memg, F.H. Zinc compartmentation in root, transport into xylem, and absorption into leaf cells in the hyperaccumulating species of *Sedum alfredii* Hance. *Planta* **2006**, *224*, 185–195, doi:10.1007/s00425-005-0194-8.

59.  Richau, K.H.; Kozhevnikova, A.D.; Seregin, I.V.; Voojis, R.; Koevoets, P.L.M.; Snith, J.A.C.; Ivanov, V.B.; Schat, H. Chelation by histidine inhibits the vacuolar sequestration of nickel in roots of the hyperaccumulator *Thlaspi caerulescens*. *New Phytol.* **2009**, *183*, 106–116, doi:10.1111/j.1469-8137.2009.02826.x.

60.  Page, V.; Weisskopf, L.; Feller, U. Heavy metals in white lupin: Uptake, root-to-shoot transfer and redistribution within the plant. *New Phytol.* **2006**, *171*, 329–341, doi:10.1111/j.1469-8137.2006.01756.x.

61.  Page, V.; Feller, U. Selective transport of zinc, manganese, nickel, cobalt and cadmium in the root system and transfer to the leaves in young wheat plants. *Ann. Bot.* **2005**, *96*, 425–434, doi:10.1093/aob/mci189.

62.  Bhatia, N.P.; Walsh, K.B.; Baker, A.J.M. Detection and quantification of ligands involved in nickel detoxification in a herbaceous Ni hyperaccumulator *Stackhousia tryonii* Bailey. *J. Exp. Bot.* **2005**, *56*, 1343–1349, doi:10.1093/jxb/eri135.

63.  Sagardoy, R.; Morales, F.; Rellen-Alvarez, R.; Abadia, A.; Abadia, J.; Lopez-Millan, A.F. Carboxylate metabolism in sugar beet plants grown with excess Zn. *J. Plant Physiol.* **2011**, *168*, 730–733, doi:10.1016/j.jplph.2010.10.012.

64.  Stolt, J.P.; Sneller, F.E.C.; Bryngellson, T.; Lundborg, T.; Schat, H. Phytochelatin and cadmium accumulation in wheat. *Environ. Exp. Bot.* **2003**, *49*, 21–28, doi:10.1016/S0098-8472(02)00045-X.

65.  Mari, S.; Gendre, D.; Pianelli, K.; Ouerdane, L.; Lobinski, R.; Briat, J.-F. Root-to-shoot long-distance circulation of nicotianamine and nicotianamine-nickel chelates in the metal hyperaccumulator *Thlaspi caerulescens*. *J. Exp. Bot.* **2006**, *57*, 4111–4122, doi:10.1093/jxb/erl184.

66.  Miyadate, H.; Adachi, S.; Hiraizumi, A.; Tezuka, K.; Nakazawa, N.; Kawamoto, T.; Katou, K.; Kodama, I.; Sakurai, K.; Takahashi, H.; *et al.* OsHMA3, a P-1B-type of ATPase affects root-to-shoot cadmium translocation in rice by mediating efflux into vacuoles. *New Phytol.* **2011**, *189*, 190–199, doi:10.1111/j.1469-8137.2010.03459.x.

67.  Page, V.; Blösch, R.M.; Feller, U. Regulation of shoot growth, root development and manganese allocation in wheat (*Triticum aestivum*) genotypes by light intensity. *Plant Growth Regul.* **2012**, *67*, 209–215, doi:10.1007/s10725-012-9679-1.

68.  Herren, T.; Feller, U. Transfer of zinc from xylem to phloem in the peduncle of wheat. *J. Plant Nutr.* **1994**, *17*, 1587–1598, doi:10.1080/01904169409364831.

69.  Riesen, O.; Feller, U. Redistribution of nickel, cobalt, manganese, zinc and cadmium via the phloem in young and maturing wheat. *J. Plant Nutr.* **2005**, *28*, 421–430, doi:10.1081/PLN-20049153.

70.  Stieger, P.A.; Feller, U. Nutrient accumulation and translocation in maturing wheat plants grown on waterlogged soil. *Plant Soil* **1994**, *160*, 87–95, doi:10.1007/BF00150349.

71.  Van Bel, A.J.E. The phloem, a miracle of ingenuity. *Plant Cell Environ.* **2003**, *26*, 125–149, doi:10.1046/j.1365-3040.2003.00963.x.

72.  Van Bel, A.J.E.; Gamalei, Y.V. Ecophysiology of phloem loading in source leaves. *Plant Cell Environ.* **1992**, *15*, 265–270.

73.  Turgeon, R.; Wolf, S. Phloem transport: Cellular pathways and molecular trafficking. *Annu. Rev. Plant Biol.* **2009**, *60*, 207–221, doi:10.1146/annurev.arplant.043008.092045.

74.  Zeller, S.; Feller, U. Long-distance transport of cobalt and nickel in maturing wheat. *Eur. J. Agron.* **1999**, *10*, 91–98, doi:10.1016/S1161-0301(98)00060-4.

75. Stephan, U.W.; Scholz, G. Nicotianamine—Mediator of transport of iron in the phloem. *Physiol. Plant.* **1993**, *88*, 522–529, doi:10.1034/j.1399-3054.1993.880318.x.

76. Hazama, K.; Nagata, S.; Fujimori, T.; Yanagisawa, S.; Yoeneyama, T. Concentrations of metals and potential metal-binding compounds and speciation of Cd, Zn and Cu in phloem and xylem saps from castor bean plants (*Ricinus communis*) treated with four levels of cadmium. *Physiol. Plant.* **2015**, *154*, 243–255, doi:10.1111/ppl.12309.

77. Mendoza-Cozatl, D.G.; Butko, E.; Springer, F.; Torpey, J.W.; Komives, E.A.; Kehr, J.; Schroeder, J.I. Identification of high levels of phytochelatins, glutathione and cadmium in the phloem sap of *Brassica napus*. A role for thiol-peptides in the long-distance transport of cadmium and the effect of cadmium on iron translocation. *Plant J.* **2008**, *54*, 249–259, doi:10.1111/j.1365-313X.2008.03410.x.

78. Zheng, L.Q.; Yamaji, N.; Yokosho, K.; Ma, J.F. YSL16 is a phloem-localized transporter of the copper-nicotianamine complex that is responsible for copper distribution in rice. *Plant Cell* **2012**, *24*, 3767–3782, doi:10.1105/tpc.112.103820.

79. Herren, T.; Feller, U. Effect of locally increased zinc contents on zinc transport from the flag leaf lamina to the maturing grains of wheat. *J. Plant Nutr.* **1996**, *19*, 379–387, doi:10.1080/01904169609365128.

80. Zeller, S.; Feller, U. Redistribution of cobalt and nickel in detached wheat shoots: effects of steam-girdling and of cobalt and nickel supply. *Biol. Plant.* **1998**, *41*, 427–434, doi: 10.1023/A:1001858728977.

81. Herren, T.; Feller, U. Influence of increased zinc levels on phloem transport in wheat shoots. *J. Plant Physiol.* **1997**, *150*, 228–231.

82. Fahn, A. *Plant Anatomy*, 3rd ed.; Pergamon Press: Oxford, UK, 1982; pp. 46–73.

83. Van Be, A.J.E. Xylem-phloem exchange via the rays—The undervalued route of transport. *J. Exp. Bot.* **1990**, *41*, 631–644, doi:10.1093/jxb/41.6.631.

84. McNeil, D.L.; Atkins, C.A.; Pate, J.S. Uptake and utilization of xylem-borne amino-compounds by shoot organs of a legume. *Plant Physiol.* **1979**, *63*, 1076–1081, doi:10.1104/pp.63.6.1076.

85. Offler, C.E.; McCurdy, D.W.; Patrick, J.W.; Talbot, M.J. Transfer cells: Cells specialized for a special purpose. *Annu. Rev. Plant Biol.* **2003**, *54*, 431–454, doi:10.1146/annurev.arplant.54.031902.134812.

86. Feller, U.; Anders, I.; Wei, S. Effects of PEG-Induced Water Deficit in *Solanum nigrum* on Zn and Ni Uptake and Translocation in Split Root Systems. *Plants* **2015**, *4*, 284–297, doi:10.3390/plants4020284.

87. Graham, R.D.; Knez, M.; Welch, R.M. How much nutritional iron deficiency in humans globally is due to an underlying zinc deficiency? *Adv. Agron.* **2012**, *115*, 1–40, doi:10.1016/B978-0-12-394276-0.00001-9.

88. Bouis, H.E.; Welch, R.M. Biofortification-A sustainable agricultural strategy for reducing micronutrient malnutrition in the global south. *Crop Sci.* **2010**, *50*, S20–S32, doi:10.2135/cropsci2009.09.0531.

89. Welch, R.M.; Graham, R.D. Agriculture: The real nexus for enhancing bioavailable micronutrients in food crops. *J. Trace Elem. Med. Biol.* **2005**, *18*, 299–307, doi:10.1016/j.jtemb.2005.03.001.

90. Welch, R.M.; Graham, R.D. Breeding for micronutrients in staple food crops from a human nutrition perspective. *J. Exp. Bot.* **2004**, *55*, 353–364, doi:10.1093/jxb/erh064.

91.  White, P.J.; Broadley, M.R. Biofortification of crops with seven mineral elements often lacking in human diets—Iron, zinc, copper, calcium, magnesium, selenium and iodine. *New Phytol.* **2009**, *182*, 49–84, doi:10.1111/j.1469-8137.2008.02738.x.

92.  Khan, M.A.; Castro-Guerrero, N.; Medoza-Cozatl, D.G. Moving toward a precise nutrition: Preferential loading of seeds with essential nutrients over non-essential toxic elements. *Front. Plant Sci.* **2014**, *5*, 51, doi:10.3389/fpls.2014.00051.

93.  Stasinos, S.; Nasopoulou, C.; Tsikrika, C.; Zabetakis, I. The bioaccumulation and physiological effects of heavy metals in carrots, onions, and potatoes and dietary implications for Cr and Ni: A review. *J. Food Sci.* **2014**, *79*, R765–R780, doi:10.1111/1750-3841.12433.

94.  Demirezen, D.; Aksoy, A. Heavy metal levels in vegetables in Turkey are within safe limits for Cu, Zn, Ni and exceeded for Cd and Pb. *J. Food Qual.* **2006**, *29*, 252–265, doi:10.1111/j.1745-4557.2006.00072.x.

95.  Murtaza, G.; Ghafoor, A.; Qadir, M.; Owens, G.; Aziz, M.A.; Zia, M.H.; Saifullah. Disposal and use of sewage on agricultural lands in Pakistan: A review. *Pedosphere* **2010**, *20*, 23–34.

96.  Jamil, M.; Zia, M.S.; Qasim, M. Contamination of agro-ecosystem and human health hazards from wastewater used for irrigation. *J. Chem. Soc. Pak.* 2010, 32, 370–378.

97.  He, Z.L.L.; Yang, X.E.; Stoffella, P.J. Trace elements in agroecosystems and impacts on the environment. *J. Trace Elem. Med. Biol.* **2005**, *19*, 125–140, doi:10.1016/j.jtemb.2005.02.010.

98.  Albering, H.J.; van Leusen, S.M.; Moonen, E.J.C.; Hoogewerff, J.A.; Kleinjans, J.C.S. Human health risk assessment: A case study involving heavy metal soil contamination after the flooding of the river Meuse during the winter of 1993–1994. *Environ. Health Perspect.* **1999**, *107*, 37–43.

99.  Kramer, U. Metal hyperaccumulation in plants. *Annu. Rev. Plant Biol.* **2010**, *61*, 517–534, doi:10.1146/annurev-arplant-042809-112156.

100. Memon, A.R.; Schroder, P. Implications of metal accumulation mechanisms to phytoremediation. *Environ. Sci. Pollut. Res.* **2009**, *16*, 162–175, doi:10.1007/s11356-008-0079-z.

101. Salt, D.E.; Smith, R.D.; Raskin, I. Phytoremediation. *Annu. Rev. Plant Physiol. Plant Mol. Biol.* **1998**, *49*, 643–668, doi:10.1146/annurev.arplant.49.1.643.

102. Clemens, S. Toxic metal accumulation, responses to exposure and mechanisms of tolerance in plants. *Biochimie* **2006**, *88*, 1707–1719, doi:10.1016/j.biochi.2006.07.003.

103. Wei, S.; Anders, I.; Feller, U. Selective uptake, distribution, and redistribution of Cd-109, Co-57, Zn-65, Ni-63, and Cs-134 via xylem and phloem in the heavy metal hyperaccumulator *Solanum nigrum* L. *Environ. Sci. Pollut. Res.* **2014**, *21*, 7624–7630, doi:10.1007/s11356-014-2636-y.

104. Kupper, H.; Lombi, E.; Zhao, F.J.; McGrath, S.P. Cellular compartmentation of cadmium and zinc in relation to other elements in the hyperaccumulator *Arabidopsis halleri*. *Planta* **2000**, *212*, 75–84, doi:10.1007/s004250000366.

105. Kramer, U.; Chardonnens, A.N. The use of transgenic plants in the bioremediation of soils contaminated with trace elements. *Appl. Microbiol. Biotechnol.* **2001**, *55*, 661–672.

106. Ferraz, P.; Fidalgo, F.; Almeida, A.; Teixeira, J. Phytostabilization of nickel by the zinc and cadmium hyperaccumulator *Solanum nigrum* L. Are metallothioneins involved? *Plant Physiol. Biochem.* **2012**, *57*, 254–260, doi:10.1016/j.plaphy.2012.05.025.

107. Rascio, N.; Navari-Izzo, F. Heavy metal hyperaccumulating plants: How and why do they do it? And what makes them so interesting? *Plant Sci.* **2011**, *180*, 169–181, doi:10.1016/j.plantsci.2010.08.016.

108. Pence, N.S.; Larsen, P.B.; Ebbs, S.D.; Letham, D.L.D.; Lasat, M.M.; Garvin, D.F.; Eide, D.; Kochian, L.V. The molecular physiology of heavy metal transport in the Zn/Cd hyperaccumulator *Thlaspi caerulescens. Proc. Natl. Acad. Sci. USA* **2000**, *97*, 4956–4960, doi:10.1073/pnas.97.9.4956.

109. Yamaji, N.; Xia, J.X.; Mitani-Ueno, N.; Yokosho, K.; Ma, J.F. Preferential delivery of zinc to developing tissues in rice is mediated by P-type heavy metal ATPase OsHMA2. *Plant Physiol.* **2013**, *162*, 927–939, doi:10.1104/pp.113.216564.

110. Kobayashi, T.; Ita, R.N.; Nishizawa, N.K. Iron deficiency responses in rice roots. *Rice* **2014**, *7*, 27, doi:10.1186/s12284-014-0027-0.

111. Colangelo, E.P.; Guerinot, M.L. Put the metal to the petal: Metal uptake and transport throughout plants. *Curr. Opin. Plant Biol.* **2006**, *9*, 322–330, doi:10.1016/j.pbi.2006.03.015.

112. Feller, U.; Vaseva, I.I. Extreme climatic events: Impacts of drought and high temperature on physiological processes in agronomically important plants. *Front. Environ. Sci.* **2014**, *2*, 39, doi:10.3389/fenvs.2014.00039.

113. Sharma, R.K.; Agrawal, M. Biological effects of heavy metals: An overview. *J. Environ. Biol.* **2005**, *26*, 301–313.

114. Welch, R.M.; Graham, R.D. Breeding crops for enhanced micronutrient content. *Plant Soil* **2002**, *245*, 205–214, doi:10.1023/A:1020668100330.

115. Yang, X.; Feng, Y.; He, Z.L.; Stoffella, P.J. Molecular mechanisms of heavy metal hyperaccumulation and phytoremediation. *J. Trace Elem. Med. Biol.* **2005**, *18*, 339–353, doi:10.1016/j.jtemb.2005.02.007.

# 4

# Response of Table Grape to Irrigation Water in the Aconcagua Valley, Chile

**Carlos Zúñiga-Espinoza [1], Cristina Aspillaga [2], Raúl Ferreyra [1] and Gabriel Selles [1,\*]**

[1] Instituto de Investigaciones Agropecuarias, INIA La Platina, Santa Rosa 1161, Santiago, Chile;
E-Mails: czuniga@inia.cl (C.Z.-E); rferreyr@inia.cl (R.F.)

[2] Roma N° 90, San Esteban, Los Andes 2120000, Chile;
E-Mail: cristina.aspillaga@gmail.com

\* Author to whom correspondence should be addressed; E-Mail: gselles@inia.cl

Academic Editor: Yantai Gan

**Abstract**: The irrigation water available for agriculture will be scarce in the future due to increased competition for water with other sectors, and the issue may become more serious due to climate change. In Chile, the table grape is only cultivated under irrigation. A five-year research program (2007–2012) was carried out in the Aconcagua Valley, the central area of grapes in Chile, to evaluate the response of table grape vines (*Vitis vinifera* L., *cv* Thompson Seedless) to different volumes of irrigation water. Four irrigation treatments were applied: 60, 88, 120 and 157% of crop evapotranspiration (ETc) during the first four years, and 40, 54, 92 and 108% of ETc in the last year. Irrigation over 90%–100% of ETc did not increase fruit yield, whereas the application of water below 90% ETc decreased exportable yield, berry size and pruning weight. For example, 60% ETc applied water reduced exportable yield by 20%, and only 40% of the berries were in the extra and large category size, while pruning weight was 30% lower in comparison to the treatment receiving more water.

**Key words:** table grape; water production function; berry size; grapevine irrigation

# 1. Introduction

Chile is one of the main exporters of table grapes in the world. There are 52,234 hectares dedicated to table grape cultivation, from the Atacama Region (30° S Latitude) to the Maule Region (36° S). The annual production of table grape was 725,000 tons in 2013/2014 [1]. The wide territorial extension of table grape cultivation means that it grows under different climatic conditions, ranging from desert in the north (30° S Lat.) to the Mediterranean climate between 30° and 40° S [2]. Annual mean precipitation varies from 22.8 mm in the north to 735 mm in the south; rainfall is concentrated mainly in the winter months [2]. Therefore, the table grape in the north must be grown under irrigation and the productivity depends upon the availability of water in spring and summer months. The Aconcagua Valley in Chile is one of the most important zones in the production of table grapes in the country, with 10,770 ha in full production annually. The most important cultivars are Thompson Seedless, Flame Seedless, Crimson Seedless and Red Globe [1]. Villagra et al. [3] measured a seasonal ETc (September to March) around 800 mm using the Eddy covariance technique. However, local farmers use a wide range of water volumes to irrigate table grapes, above or below 800 mm per growing season.

The availability of water for agriculture will be scarce in the future, due to increased competition for water with other sectors of the economy, and climatic change may lead to more recurrent drought situations [4,5]. In the area of table grape production in Chile (30° to 36° South latitude), there is evidence that precipitation has decreased in the last century; rainfall is predicted to decline 25%–35% by 2040–2070, and average temperature to increase by 2–4 °C [6,7] as a consequence of climate change. Furthermore, Chile is periodically affected by the El Niño-Southern Oscillation (ENSO), which leads to severe droughts (La Niña event) and economic losses [8]. This means less storage of water in the soil, less runoff to reservoirs and less recharging of aquifers. In addition, as a consequence of climate change the 0° isotherm will increase in altitude [6], and the area of snow reserves in the mountains for river water flow in spring and summer will decrease; therefore, irrigation water in the period of maximum crop development will be limited.

Crop production must be more efficient in the use of water [5]. Strategies are required to determine crop evapotranspiration (ETc) and irrigation needs [3], and using improved irrigation practices to reduce the quantity of water applied to crops without affecting yields or product quality [9]. Regulated deficit irrigation (RDI) and sustained deficit irrigation (SDI) have been used as strategies to reduce the volume of water applied without affecting yields [9]. With RDI, water is applied to crops below the ETc in specific phenological periods, and this technique has been shown to be successful in crops such as peach [10], olive [11,12] and wine grape [13,14].

With SDI technique, reduced water is applied to crops during the entire development period, independent of the plant physiological stage [9]. In some fruit crops, this technique has produced better results than RDI in terms of crop production and water saving [15]. There is sufficient evidence that supplying the full ETc requirements to tree crops and vines may not be necessary in many situations [16]. However, most of the experiments in vine grape have been on wine grapes and little on table grapes. Berry quality variables of table grapes differ from those of wine grapes. In table grape, berry size, firmness, color, acidity and total soluble solids are important quality parameters [17]. There are a few RDI studies with table grapes where an irrigation restriction was imposed after veraison when berries have almost reached full size [18–20]. El-Ansari et al. [19] showed that the cultivar "Muscat of

Alexandria" decreased firmness and acidity and increased total soluble solids of the berries under RDI. Ezzahouani and Williams [20] found that under different irrigation treatments the cultivar "Danlas" obtained the highest yield and berry weight under well irrigated treatments. Williams *et al.* [17,21] studied the effect of SDI on Thompson Seedless for raisin production, and concluded that application of water above 80% of the crop evapotranspiration (ETc) did not increase fruit yield, whereas below 60% ETc decreased berry yield and weight but increased soluble solids.

The aim of this study was to determine the effect of different amounts of irrigation on the yield and fruit quality of table grape in the Aconcagua Valley of Chile.

## 2. Materials and Methods

### 2.1. Experimental Site and Irrigation Treatments

The experimental site was located in a commercial table grape vineyard in the Aconcagua Valley, Valparaiso Region, Chile (70°41′23″ W, 32°47′21″ S). The soil is a Fluventic Haploxerolls, 1 m depth, with a clay loam texture in all depths. Annual rainfall and reference evapotranspiration (ETo) during the study period are presented in Table 1. The cultivar was Thompson Seedless on Freedom rootstock, trained as overhead trellis system and irrigated by drip (double line). The vineyards were planted in 2003, with a plant spacing of 3 × 2.5 m.

**Table 1.** Seasonal precipitation, seasonal reference evapotranspiration (ETo), seasonal crop evapotranspiration (ETc), applied water and percentage of ETc in each experimental year.

| Season | Annual Rain | Season ETo | Season ETc | Applied water ($m^3ha^{-1}$) | | | | Percent ETc (%) | | | |
|---|---|---|---|---|---|---|---|---|---|---|---|
| | | | | T1 | T2 | T3 | T4 | T1 | T2 | T3 | T4 |
| | (mm) | (mm) | (mm) | | | | | | | | |
| 2007/08 | 116.3 | 845.2 | 799.2 | 5279 | 7647 | 9705 | 11796 | 66 | 96 | 121 | 148 |
| 2008/09 | 242.9 | 876.4 | 741.4 | 4717 | 6388 | 9397 | 11217 | 64 | 86 | 127 | 151 |
| 2009/10 | 182.4 | 825.6 | 658.1 | 3597 | 5755 | 7865 | 10806 | 55 | 87 | 120 | 164 |
| 2010/11 | 141.8 | 870.18 | 690.18 | 3992 | 5782 | 8395 | 11498 | 58 | 84 | 122 | 167 |
| 2011/12 | 111.1 | 962.3 | 674.12 | 2663 | 3615 | 6171 | 7293 | 39 | 54 | 92 | 108 |

The experiment was performed during five years; in the four first seasons (2007/08 to 2010/11) four irrigation treatments were applied during the entire season; T1: 60% of crop evapotranspiration (ETc), T2: 88% ETc, T3: 120% ETc and T4: 157% ETc. In the last season (2011/12) less water was applied in all the treatments; 40, 54, 92 and 108% of ETc for T1, T2, T3 and T4, respectively. Each season, irrigation treatments were started on 1 October and finished on 31 March. The water applied each season and the resulting percentages of ETc are presented in Table 1. Each treatment was replicated four times in a randomized block design, each elementary plot contained 16 vines, and measurements were done only in the four central plants to avoid border effects. ETc was calculated as ETo × kc, where kc is the crop coefficient [22]. ETo was estimated by the Penman-Montheith method [22], using climatic data from an automatic weather station near the field experiment (www.agroclima.cl network). Crop coefficient (kc) was estimated following the methodology proposed by Villagra *et al.* [3] and Williams and Ayars [23]. Irrigation was scheduled in a low frequency regime as recommended by Selles *et al.* [24] for the fine-textured soil of the Aconcagua valley.

## 2.2. Soil and Plant Water Status

Soil water content and stem water potential (SWP) were measured throughout the entire season. Soil water content was measured daily with a capacitive probe (Diviner 2000, Sentek Inc., Sidney, Australia) with 10 cm increment down to the depth of 1 m. Seven access tubes were used in each treatment, placed 30 cm away from the plant row. The readings were expressed as total soil available water (SAW). To express soil water content measured with the capacitive probe as SAW, the soil was irrigated around the access tubes until field capacity (FC) was reached as proposed by Cassel and Nielsel [25]. After that, soil water content was measured with the probe and the value obtained was established as FC. In addition, the permanent wilting point was estimated as half of FC [26,27].

Midday stem water potential (SWP) was measured at midday (2–4 PM, solar time) every other week, before an irrigation event, using a pressure chamber technique [28]. Three leaves were used per replicate; the leaves were covered with an aluminized plastic bag one hour before being measured [29].

## 2.3. Vegetative Growth and Fruit Production

Each season, pruning weight was determined on four central plants per replicate (16 per treatment) and expressed as pruning dry matter. A sample of fresh pruned branches was dried at 70 °C in a forced-air oven for 48 h to determine the water content of the sample.

During each season, the intercepted solar radiation (ISR) by the vines was measured from bud break to harvest. At midday, the flux density of photosynthetically active incident radiation ($PAR_i$) over and under the orchard ($PAR_{bd}$) was measured with a ceptometer (AccuPAR, Decagon Devices, Washington, DC, USA). Data were measured in each replicate in one quadrant of four plants each. Fifteen measurements were made per quadrant; three in each row and five between rows. Mean of ISR ($\mu mol\ m^{-2}\ s^{-1}$) were expressed as percentage using:

$$ISR = \left[1 - \left(\frac{PAR_{bd}}{PAR_i}\right)\right] \times 100 \tag{1}$$

After fruit set, the number of bunches per vine and the number of berries per bunch were defined as in normal commercial table grape management. ($40 \pm 2$ bunches per vine and $113 \pm 10$ berries per bunch). At harvest, exportable fruit production was measured in the four central plants per replicate. All harvested export bunches were weighed, and a random sample of 100 berries per replicate (400 per treatment) were weighed individually. A sample of bunches was commercially packed and stored at 0 °C and 90% relative humidity for laboratory analysis; berry firmness was measured with 200 berries with attached pedicel per treatment using a FirmTech 2 apparatus (BioWorks Wamego, KS, USA), along with soluble solids, juice acidity and shatter.

*2.4. Statistical Analysis*

Data were subjected to analysis of variance using MIXED model, and mean separation was performed by the LSD method or Duncan's multiple range test where appropriate (SAS Institute Inc., Cary, NC, USA).

## 3. Results and Discussion

The volume of water applied in each treatment from bud break until the end of maturity is shown in Table 1. Winter precipitation was sufficient to maintain the soil available water (SAW) close to field capacity (FC) until bud break time each year. The irrigation treatments produced a reduction of SAW during the season (Figure 1) in the treatments which received less water (T1 and T2). Accordingly, a moderate water deficit was produced and reflected in SAW (Table 2) and SWP (Table 3).

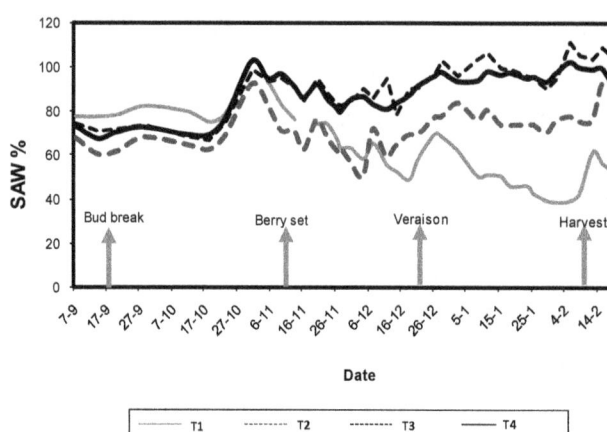

**Figure 1.** Typical variation of soil available water (SAW%), during the 2007/08 season. Arrows indicate different phenological stages.

**Table 2.** Average soil available water (SAW%) in each experimental year.

| Irrigation Treatment | 2007/08 | | 2008/09 | | 2009/10 | | 2010/11 | | 2011/12 | |
|---|---|---|---|---|---|---|---|---|---|---|
| | S-V* | V-H* | S-V | V-H | S-V | V-H | S-V | V-H | S-V | V-H |
| T1 | 63.73 | 64.4 | 60.32 | 54.45 | 86.2 | 56.21 | 90.91 | 66.68 | 83.88 | 41.77 |
| T2 | 70.9 | 73.8 | 67.29 | 67.93 | 70.9 | 63.81 | 91.55 | 72.16 | 75.65 | 48.15 |
| T3 | 79.35 | 81.75 | 75.07 | 77.69 | 85.95 | 73.91 | 80.48 | 79.14 | 84.8 | 88.85 |
| T4 | 81.22 | 93.17 | 78.36 | 81.73 | 98.46 | 88.73 | 89.59 | 84.68 | 85.73 | 82.9 |

*(S-V bud break to veraison, V-H, veraison to harvest).

**Table 3.** Average stem water potential (SWP, MPa) in each experimental season.

| Irrigation treatment | Fruit set-Veraison (MPa) | | | | | Veraison-Harvest (MPa) | | | | |
|---|---|---|---|---|---|---|---|---|---|---|
| | 2007/08 | 2008/09 | 2009/10 | 2010/11 | 2011/12 | 2007/08 | 2008/09 | 2009/10 | 2010/11 | 2011/12 |
| T1 | −0.64 c | −0.78 b | −0.73 b | −0.63 | −0.88 c | −0.71 | −1.00 b | −0.96 b | −0.83 | −1.16 c |
| T2 | −0.62 bc | −0.76 a | −0.73 b | −0.64 | −0.79 b | −0.67 | −0.88 ab | −0.83 a | −0.86 | −0.98 b |
| T3 | −0.59 ab | −0.76 a | −0.63 a | −0.62 | −0.72 a | −0.66 | −0.87 ab | −0.80 a | −0.77 | −0.81 a |
| T4 | −0.53 a | −0.68 a | −0.63 a | −0.60 | −0.68 a | −0.63 | −0.82 a | −0.77 a | −0.83 | −0.80 a |

Means followed by a different letter within a given year are significantly different at $P < 0.05$

Allen *et al.* [22] established that soil water depletion greater than 30% of SAW (<70% SAW in the soil) is a critical point for table grapes. In this study, only T1 and T2 presented SAW below 70% in the soil, mostly from veraison to harvest. In Thompson Seedless, SWP at midday below –0.9 MPa was defined as moderate water stress by Selles *et al.* [24]. Grimes and Williams [30] consider –1 Mpa as the threshold value. From this point of view, only T1 was subjected to a moderate water stress between veraison and harvest. The average water received by T1 was only 60% of ETc (2007/2008 to 2010/11) and 39% of ETc in the last season (2011/12); a severe water stress was not observed. As all treatments in each season began with the SAW close to FC (Table 2), part of the water used by plants in the T1 treatment came from the soil, preventing severe plant water stress during the season (Figure 1).

Irrigation treatments had an effect on winter pruning weight; the differences between T1 and T4 were significant in three out of the five experiment years; plants which received less water showed lower pruning weight (Table 4). A linear relationship was found between applied water (% ETc) and the relative pruning weight of the vines ($r^2$ = 0.67, Figure 2). For the same cultivar, Williams *et al.* [21] found also a linear relationship between pruning weight and SWP at midday. This relationship shows that vine pruning weight is sensitive to moderate water stress.

**Table 4.** Winter pruning dry weight (kg plant$^{-1}$) and average solar radiation intercepted by the vines (ISR, %) from veraison to harvest.

| Irrigation treatment | Pruning dry weight (kg plant$^{-1}$) | | | | | ISR (%) from veraison to harvest | | | | |
|---|---|---|---|---|---|---|---|---|---|---|
| | 2007/08 | 2008/09 | 2009/10 | 2010/11 | 2011/12 | 2007/08 | 2008/09 | 2009/10 | 2010/11 | 2011/12 |
| T1 | 1.82 b | 1.61 b | 2.47 | 2.16 | 2.11 b | 80.55 | 87.53 | 85.35 | 84.47 a | 81.7 c |
| T2 | 2.04 ab | 2.01 ab | 2.51 | 2.07 | 2.08 b | 76.93 | 84.13 | 84.34 | 83.52 a | 85.72 bc |
| T3 | 2.34 ab | 2.20 ab | 2.97 | 2.24 | 2.92 a | 84.2 | 89.58 | 89.95 | 90.54 ab | 90.54 ab |
| T4 | 2.89 a | 2.46 a | 3.18 | 2.5 | 3.17 a | 83.2 | 86.53 | 88.26 | 81.13 a | 92.53 a |

Means followed by a different letter within a given year are significantly different at *P* < 0.05.

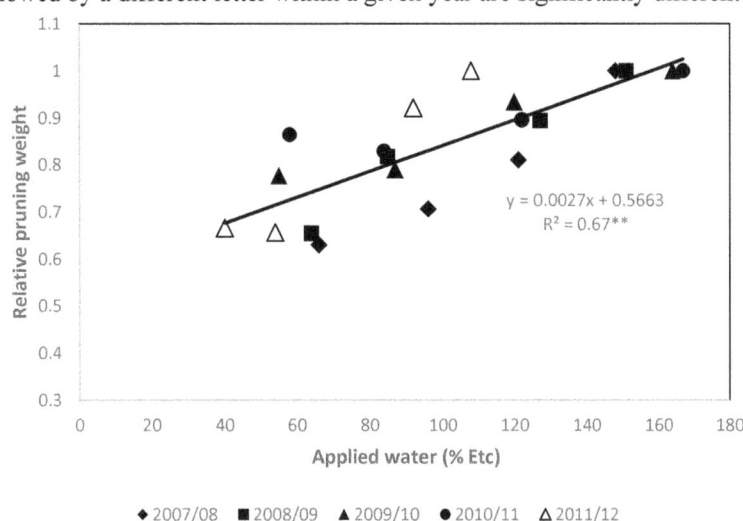

**Figure 2.** Relationship between applied water (% ETc) and relative winter pruning weight in the different experimental seasons.

ISR was similar in all treatments in four of the five years of the experiment, but in 2011/12, T1 treatment which received only 39% ETc, presented significant differences in ISR compared to the other treatments (Table 4). That year the SAW and SWP of T1 were lower than in other years between veraison and harvest (Tables 2 and 3), which clearly affected vine vegetative growth (Table 4). Similar results were found also in Thomson Seedless [21], where sustained deficit irrigation reduced leaf area and pruning weight per vine compared to vines irrigated at 100% ETc.

In 2011/2012, SWP decreased below −1 MPa, showing a moderate water stress in T1. Selles *et al.* [31] also showed that vegetative growth is affected by the amount of water applied in Crimson Seedless cultivar growing in the Aconcagua Valley. That study also reported that water applied affected not only pruning weight but also trunk growth. Willians *et al.* [21] also showed that vegetative growth (shoot length, pruning weight and leaf area) of Thomson Seedless is affected by the amount of water applied.

The volumes of water applied produced a significant decrease in the mean bunch weight in four of the five years; T1 average bunch weight was less than T3 and T4. Berry weight was also affected by the amount of water (Table 5). The percentage of extra and large berries (>5.2g) increased with increasing water applied (Figure 3).

**Table 5.** Bunch and berry weight at harvest (g) in each treatment, in five experimental seasons.

| Irrigation treatment | Bunch weight (g) | | | | | Berry weight (g) | | | | |
|---|---|---|---|---|---|---|---|---|---|---|
| | 2007/08 | 2008/09 | 2009/10 | 2010/11 | 2011/12 | 2007/08 | 2008/09 | 2009/10 | 2010/11 | 2011/12 |
| T1 | 607.5 b | 623.2 b | 578.5 b | 560.7 | 511.1 b | 4.76 c | 4.86 b | 5.19 b | 5.28 | 5.5 b |
| T2 | 650.9 ab | 676.9 ab | 618.8 ab | 625.4 | 528.0 ab | 5.08 bc | 5.02 b | 5.17 b | 5.39 | 6.0 ab |
| T3 | 674.6 ab | 729.7 a | 661.7 a | 671.8 | 549.4 ab | 5.67 ab | 5.49 a | 5.56 ab | 5.48 | 6.11 ab |
| T4 | 714.4 a | 723.2 a | 682.9 a | 631.9 | 594.0 a | 5.85 a | 5.64 a | 5.74 a | 5.44 | 6.23 a |

Means followed by a different letter within a given year are significantly different at $P < 0.05$.

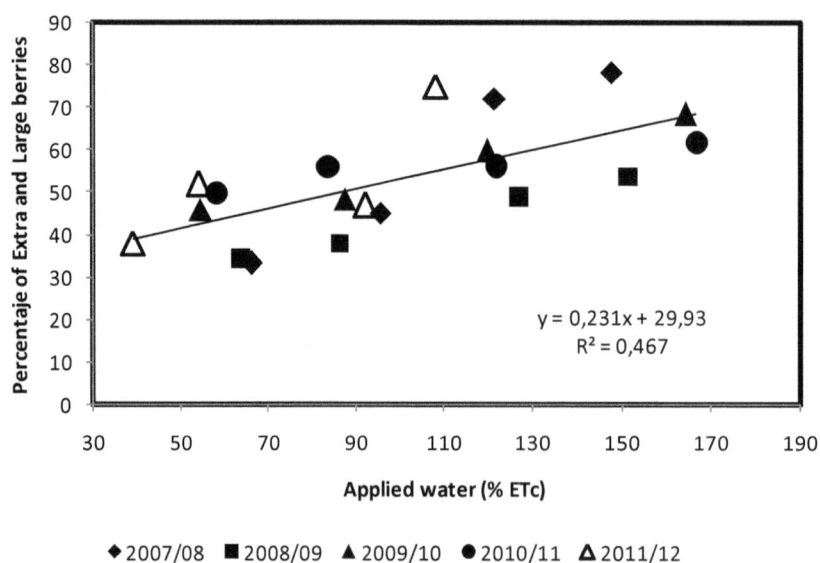

y = 0,231x + 29,93
R² = 0,467

◆ 2007/08   ■ 2008/09   ▲ 2009/10   ● 2010/11   △ 2011/12

**Figure 3.** Relationship between percentage of Extra and Large berries and percentage ETc of water applied in the different experimental seasons.

Williams *et al.* [17], using Thompson Seedless cultivar grown for raisins, found a linear relationship between SWP and berry weight at harvest; berry weight increased with increased water applied up to 80% ETc, while more water beyond 80% ETc did not produce greater berry weight. This relationship was also linear in the present study, even for greater amounts of water. The difference may be due to the fact that there are fewer berries per bunch in table grape than in raisin production, thus the berries may grow more when there is less competition within the same bunch. In table grape, berry size is a very important commercial quality component; extra and large sizes have better market prices and it is very important for the grower that most of the bunches have these berry sizes. Other quality parameters are: color (green color in the case of Thompson Seedless), berry firmness, sugar content and juice acidity. In this study, the application of less water (e.g., 40% ETc in 2011/2012) affected the percentage of green berries in bunches due to more solar light received by bunches in T1, with lower intercepted solar radiation (Figure 4). This agrees with the results of Selles *et al.* [32], who found that with less than 80% ISR there was a predominance of yellow color in this cultivar.

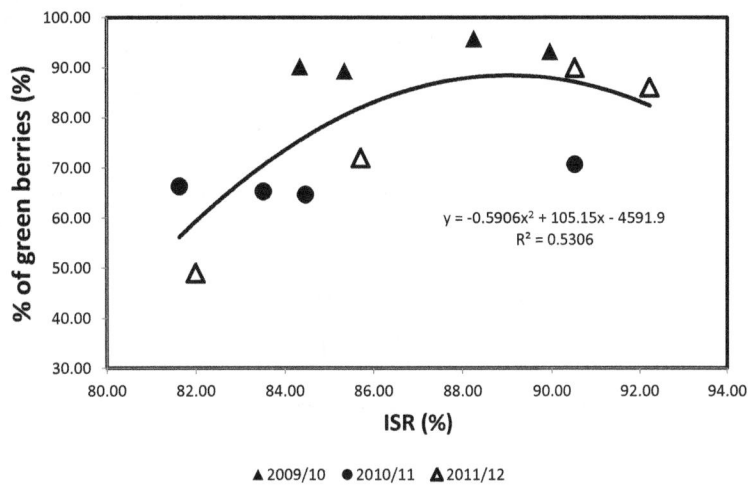

The equation shown on the chart: $y = -0.5906x^2 + 105.15x - 4591.9$, $R^2 = 0.5306$

Legend: ▲ 2009/10  ● 2010/11  △ 2011/12

**Figure 4.** Percentage of green commercial berries per bunch as a function of intercepted solar radiation (ISR, %) in the different experimental growing seasons.

The quality parameters firmness (Table 6) and sugar content (Table 7) were not affected by the irrigation treatment in any experimental year, and were in the normal range for this cultivar [33]. Williams *et al.* [17] found an increase in berry sugar with water application under 60% of ETc in Thompson Seedless for raisins. However, juice acidity was between 0.7 and 0.8 mg of tartaric acid/100 mL juice in all years and treatments; these are considered normal values for this variety [33]. Also, shattering was very low (less than 1.7%) in all treatments and all years. In summary, the only quality parameters affected by irrigation treatment were berry size and berry color (lower percentage of green berries in the bunch).

**Table 6.** Berry firmness at harvest (g mm$^{-2}$) in each treatment, in five experimental years.

| Irrigation treatment | Berry firmness (g/mm$^2$) | | | | |
|:---:|:---:|:---:|:---:|:---:|:---:|
| | 2007/08 | 2008/09 | 2009/10 | 2010/11 | 2011/12 |
| T1 | 282.1 b | 235.5 b | 262.9 | 283.3 ab | 258.2 |
| T2 | 298.4 ab | 242.2 b | 275.6 | 279.7 b | 274.7 |
| T3 | 304.6 ab | 261.8 ab | 290.4 | 320.5 a | 272.7 |
| T4 | 327.3 a | 275.3 a | 294.2 | 313.5 ab | 275.1 |

Means followed by a different letter within a given year are significantly different at $P < 0.05$.

**Table 7.** Berry sugar content at harvest (°Brix), in each treatment, in five experimental seasons.

| Irrigation treatment | Sugar content (°Brix) | | | | |
|:---:|:---:|:---:|:---:|:---:|:---:|
| | 2007/08 | 2008/09 | 2009/10 | 2010/11 | 2011/12 |
| T1 | 18.97 | 20.5 | 17.39 | 21.45 | 21.08 a |
| T2 | 18.21 | 18.23 | 17.34 | 21.92 | 21.06 a |
| T3 | 18.38 | 18.98 | 17.61 | 21.31 | 20.95 ab |
| T4 | 18.84 | 19.35 | 17.43 | 21.79 | 20.43 b |

Means followed by a different letter within a given year are significantly different at $P < 0.05$.

Finally, a production function was established correlating relative yield (actual treatment yield/maximum yield) to water applied in terms of percentage of ETc (Figure 5).

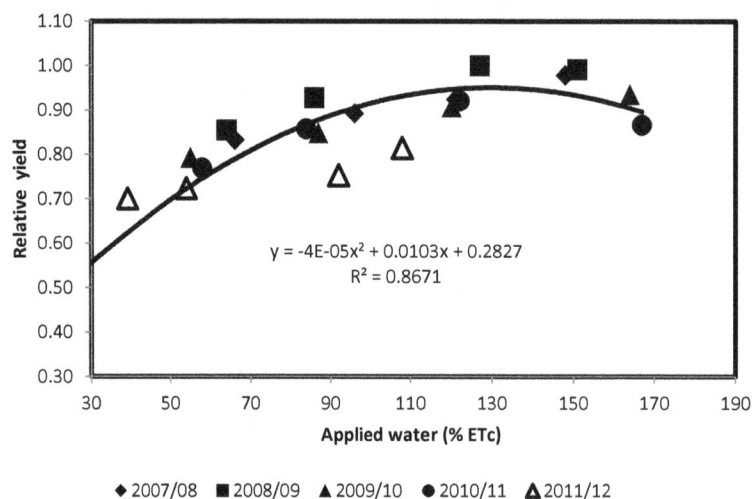

$$y = -4E\text{-}05x^2 + 0.0103x + 0.2827$$
$$R^2 = 0.8671$$

◆ 2007/08   ■ 2008/09   ▲ 2009/10   ● 2010/11   △ 2011/12

**Figure 5.** Relative yield as a function of applied water (% ETc). Relative yields represent the yield of each treatment divided by the highest yield recorded.

Relative yield increased by 30% when applied water increased from 40%–100% of ETc, and above this amount relative yield did not increase (Figure 5). Williams *et al.* [17], in Thompson Seedless destined to raisin production, found that yield increased when water applied was increased up to 80% ETc. A linear relationship between water applied and relative yield of Crimson Seedless cultivar in the Aconcagua Valley was reported by Ferreyra *et al.* [34]; the exportable production increased by 22%

when the water applied was increased from 40%–100% of ETc. Netzer *et al.* [35] in Superior cultivar in Israel found that yield decreased by 29% when water application was reduced from 100%–40% of ETc. Similar results were found by Vita *et al.* [36] in Argentina, also in the Superior cultivar. In our case, it is interesting also to consider that berry size decreased as applied water was reduced (Figure 3). That means that a reduced amount of water not only decreases total yield but the berry size, affecting commercial quality.

Water use efficiency (WUE), defined as kilograms of fresh fruit per cubic meter of applied water, increased as applied water decreased, from 2.3 (160% ETc) to 7 kg m a$^{-3}$ (40% ETc) (Figure 6). Deficit irrigation (RDI or SDI) has been proposed as one way to increase water use efficiency particularly for woody perennial crops [9]. However, in our case increasing WUE over 3.7 kg m$^{-3}$ the fruit commercial quality diminished. Thompson Seedless for table grape production has high sensitivity to water deficit when quality standards are considered. SDI below 80%–90% ETc is not recommended, at least before harvest.

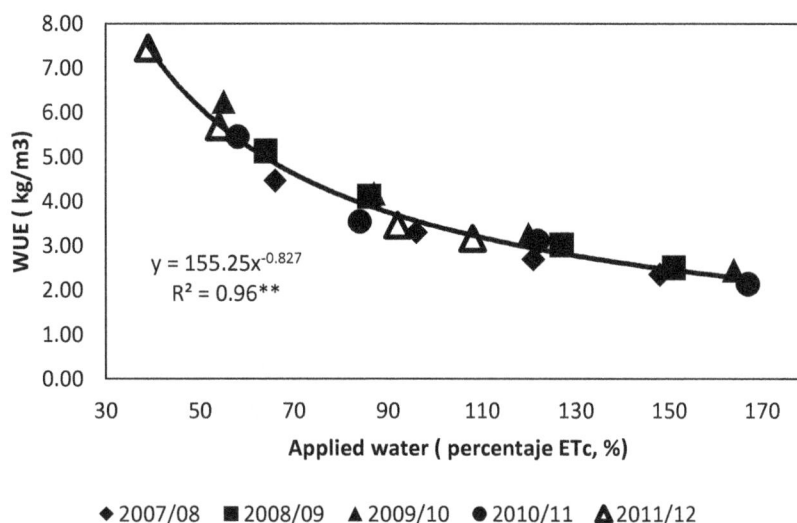

$$y = 155.25x^{-0.827}$$
$$R^2 = 0.96**$$

◆ 2007/08   ■ 2008/09   ▲ 2009/10   ● 2010/11   ▲ 2011/12

**Figure 6.** Water use efficiency (kg of fresh fruit per cubic meter of applied water, kg m$^{-3}$) as a function of applied water (% ETc).

## 4. Conclusions

Irrigation water above 90%–100% ETc did not increase fruit yield in table grapes, whereas the application of water below 90% ETc decreased exportable yield and fruit quality as reflected by smaller berry size and a greater proportion of yellow fruit. Irrigation amounts did not have a significant effect on the other quality parameters such as firmness, sugar content and juice acidity. Pruning weight was also affected when less water is applied, reducing shoot wood for future vine fructification, compromising sustainable grape production. The SDI technique on table grapes could be used as a short term strategy to avoid water scarcity, but not as a permanent or long term strategy, at least in the Thomson Seedless cultivar. In this cultivar, it is better to irrigate a smaller surface with adequate amounts of water, so the yield and quality of the fruit are not affected.

**Acknowledgements**

The authors acknowledge INNOVA-CORFO, whom financed this research (project 05-CR11PAT-11). Also the authors acknowledge Agricola El Maitenal S.A., where the research was done.

**Authors Contribution**

Gabriel Selles was responsible of the all research project and with Raúl Ferreyra were responsible for the interpretation of results and manuscript preparation. Carlos Zuñiga and Cristina Aspillaga were responsible for technician supervision and all phases of field operations, measurements and statistical analysis

**Conflicts of Interest**

The authors declare no conflict of interest.

**References**

1. CIREN. Catastro Frutícola. In *Principales Resultados Región de Valparaíso*; Centro de Información de Recursos Naturales (CIREN): Santiago, Chile, 2014; p. 44.
2. Novoa, R. and Villaseca, S. (Eds.) *Mapa Agroclimático de Chile*; Instituto de Investigaciones Agropecuarias: Santiago, Chile, 1989; p. 221.
3. Villagra, P.; García de Cortázar, V.; Ferreyra, R.; Aspillaga, C.; Zuñiga, C.; Ortega-Farias, S.; Selles, G. Estimation of water requirements and Kc values of "Thompson Seedless" table grapes grown in the overhead trellis system, using the Eddy covariance method. *Chill. J. Agric. Res.* **2014**, *74*, 213–218.
4. Laraus, J.L. The problems of sustainable water use in the Mediterranean and research requirements for agriculture. *Ann. Appl. Biol.* **2004**, *144*, 259–272.
5. Morison J.I.L.; Baker, N.R.; Mullineaux M.P.; Davies, W.J. Improving water use in crop production. *Philos. Trans. R. Soc. B* **2008**, *363*, 639–665.
6. CONAMA. *Estudio de Variabilidad Climática en Chile para el Siglo XXI*; Departamento de Geofísica, Facultad de Ciencias; Físicas y Matemáticas, Universidad de Chile: Santiago, Chile, 2006; p. 63.
7. AGRIMED. Análisis de vulnerabilidad del sector silvoagropecuario, recursos hídricos y edáficos de Chile frente a escenarios de Cambio Climático. In *Capítulo I: Impactos Productivos en el Sector Silvoagropecuario de Chile Frente a Escenarios de Cambio Climático*; Conama y Ministerio de Agricultura: Santiago, Chile, 2008; p. 181.
8. Bates, B.C.; Kundzewicz, Z.W.; Wu, S.; Palutikof, J.P. *Climate Change and Water. IPCC Technical Paper VI*; IPCC Secretariat: Geneva, Switzerland, 2008; p. 210.
9. Fereres, E.; Soriano, M.A. Deficit irrigation for reducing agricultural water use. *J. Exp. Bot.* **2007**, *58*, 147–159.

10. Ferreyra, E.R.; Selles, V.G.Y.; Lemus, S.G. Efecto del estrés hídrico durante la fase II del crecimiento del fruto del duraznero cv. Kakamas en el rendimiento y estado hídrico de las plantas. *Agric. Téc. (Chile)* **2002**, *62*, 565–573.

11. Selles, G.; Ferreyra, R.; Selles, I.; Lemus, G. Efecto de diferentes regímenes de riego sobre la carga frutal, tamaño de fruta y rendimiento del olivo cv Sevillana. *Agric. Téc. (Chile)* **2006**, *66*, 48–56.

12. Motilva, M.J.; Tovar, M.J.; Romero, M.P.; Alegre, S.; Girona, J. Influence of regulated deficit irrigation strategies applied to olive trees (Arbequina cultivar) on oil yield and oil composition during the fruit ripening period. *J. Sci. Food Agric.* **2000**, *80*, 2037–2043.

13. Ferreyra, R.; Selles, G.; Peralta, J.; Burgos, L.; Valenzuela, J. Efecto de la restricción del riego en distintos períodos de desarrollo de la vid cv. Cabernet Sauvignon sobre producción y calidad del vino. *Agric. Téc. (Chile)* **2002**, *62*, 406–417.

14. Girona, J.; Mata, M.; del Campo, J.; Arbonés, A.; Bartra E.; Marsal, J. The use of midday leaf water potential for scheduling deficit irrigation in vineyards. *Irrig. Sci.* **2006**, *24*, 115–127.

15. Goldhamer, D.A.; Viveros, M.; Salinas M. Regulated deficit irrigation in almonds: Effects of variations in applied water and stress timing on yield and yield components. *Irrig. Sci.* **2006**, *24*, 101–114.

16. Fereres, E.; Evans, R.G. Irrigation of fruit trees and vines: An introduction. *Irrig. Sci.* **2006**, *24*, 55–57.

17. Williams, L.E.; Grimes, D.; Phene, C. The effects of applied water at various fractions of measured evapotranspiration on reproductive growth and water productivity of Thompson Seedless grapevines. *Irrig. Sci.* **2010**, *28*, 233–243.

18. Blanco, O.; Facil, J.M.; Negueroles, J. Response of table grape cultivar "Autumn Royal" to regulated deficit irrigation applied in post-veraison period. *Span. J. Agric. Res.* **2010**, *8*, 76–85.

19. El-Ansari, D.O.; Nakayama, S.; Hirano, K.; Okamoto, G. Response of "Muscat" table grapes to post-veraison regulated deficit irrigation in Japan. *Vitis* **2005**, *44*, 5–9.

20. Ezzahouani, A.; Williams, L.E. Effect of irrigation amount and preharvest irrigation cutoff date on vine water status and productivity of "Danlas" grapevines. *Am. J. Enol. Vitic.* **2007**, *58*, 333–340.

21. Williams, L.E.; Grimes, D.; Phene, C. The effects of applied water at various fractions of measured evapotranspiration on water relations and vegetative growth of Thompson Seedless grapevines. *Irrig. Sci.* **2010**, *28*, 221–232.

22. Allen, R.; Pereira, L.; Raes, D.Y.; Smith, M. *Crop Evapotranspiration. Guidelines for Computing Crop Water Requirements*; FAO Irrigation and Drainage Paper No 56; FAO: Rome, Italy, 1998; p. 300.

23. Williams, L.E.; Ayars, J. Grapevine water use and the crop coefficient are linear functions of the shaded area measured beneath the canopy. *Agric. For. Meteorol.* **2005**, *132*, 201–211.

24. Selles, G.; Ferreyra, R.; Contreras, G.; Ahumada, R.; Valenzuela, J.; Bravo, R. Manejo del riego por goteo en uva de mesa cv. Thompson Seedless cultivada en suelos de textura fina. *Agric. Téc. (Chile)* **2003**, *63*, 180–192.

25. Cassel, D.; Nielsen, D. Field capacity and available water capacity. In *Methods of Soil Analysis, Physical and Mineralogical Methods*; Klute, A., Ed.; American Society of Agronomy: Wisconsin, WI, USA, 1986; pp. 901–924.

26. Israelsen, O.W.; Hansen, V.E. *Irrigation Principles and Practices*, 3rd ed.; John Wiley and Sons: New York, NY, USA, 1962; p. 447.

27. Saxton, K.E.; Rawls, W.J. Soil water characteristic estimates by texture and organic matter for hydrologic solutions. *Soil Sci. Soc. Am. J.* **2006**, *70*, 1569–1578.

28. Schackel, K.A.; Ahmadi, H.; Biasi, W.; Buchner, R.; Goldhamer, D.; Gurusinghe, S.; Hasey, J.; Kester, D.; Krueger, B.; Lampinen, B., *et al.* Plant water status as an index of irrigation needs in deciduous fruit trees. *Hort Technol.* **1997**, *7*, 23–29.

29. Meyer, W.S.; Reicosky, D.C. Enclosing leaves for water potential measurements and its effect on interpreting soil-induced water stress. *Agric. For. Meteorol.* **1985**, *35*, 187–192.

30. Grimes, D.W.; Williams, L.E. Irrigation effects on plant water relations and productivity of Thompson Seedless grapevines. *Crop Sci.* **1990**, *30*, 255–260.

31. Selles, G.; Ferreyra, R.; Ahumada, R.; Muñoz, I.; Silva, H. Effect of soil water content and berry phenological stages on trunk diameter variations in table grape. *Acta Hortic.* **2008**, *792*, 573–580.

32. Selles, G.; Ruiz, R.; Aspillaga, C.; Lira, W. Efecto del sombreamiento del parronal sobre la acumulación de sólidos solubles, color y golpe de sol en uva de mesa var. Thompson Seedless. Aconex **2010**, *106*, 22–25.

33. Muñoz-Robredo, P.; Robledo, P.; Manríquez, D.; Molina, R.; Defilippi, B. Characterization of sugars and organic acids in commercial varieties of table grape. *Chil. J. Agric. Res.* **2011**, *71*, 452–458.

34. Ferreyra, R.; Selles, G.; Silva, H.; Ahumada, R.; Muñoz, I.; Muñoz, V. Efecto del agua aplicada en las relaciones hídricas y productividad de la vid "Crimson Seedless". *Pesqui. Agropecu. Bras.* **2006**, *41*, 1109–1118.

35. Netzer, Y.; Yao, C.; Shenker, M.; Bravdo, B.A.; Schwartz, A. Water use and the development of seasonal crop coefficient for Superior Seedless grapevines trained to an open-gable trellis system. *Irrig. Sci.* **2009**, *27*, 109–120.

36. Vita, F.; Liotta, M.; Parera, C. Effects of irrigation deficit on table grape cv. Superior Seedless production. *Acta Hortic.* **2004**, *646*, 183–186.

# 5

# Biological Control of Spreading Dayflower (*Commelina diffusa*) with the Fungal Pathogen *Phoma commelinicola*

**Clyde D. Boyette [1],\*, Robert E. Hoagland [2] and Kenneth C. Stetina [1]**

[1] USDA-ARS, Biological Control of Pests Research Unit, Stoneville, MS 38776, USA;
E-Mail: kenneth.stetina@ars.usda.gov

[2] USDA-ARS, Crop Production Systems Research Unit, Stoneville, MS 38776, USA;
E-Mail: bob.hoagland@ars.usda.gov

\* Author to whom correspondence should be addressed; E-Mail: doug.boyette@ars.usda.gov

Academic Editor: Rakesh S. Chandran

**Abstract:** Greenhouse and field experiments showed that conidia of the fungal pathogen, *Phoma commelinicola,* exhibited bioherbicidal activity against spreading dayflower (*Commelina diffusa*) seedlings when applied at concentrations of $10^6$ to $10^9$ conidia·mL$^{-1}$. Greenhouse tests determined an optimal temperature for conidial germination of 25 °C–30 °C, and that sporulation occurred on several solid growth media. A dew period of $\geq$ 12 h was required to achieve 60% control of cotyledonary-first leaf growth stage seedlings when applications of $10^8$ conidia·mL$^{-1}$ were applied. Maximal control (80%) required longer dew periods (21 h) and 90% plant dry weight reduction occurred at this dew period duration. More efficacious control occurred on younger plants (cotyledonary-first leaf growth stage) than older, larger plants. Mortality and dry weight reduction values in field experiments were ~70% and >80%, respectively, when cotyledonary-third leaf growth stage seedlings were sprayed with $10^8$ or $10^9$ conidia·mL$^{-1}$. These results indicate that this fungus has potential as a biological control agent for controlling this problematic weed that is tolerant to the herbicide glyphosate.

**Keywords:** bioherbicide; biocontrol; dayflower; fungal pathogen; weed control

## 1. Introduction

Spreading dayflower (*Commelina diffusa* Burm. f.) is a perennial, monocotyledenous weed occurring worldwide in tropical and subtropical areas, and an annual weed in temperate climates. It spreads diffusely, creeping along the ground, branching heavily and rooting at the nodes, obtaining stem lengths up to 1 m [1]. *C. diffusa* can reproduce vegetatively and by seed, and cut stems root readily in moist ground. This weed prefers moist, fertile soil (e.g., gardens, cultivated fields), but will also grow on roadsides and in non-crop areas. It has a sprawling growth habit, and its long stems can create a tangled web in gardens and flower beds. It is related to several houseplant species e.g., wandering jew (*Tradescantia zebrine* (Schinz) D.R. Hunt) and perennial spiderwort (*Tradescantia virginiana* L.). *Commelina* spp. have been used as a ground cover to reduce soil erosion [2], which may have contributed to their spreading. Its potential as a fodder crop may be useful to provide protein to ruminants on smallholder farms [3].

When growing in rice and other lowland crops, this weed may act as a quasi-aquatic plant that can withstand flooding, and it readily infests cultivated lands, roadsides, pastures and wastelands [1]. *C. diffusa* is problematic, primarily in young crops (2–5 weeks old), but can also be a problem in mature crops due to its sprawling behavior [4]. It is a troublesome weed of cotton, rice and soybean in warm temperate areas of the U.S. and other countries [5–7]. It is also reported as a major weed of bananas in Mexico and Hawaii; beans, oranges, lemons, grapes, apricots, coffee and cotton in Mexico; papaya in Hawaii; sugarcane in Puerto Rico, and sorghum in Thailand [2]. It is also a weed in maize and vegetables in Mexico; bananas, papayas, and pineapples in the Philippines; rice in Colombia; sugarcane in Mexico and Trinidad; taro and pastures in Hawaii and coffee in Costa Rica [2].

*C. diffusa* is a host of the root-burrowing (*Radophilus similis*) [8], reform (*Rotylenchulus reniformis*), banana lesion (*Pratylenchus goodeyi*) [9] and root-knot (*Meloidogyne exigua*) [10] nematodes. Severe outbreaks of cucumber mosaic virus have been correlated with high densities of *C. diffusa* serving as a reservoir of virus and aphid vectors [11].

Worldwide, there are about 170 species of *Commelina* and generally, most are difficult to control. Several *Commelinia* spp. exhibit resistance or tolerance to several chemical herbicides. *C. diffusa* was one of the first plants reported as being resistant to 2.4-dichlorophenoxy-acetic acid [12]. Herbicidal control of *C. diffusa* can be variable depending on the herbicide, growth stage, environmental parameters, *etc.*, and various herbicides and combinations of herbicides have exhibited a range of efficacy for control of *C. diffusa* and other related species as summarized [4,13–15]. Recent guidelines for control of *C. diffusa* in rice in Mississippi (USA) indicate that only ~50% of the 42 single herbicide or herbicide combination treatments provided good to excellent control, while 26% gave fair control and the remainder gave zero to poor control [16]. Some alternative herbicide options can be used to control this weed, alone or in combination with other modes of action in rice, during early post-emergence applications prior to flooding [17].

*C. diffusa*, Benghal dayflower (*C. benghalensis* L.), and Asiatic dayflower (*C. communis* L.) have been reported to be difficult to control with glyphosate in genetically-modified crops [18–24]. The ecological, biological and physiological factors related to glyphosate-tolerant *C. communis* in agronomic systems in Iowa have recently been studied [25]. Because of the increasing importance of *C. diffusa*, and its resistance or tolerance to many herbicides, alterative weed control measures may be required. The use of bioherbicides has been recognized as a potential technological alternative to chemical herbicides in certain situations, and global interest exists in the bioherbicide concept, with active research and development projects established by commercial entities in the U.S., Canada, Europe, Australia, Japan, and other countries [26–30].

A leaf-spot disease (oblong lesions, *ca.* 1.3–1.8 cm) was observed on *C. diffusa* in a flooded rice field near Stuttgart, AR, USA (Figure 1). Infected leaf and stem tissues were collected and a fungal pathogen was isolated from this diseased tissue. This fungus was provisionally identified as *Phyllosticta commelinicola* E. Young, a synonym of *Phoma commelinicola* (E. Young) Gruyter [31]. The objectives of these studies were to isolate and examine this pathogen with respect to its growth and germination on various growth media, correlate inoculum concentration and bioherbicidal activity (inundative application) with plant growth stage, develop time courses for weed control and disease progression on *C. diffusa*, and evaluate weed control under field conditions. Knowledge of these basic parameters is essential for evaluating a plant pathogen as a bioherbicide for weed control [32].

**Figure 1.** *Commelina diffusa* photographs. (**A**): flowering plant in the field; (**B**): pressed/dried specimen exhibiting leaf spotting incited by *P. commelinicola*.

## 2. Materials and Methods

### 2.1. Seed Sources, Test Plant Propagation

*C. diffusa* seeds were collected near Stuttgart AR, USA, planted in a 2:1 potting mix of Jiffy mix:sandy soil (Jiffy Mix, Jiffy Products of America, Inc., Batavia, IL, USA) contained in plastic trays (25 × 52 cm) and allowed to germinate. Germinated seedlings were transplanted into 10-cm$^2$ plastic pots (1 plant per pot) containing the soil mixture above, and grown under greenhouse conditions (28 °C to 32 °C, 40 to 60% relative humidity (RH), ~14 h day length, and 1650 $\mu E \cdot m^{-2} \cdot s^{-1}$ photosynthetically active radiation (PAR) measured at midday).

### 2.2. Isolation and Culture of Phoma commelinicola

Several isolates of the fungus were isolated from diseased *C. diffusa* tissue by surface sterilizing sections of diseased tissue in 0.05% NaOCl for 1 min, rinsing in sterile distilled water and then placing the sections on autoclaved (121 C, 15 min; at 103.42 kPa) potato-dextrose agar (PDA, Difco, Detroit, MI, USA) plates amended with the antibiotics chloramphenicol (0.75 mg·mL$^{-1}$) and streptomycin sulfate (1.25 mg·mL$^{-1}$). The plates were incubated for 48 h at 25 °C and then advancing edges of fungal colonies were transferred to PDA plates followed by incubation for 5 days at 25 °C under alternating 12-h light (cool, white fluorescent bulbs)/12-h dark regimens. Tests of these isolates indicated very similar virulence on the host plant. We chose an isolate with the highest virulence (SFN-73) and used it in further studies. When re-inoculated onto healthy seedlings at $1 \times 10^8$, the fungus (SFN-73) was highly virulent and killed all inoculated plants within 5 days, while the controls remained healthy, thus fulfilling Koch's postulates (data not shown). The fungus was then sub-cultured on PDA without antibiotics, and preserved under refrigeration (4 °C to 5 °C) on sterilized sandy loam soil (25% water holding capacity), or on sterile silica gel containing skim milk [33].

Several media were examined for growth and conidial production of the fungus: water agar, 2.0% (WA), potato dextrose agar (PDA), yeast extract agar (YEA) and Czapek-Dox agar (CDA) from Difco (Detroit, MI, USA), V8 agar (V8A) [33] from Campbell Soup Co. (Camden, NJ, USA) and dayflower decoction agar (DFA). DFA was prepared by adding 100 g of finely ground, air-dried dayflower leaf and stem tissue to 2.0% WA to yield a 10% (w:v) product.

### 2.3. Effect of Temperature on Conidial Germination and Radial Growth Rate

Conidial germination was measured by spreading 100 μL of a suspension ($1.0 \times 10^6$ conidia·mL$^{-1}$) prepared in sterile distilled water on PDA plates, and incubating them at 10 °C, 15 °C, 20 °C, 25 °C, 30 °C, or 35 °C on open-mesh wire shelves of an incubator (Precision Scientific Inc., Chicago, IL, USA). A 12-h photoperiod was provided by two 20 W, cool-white fluorescent lamps positioned in the incubator door. The light intensity at the plate level was 200 μE·m$^{-2}$·s$^{-1}$ PAR as measured with a light meter (LI-COR Inc., Lincoln, NE, USA). Germinated conidia (500 plate$^{-1}$) were counted after 16 h using a haemocytomer.

For radial growth studies, 5-mm plugs were taken from the advancing margins of 7-day-old colonies of the fungus and placed in the centers of PDA plates. The plates were incubated at temperatures of 10 °C, 15 °C, 20 °C, 25 °C, 30 °C, or 35 °C as described above for the conidial germination studies. Colony diameters (fungal growth) were measured after 7 days of incubation. In each experiment, five replicate plates for each temperature were utilized. Both experiments were conducted twice, and the results of each experiment were pooled following testing for homogeneity.

### 2.4. Effect of Dew Period Duration on Weed Control and Dry Weight Reduction of C. diffusa Seedlings

*C. diffusa* seedlings (cotyledonary to first leaf growth stage) were sprayed (hand held sprayer; Spray-Tool, Aervoe Industries, Gardnerville, NV, USA) until runoff (*ca.* 100 L·ha$^{-1}$) occurred with a spray mixture containing $1.0 \times 10^8$ conidia·mL$^{-1}$ in distilled water. Control plants were sprayed with distilled water. The inoculated plants were then placed in darkened dew chambers at 25 °C and 100% RH for periods of 3, 6, 9, 12, 15, 18, 21 or 24 h. Following this dew treatment, the plants were placed on sub-irrigated trays in the greenhouse as described above. Weed control and dry weight reductions were recorded 14 days after treatment (DAT). For dry weight determinations, the above ground biomass was harvested, oven-dried (48 h, 85 °C), weighed, and the percentage biomass reduction (compared with untreated control plants) was determined. The experiment was conducted twice with 3 sets of 10 plants for each experiment.

## 2.5. Effect of Inoculum Concentration and Plant Growth Stage

C. diffusa plants (cotyledonary, 1 to 2 true-leaf, 3 to 4 true-leaf and 5 to 7 true-leaf growth stages) were sprayed with conidial suspensions of $1.0 \times 10^6$ to $1.0 \times 10^9$ conidia·mL$^{-1}$ and held in a dew chamber for 16 h at 25 °C. Control plants were sprayed with distilled water only. Plants were moved to the greenhouse, and mortality and dry weight reductions were recorded 14 DAT. Experiments were conducted twice with 3 sets of 10 plants for each experiment. The experiment was conducted twice with 3 sets of 10 plants for each experiment.

## 2.6. Effects of Phoma commelinicola on Crop Seedlings

Greenhouse tests were conducted on seedlings of several crops to access the possible detrimental effects of this pathogen. Rice (Oryza sativa L.), soybean (Glycine max (L.) Merr.), cotton (Gossypium hirsutum L.) and corn (Zea mays L.) plants were grown from seeds in the greenhouse under the conditions described above. After 10 to 14 days (when seedlings were ~3–7 cm tall), plants were sprayed with P. commelinicola inoculum concentrations of $1.0 \times 10^6$ or $1.0 \times 10^8$ conidia·mL$^{-1}$. Visual disease symptomatology and dry weight analyses of fungal-inoculated plants were monitored 14 DAT. The experiment was conducted twice with 3 sets of 10 plants for each experiment.

## 2.7. Field Experiments

Field experiments were conducted in 1998 and 1999 at the University of Arkansas, Rice Research and Extension Center, Stuttgart, AR, USA, on Crowley silt loam (fine montmorillonitic, thermic Typic Albaqualfs), pH 6.2 to 6.5, with an organic matter content of ~1.0%. The experiments were established in an irrigated rice field, divided into $1.0 \times 1.0$ m micro-plots ($1.0 \times 10^{-5}$ ha). The field was naturally infested with dayflower seedlings (avg. 75 seedlings per plot) that were in the cotyledonary to first leaf growth stages. Within each plot, 18–20 test plants were randomly selected and marked using wooden stakes (7.6 cm long). Treatments consisted of P. commelinicola conidia applied at either 0.0 conidia·mL$^{-1}$ (water control) or conidia in water applied at several concentrations from $1.0 \times 10^6$ to $1.0 \times 10^9$ conidia·mL$^{-1}$. The selected plants were monitored for disease development at 3-day intervals for 21 days. All treatments were replicated 4 times and the experiment was repeated.

## 2.8. Statistical Procedures

The greenhouse and field experiments were arranged as randomized complete block factorial designs with three and four replications, respectively. Data collected over the 2-year field testing period were examined for homogeneity of variance [34], combined, and analyzed using ANOVA. Field data from both years were pooled following subjection to Bartlett's test for homogeneity, and analyzed using analysis of variance. Because arcsine and square-root transformation of the data did not alter the interpretation, non-transformed data are presented. When significant differences were detected by the F-test, means were separated with Fisher's protected LSD test at the 0.05 probability level. Disease progression was based on a modified Horsfall and Barratt [35] rating scale of 0 to 5.0, assigning symptom expression as 0 represents unaffected, and 1.0, 2.0, 3.0, 4.0, and 5.0 represents 20, 40, 60, 80 and 100% leaf and stem injury (or dead plants), respectively. Percentage weed control was determined by dividing the number of dead and severely injured plants (symptom expression ratings of 4.0–5.0) by the total number of plants treated × 100. Data were analyzed using standard mean errors and best-fit regression analysis. All data were analyzed using SAS (Version 9.1, SAS Institute, Inc., Cary, NC, USA) statistical software.

## 3. Results and Discussion

### 3.1. Isolation of Phoma commelinicola

The fungus was readily isolated from diseased tissue and observed to sporulate abundantly on PDA. Several isolates were collected and found to exhibit similar virulence. We chose an isolate with the highest virulence (SFN-73) and used it in these studies. When re-inoculated onto healthy seedlings at $1.0 \times 10^8$, the fungus (SFN-73) was highly virulent and killed all inoculated plants within 5 DAT, while the controls remained healthy, thus fulfilling Koch's postulates. The organism produced disease symptomatology typical of other diseases incited by other *Phoma* spp. (*i.e.*, lesions on leaves and stems) (Figure 1). The organism produced typical *Phoma* lesions on leaves and stems, with pycnidia scattered throughout the lesions. Under moist conditions, slimy masses of conidia accumulated on the upper surface of the leaf, breaking the epidermal layer and cuticle. Conidia were unicellular, hyaline (2.0 to 4.0 × 1.5 to 2.5 μm) extruded through ostioles contained in black pycnidia (60 to 165 × 45 to 140 μm), ostiolate, protruding into plant tissues or agar surfaces.

### 3.2. Germination and Growth of Phoma commelinicola

Germination of conidia on PDA occurred at 10 to 35 °C with optimal germination at 20 to 30 °C and maximum (85%) at 25 °C (Figure 2). The fungus grew at all temperatures tested (10 to 35 °C), but growth was significantly reduced at 10, 15 and 35°C (Figure 3). The fungus also grew and sporulated prolifically on several different solid substrate media. Dayflower decoction agar (DFA) and PDA produced the most abundant conidia ($6.0 \times 10^8$ and $5.0 \times 10^8$, respectively) (Table 1). The lowest growth rate occurred on Czapek-Dox agar (4.1 mm·day$^{-1}$) and the highest rate of growth was found on DFA decoction agar (10.5 mm·day$^{-1}$). Due to the commercial availability of PDA, it was used in all other greenhouse and field experiments. This growth is comparable to conidial yields produced by other *Phoma* spp. that have been evaluated as bioherbicides [36,37].

**Table 1.** Effect of various growth media [a] on radial growth and conidial production of *Phoma commelinicola* under various light and dark regimes [b].

| Growth | Radial Growth (mm·day$^{-1}$) | | | Conidia ($10^8$ plate$^{-1}$) | | |
|---|---|---|---|---|---|---|
| | Light | Dark | Light/Dark | Light | Dark | Light/Dark |
| WA | - | - | - | - | - | - |
| DFA | 10.2 [a,c] | 9.0 [a] | 10.5 [a] | 4.2 [a] | 2.8 [a] | 6.0 [a] |
| PDA | 9.5 [b] | 8.0 [b] | 9.9 [b] | 3.0 [b] | 1.3 [b] | 5.0 [b] |
| V8A | 8.4 [c] | 7.0 [c] | 8.5 [c] | 2.1 [c] | 0.9 [c] | 3.9 [c] |
| YEA | 7.5 [d] | 6.9 [c] | 7.4 [d] | 1.0 [d] | 0.3 [d] | 2.3 [d] |
| CDA | 4.0 [e] | 2.8 [d] | 4.1 [e] | 1.0 [d] | 0.2 [d] | 1.2 [e] |

[a] WA, water agar; DFA, dayflower decoction agar; PDA, potato dextrose agar; V8A (V8 vegetable juice); agar; YEA, yeast-extract agar; CDA, Czapek-Dox agar; [b] Light conditions (24 h continuous light at 28 °C); Dark conditions (24 h continuous light at 28 °C); Light/dark conditions (12 h light/dark at 28 °C); [c] Means within the same column followed by the same letter do not differ at $p = 0.05$, according to FLSD.

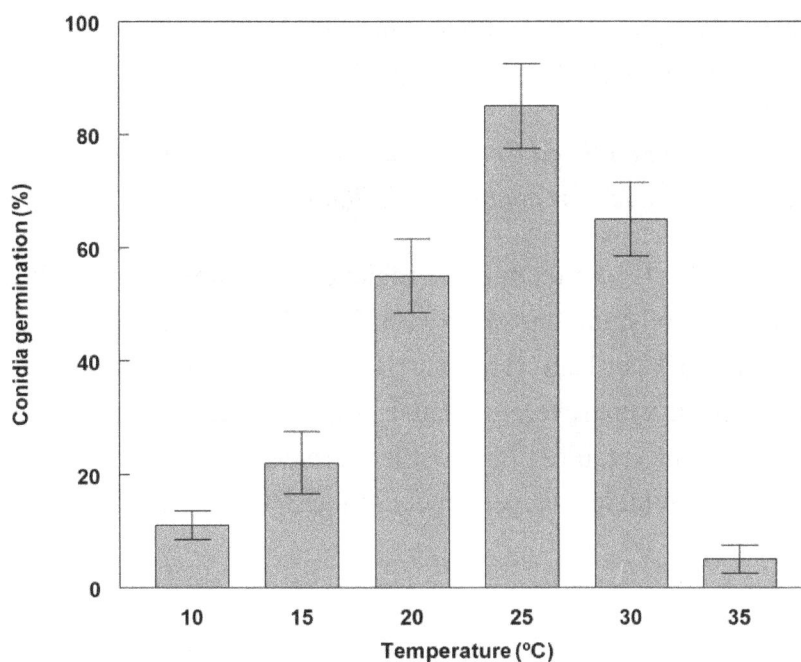

**Figure 2.** Effect of temperature on germination of conidia of *P. commelinicola* on PDA, 7 days after inoculation and growth under alternating light/dark conditions (12 h light/dark at 28 °C). Error bars represent Fisher's LSD ($p = 0.05$).

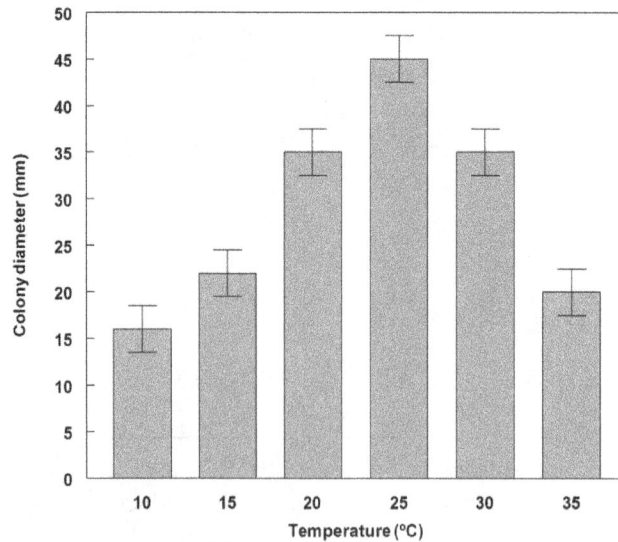

**Figure 3.** Effect of temperature on radial growth of *P. commelinicola* colonies on PDA, 7 days after inoculation and growth under alternating light/dark conditions (12 h light/dark at 28 °C). Error bars represent Fisher's LSD ($p = 0.05$).

*3.3. Effects of Dew Period Duration on Weed Control and Dry Weight Reduction of C. diffusa Inoculated with P. commelinicola under Greenhouse Conditions*

A minimum of 12 h dew at 25 °C was required to achieve ~65% weed control (Figure 4). Optimal weed control occurred at dew period durations of between 15 to 24 h. A similar trend was observed for the dry weight reduction data (~67% at 12 h) (Figure 5). Complete mortality (100%) was not achieved at any dew period, but many plants were severely stunted, which resulted in greatly reduced dry weight. Lengthy dew periods are commonly required for most bioherbicidal plant pathogens [32]. This factor has been a major constraint for commercial development of these biocontrol organisms [38].

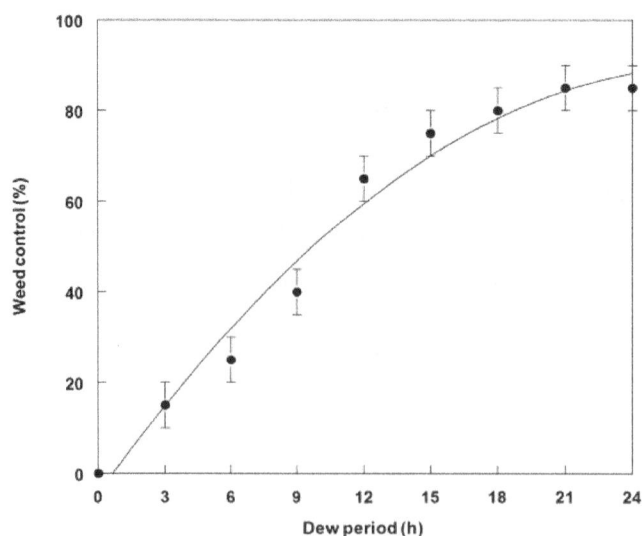

**Figure 4.** Effect of dew period duration at 25 °C on weed control of spreading dayflower inoculated with *P. commelinicola* at $1.0 \times 10^8$ under greenhouse conditions. Error bars represent Fisher's LSD ($p = 0.05$).

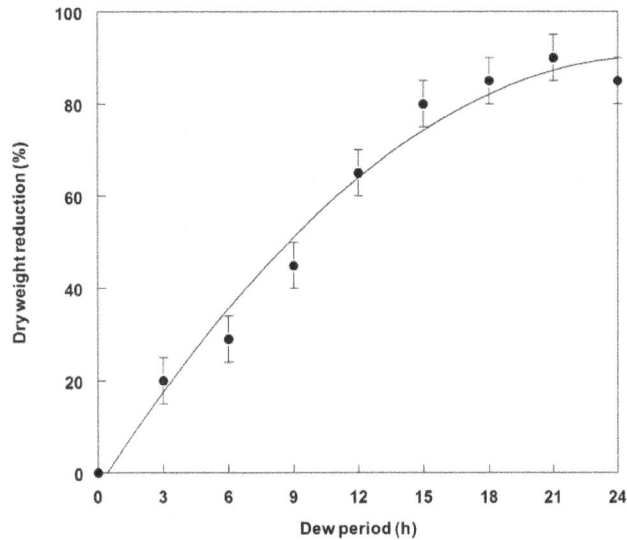

**Figure 5.** Effect of dew period duration on dry weight reduction of spreading dayflower inoculated with *P. commelinicola* at $1.0 \times 10^8$ under greenhouse conditions. Error bars represent Fisher's LSD ($p = 0.05$).

### 3.4. Effects of P. commelinicola Inoculum Concentration on Weed Control and Dry Weight Reduction of C. diffusa under Greenhouse Conditions

Generally, weed control on *C. diffusa* plants under greenhouse conditions was significantly increased at all growth stages as the fungal inoculum concentration increased (Figure 6). For example, at low inoculum concentration ($0.001 \times 10^9$ conidia·mL$^{-1}$), the youngest plants were controlled about 5%, while the highest concentration ($1.0 \times 10^9$ conidia·mL$^{-1}$) provided ~85% control. Plants in the 6- to 7- and 8- to 9-leaf stages were more resistant to infection than younger plants, *i.e.*, plants in the 8- to 9-leaf growth stages were controlled at the 0 and 15% levels by $0.001 \times 10^9$ and $1.0 \times 10^9$ conidia·mL$^{-1}$, respectively. Similar results were obtained for the dry weight reductions of plants at these growth stages and conidia concentrations (Figure 7). Since risk assessment of bioherbicides on non-target plants is necessary and important, we examined the effects of this bioherbicide on several crops under greenhouse conditions. Seedlings of several crops (rice, soybean, cotton and corn), sprayed with *P. commelinicola* inoculum concentrations at $1.0 \times 10^6$ and $1.0 \times 10^8$ conidia·mL$^{-1}$, exhibited no visual disease symptomatology or dry weight reduction when evaluated 14 DAT under greenhouse conditions (data not shown).

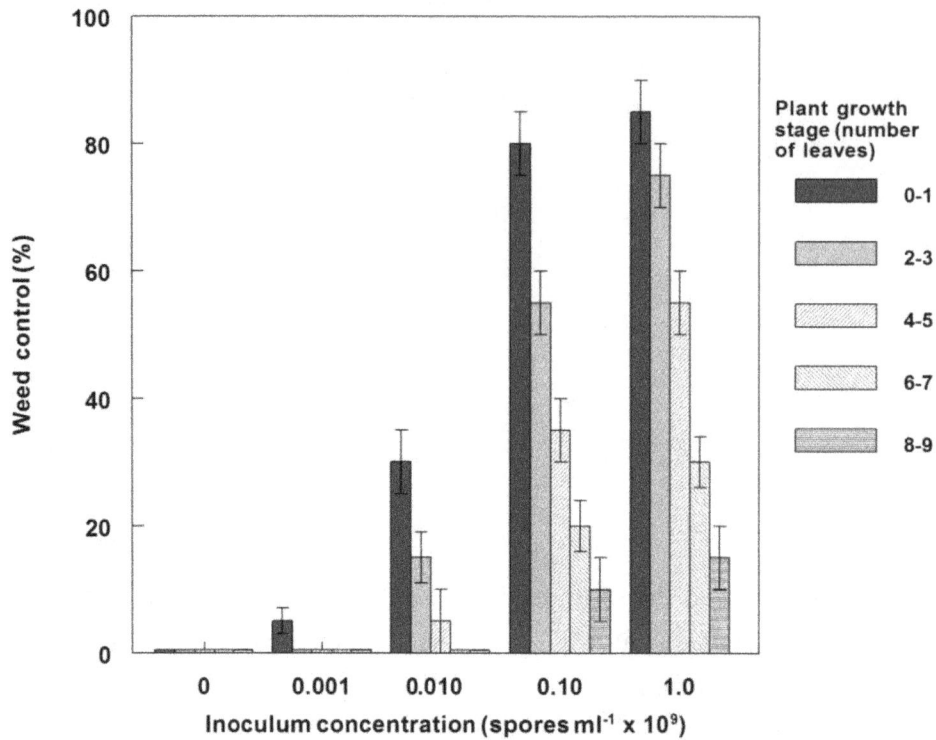

**Figure 6.** Effect of plant growth stage on weed control (mortality) of spreading dayflower inoculated with *P. commelinicola* at various inoculum concentrations under greenhouse conditions. Error bars represent Fisher's LSD ($p = 0.05$).

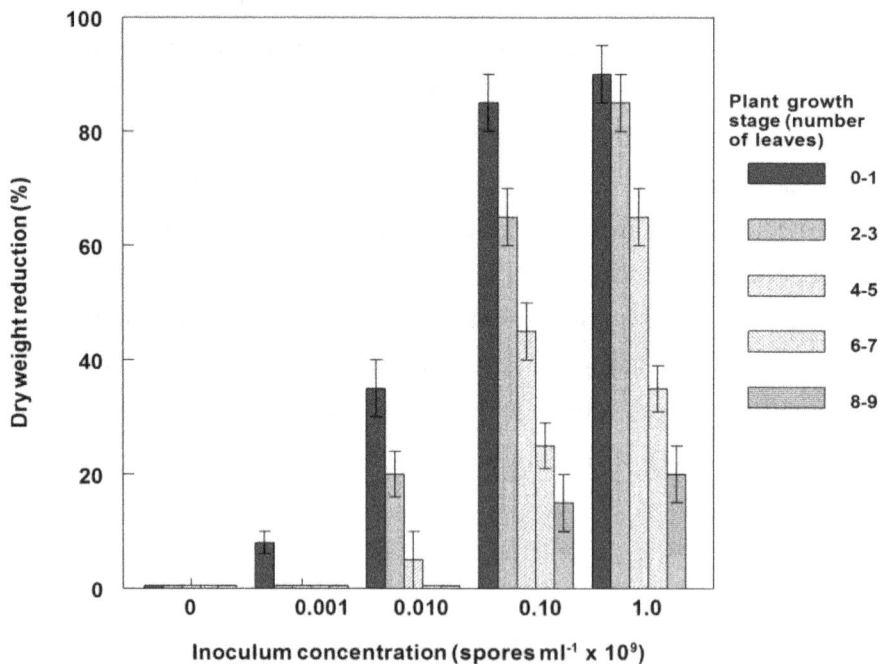

**Figure 7.** Effect of plant growth stage on dry weight reduction of spreading dayflower inoculated with *P. commelinicola* at various inoculum concentrations under greenhouse conditions. Error bars represent Fisher's LSD ($p = 0.05$).

*3.5. Disease Progression of P. commelinicola on C. diffusa under Greenhouse Conditions*

Disease on *C. diffusa* incited by *P. commelinicola* progressed in a linear fashion from 3 to 12 DAT under greenhouse conditions, with a disease rating of 3.5 occurring at 12 DAT (Figure 8). Disease progressed to 3.8 at 15 DAT and eventually increased to 4.5 at 21 DAT.

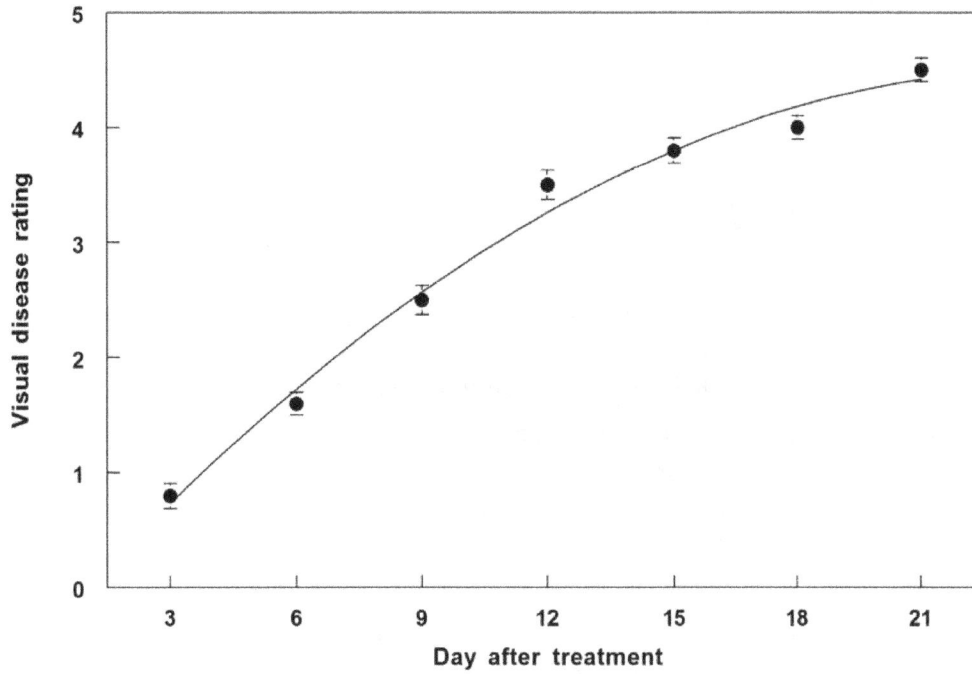

**Figure 8.** Disease progression of *P. commelinicola* on greenhouse-grown *C. diffusa* over a 21-day period after inoculation. The relationship for *P. commelinicola* disease progression is best described by the equation: $Y = -0.15 + 0.37X - 0.01X^2$, $R^2 = 0.98$. Error bars represent Fisher's LSD ($p = 0.05$).

*3.6. Effects of P. commelinicola Inoculum Concentration and C. diffusa Growth Stage on Weed Control and Dry Weight Reduction under Field Conditions*

In field experiments, the highest weed control (55 to 70%) occurred on cotyledonary to third-leaf stage plants at $1.0 \times 10^7$ to $1.0 \times 10^9$ conidia·mL$^{-1}$, 21 DAT (Figure 9). A similar trend occurred in dry weight reduction with ~80% reduction after 21 DAT at $1.0 \times 10^8$ or $10^9$ conidia·mL$^{-1}$ (Figure 10). The LD$_{50}$ and GR$_{50}$ values for weed control (Figure 9) and dry weight reduction (Figure 10) were $2.0 \times 10^7$ conidia·mL$^{-1}$ and $4.0 \times 10^6$ conidia·mL$^{-1}$, respectively.

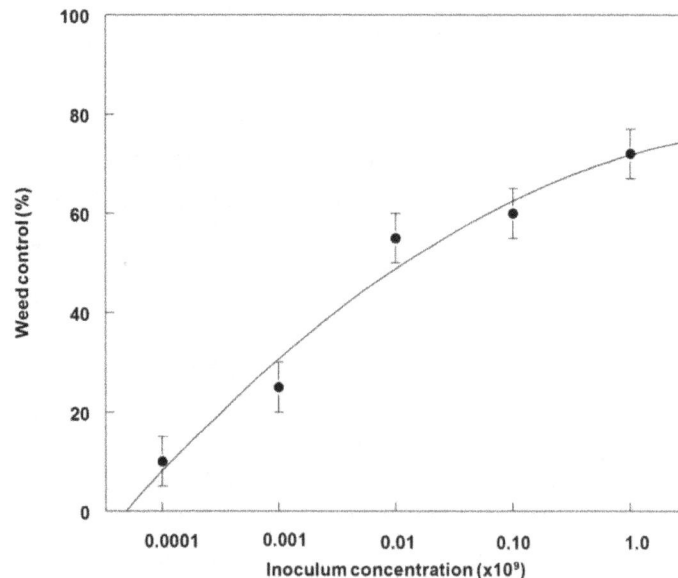

**Figure 9.** Effect of *P. commelinicola* inoculum concentration on the control of *C. diffusa* in the cotyledonary to third leaf growth stage under field conditions. $Y = -18.8 + 29.2X - 2.2X^2$, $R^2 = 0.96$.

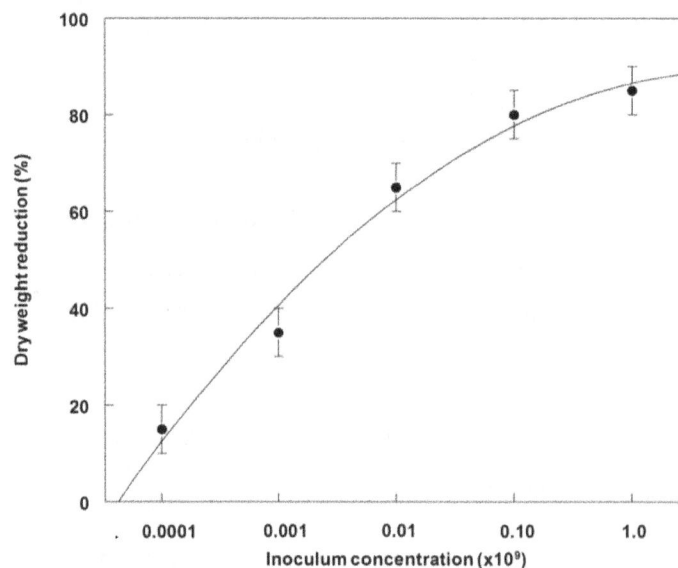

**Figure 10.** Effect of *P. commelinicola* inoculum concentration on the dry weight reduction of *C. diffusa* in the cotyledonary to third leaf growth stage under field conditions. $Y = -22.0 + 37.8X - 3.2X^2$, $R^2 = 0.98$.

*3.7. Disease Progression of P. commelinicola on C. diffusa under Field Conditions*

The disease progression of this fungus on *C. diffusa* under field conditions (Figure 11) was similar to that found under greenhouse conditions (Figure 8). However, 15 days were required to achieve a rating value of 3.5 as compared to 12 days under greenhouse conditions (Figure 8). The maximal disease rating of 3.8 occurred at 21 DAT in the field (Figure 11). No visual infectivity or injury was observed on rice plants using this formulation under field conditions (data not shown).

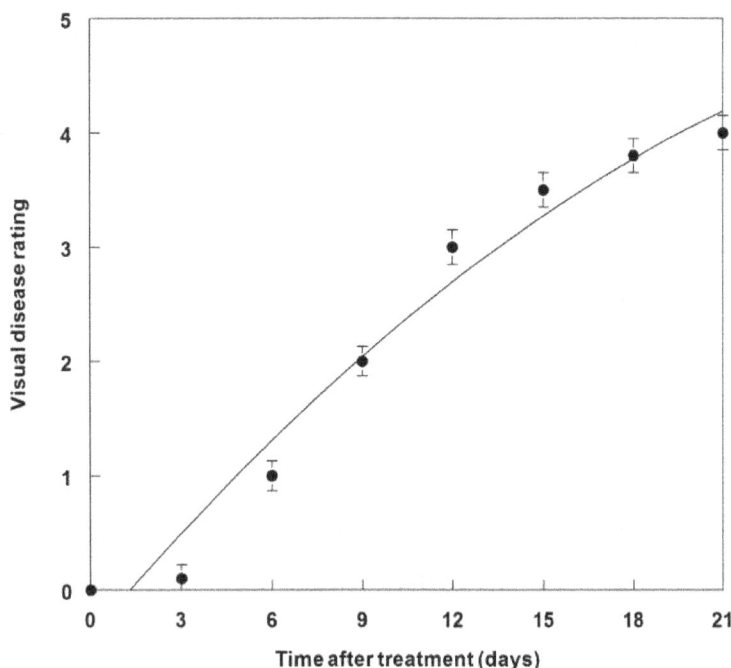

**Figure 11.** Disease progression of *P. commelinicola* on field-grown *C. diffusa* over a 21-day period after inoculation. The relationship for *P. commelinicola* disease progression is best described by the equation: $Y = -0.39 + 0.31X - 0.04X^2$, $R^2 = 0.96$. Error bars represent Fisher's LSD ($p = 0.05$).

Despite the high phytopathogenic potential shown by various *Phoma* spp. and pathovars, the bioherbicidal potential of these microbes has been largely ignored. Some attempts to evaluate this diverse genus as bioherbicides have been reported. For example, Heiny [39,40] isolated a highly host specific strain of *P. proboscis* from diseased field bindweed (*Convolvulus arvensis* L.). Heiny and Templeton [37] reported significant bioherbicidal effects when conidia of this fungus were applied to weed seedlings, under temperatures from 16–28 °C, and ≥9 h dew period and the compatibility with synthetic herbicides was investigated [40]. Studies of environmental factors on the effectiveness of a *Phoma herbarum* strain against *Commelina communis* showed that temperatures of 28–32 °C and a 48-h dew period were required for optimal control [41]. A strain of *P. herbarum* from diseased leaves of *Parthenium hysterophorus* L. was found in central India [42]. The fungus caused >90% inhibition of seed germination, seedling mortality and leaf damage, followed by a reduction in the height of this weed [42]. Three strains of *Phoma herbarum* were isolated from the diseased leaves and stem of *Lantana camara* L. [43] and all strains incited severe infection of weed seedlings [44]. Other *Phoma* spp. have also been reported from various regions of India: *P. campanulata* on *Cassia fistula* L.; *P. exigua* on *Sesamum indicum* L.; *P. eupyrena* on *Achyrenthus aspera* L.; *P. glomerata* on *Crotalaria juncea* L. and *Parthenium hysterophrous* L.; *P. lantanae* on *Lantana camara*; *P. palmarum* on *Calotropis procera* L.; *P. tridocis* on *Tridex procumbens* L.; and *P. herbarum* var. *ipomoeae* and *P. euphorbiae* on *Euphorbia hirta* [45,46]. Recently, *Phoma macrotoma* has been developed as a commercial bioherbicide (Phoma™) for controlling various broadleaf weeds in turf [47].

Fungal conidia are the predominant propagules used in bioherbicidal research [26,29,30]. However, mycelial formulations of various fungal bioherbicides, including *Phoma* spp., have shown weed control potential against some important weeds. For example, mycelial suspensions of *P. herbarum* incited significantly more disease severity than conidia on dandelion (*Taraxacum officinale* Weber) [48]. Similarly, mycelial fragments of *P. exigua* and *P. herbarum* incited significantly more disease severity of *T. officinale* than fungal conidia [49]. Zhao and Shamoun [36] found that culture growth media and age of mycelial cultures could affect the disease severity of *P. exigua* on the perennial evergreen weedy shrub, salal (*Gaultheria shallon* Pursh).

The Clearfield™ system has become the predominant rice production system in southern rice producing states [50]. The rice cultivars utilized in this system are natural mutants with tolerance to the herbicide imazethapyr (2-(4,5-dihydro-4-methyl-4-(1-methylethyl)-5-oxo-1H-imidazol-2-yl)-5-ethyl-3-pyridine-carboxylic acid) (Newpath™). Although this herbicide controls many grassy and broadleaf weeds, it fails to control some weeds, including *C. diffusa,* which can result in tremendous weed infestations if other weed control measures are not utilized [51]. Whether this isolate of *P. commelinicola* could be incorporated in this system remains to be determined.

## 4. Conclusions

*P. commelinicola* may be an effective bioherbicide for controlling *C. diffusa,* a problematic weed in rice production in the southern U.S. Fungal conidia can be readily produced on several solid substrate growth media. The fungus grows and germinates over a wide range of temperatures. Although a rather lengthy dew period is required (15 to 18 h) to achieve levels of control of 75 to 80%, respectively, this free moisture condition is met in flooded rice fields. No visual disease symptomology was observed on rice (cv, Starbonnet). However, further research is in progress to define the host range of the fungus using various rice cultivars and other economically important rice weeds. Special consideration will be given to the effects of this pathogen on other principal *Commelina* spp., especially *C. benghalensis,* an exotic, invasive weed that is resistant to herbicides in many areas of the world, including the southern U.S. [52]. We also wish to examine the effects of surfactants and other adjuvants to improve the efficacy and use of mycelial formulations of this organism for control of *Commelina* spp. Possible interactions (synergistic, additive or antagonistic) of this pathogen with herbicides will be examined since important synergistic interactions of other plant pathogens with herbicides have been discovered [32,53,54]. Furthermore, since another pathogen (*C. gloeosporioides* f. sp. *aeschynomene*, LockDown™) [55] is compatible with the Clearfield™ rice production system, additional research will also examine the compatibility of *P. commelinicola* in this rice production protocol.

## Acknowledgments

We thank Robin H. Jordan for expert technical assistance during portions of these studies.

## Conflicts of Interest

The authors declare no conflict of interest.

# References

1. Bryson, C.T.; DeFelice, M.S. *Weeds of the South*; University of Georgia: Athens, GA, USA, 2009; p. 325.

2. Isaac, W.A.; Gao, Z.; Li, M. Managing *Commelina* species: Prospects and limitations. *Herbic. Curr. Res. Case Stud. Use* **2013**, 543–561, doi:10.5772/55842.

3. Lanyasunya, T.P.; Wang, R.H.; Abdulrazak, S.A.; Mukisira, E.A. The potential of the weed, *Commelina diffusa* L., as a fodder crop for ruminants. *S. Afr. J. Anim. Sci.* **2006**, *36*, 28–32.

4. Invasive Species Compendium: *Commelina diffusa* (Spreading Dayflower). Available online: http://www.cabi.org/isc/datasheet/14979 (accessed on 2 July 2015).

5. Webster, T.M. Weed survey—Southern States 2005: Broadleaf Crops Subsection (Cotton, Peanut, Soybean, Tobacco, and Forestry). *Proc. South. Weed Sci. Soc.* **2005**, *58*, 291–306.

6. Norsworthy, J.K.; Burgos, N.R.; Scott, R.C.; Smith, K.L. Consultant perspectives on weed management needs in Arkansas rice. *Weed Technol.* **2007**, *21*, 832–839.

7. Singh, G.; Singh, Y.; Singh, V.P.; Johnson, D.E.; Mortimer, M. System level effects in weed management in rice-wheat cropping in India. In Proceedings of the British Crop Protection Conference International Congress on Crop Science and Technology (Glasgow, UK), Alton, Hampshire, UK, 2005; pp. 545–550.

8. Queneherve, P.; Chabrier, C.; Auwerkerken, A.; Topart, P.; Martiny, B.; Martie-Luce, S. Status of weeds as reservoirs of plant parasitic nematodes in banana fields in Martinique. *Crop Prot.* **2006**, *25*, 860–867.

9. Robinson, A.F.; Inserra, R.N.; Caswell-Chen, E.P.; Vovlas, N.; Troccoli, A. *Rotylenchulus* species: Identification, distribution, host ranges and crop plant resistance. *Nematropica* **1997**, *15*, 165–170.

10. Rich, J.R.; Brito, J.A.; Kaur, R.; Ferrell, J.A. Weed species as hosts of *Meloidogyne*: A review. *Nematropica* **2008**, *39*, 157–185.

11. Richard, A.; Farreyrol, K.; Rodier, B.; Leoce-Mouk-San, K.; Wong, M.; Pearson, M.; Grisoni, M. Control of virus diseases in intensively cultivated vanilla plots of French Polynesia. *Crop Protect.* **2009**, *28*, 870–877.

12. Sosnoskie, L.M.; Hanson, B. Herbicide Resistance and Its Management. Available online: http://ucanr.edu/blogs/blogcore/postdetail.cfm?postnum=12365 (accessed on 6 July 2015).

13. Isaac, W.A.P.; Brathwaite, R.A.I. *Commelina* species—A review of its weed status and possibilities for alternative weed management in the tropics. *Agro. Thesis* **2007**, *5*, 3–18.

14. Wilson, A.K. Commelinaceae—A review of the distribution, biology and control of the important weeds belonging to this family. *Int. J. Pest Manag.* **1981**, *27*, 405–418.

15. Anonymous. *CABI Crop Protection Compendium, Global Module*; CAB International: Wallingford, UK, 2007. Available online: http://www.cabi/compendia/cpc/index.htm (accessed on 21 June 2015).

16. Dodds, D.; Calcote, K.; Byrd, J. *Weed Control Guidelines for Mississippi*; Mississippi State University Extension Service and Mississippi Agricultural and Forestry Experimental Station: Starkville, MS, USA, 2015; p. 64.

17. Scott, R.C.; Boyd, J.W.; Selden, G.; Norsworthy, J.K.; Burgos, N. *Recommended Chemicals for Weed and Brush Control—MP44*; University of Arkansas Extension Publication: Fayetteville, AR, USA, 2015; p. 79.

18. Fawcett, J. Glyphosate tolerant Asiatic dayflower (*Commelina communis*) control in no-till soybeans. *Proc. North Cent. Weed Sci. Soc.* **2002**, *57*, 183.

19. Santos, L.D.T.; Meira, R.M.S.A.; Santos, I.C.; Ferreira, F.A. Effect of glyphosate on the morpho-anatomy of leaves and stems of *C. diffusa* and *C. benghalensis*. *Planta Daninha* **2004**, *22*, 101–107.

20. Webster, T.M.; Burton, M.G.; Culpepper, A.S.; York, A.C.; Prostko, E.P. Tropical spiderwort (*Commelina benghalensis*): A tropical invader threatens agroecosystems of the Southern United States. *Weed Technol.* **2005**, *19*, 501–508.

21. Culpepper, A.S. Glyphosate-induced weed shifts. *Weed Technol.* **2006**, *20*, 277–281.

22. Isaac, W.A.P.; Brathwaite, R.A.I.; Cohen, J.E.; Bekele, I. Effects of alternative weed management strategies on *Commelina diffusa* Burm. infestations in fair trade banana (*Musa* spp.) in St. Vincent and the Grenadines. *Crop Prot.* **2007**, *26*, 1219–1225.

23. Webster, T.M.; Faircloth, W.H.; Flanders, J.T.; Prostko, E.P.; Grey, T.L. The critical period of Bengal dayflower (*Commelina benghalensis*) control in peanut. *Weed Sci.* **2007**, *55*, 359–364.

24. Ulloa, S.M.; Owen, M.D.K. Response of Asiatic dayflower (*Commelina communis*) to glyphosate and alternatives in soybean. *Weed Sci.* **2009**, *57*, 74–80.

25. Gomez, J.M. Glyphosate-tolerant Asiatic Dayflower (*Commelina communis* L.): Ecological, Biological and Physiological Factors Contributing to Its Adaptation to Iowa Agronomic Systems. Master Thesis, Iowa State University, Ames, IA, USA, 2012.

26. Charudattan, R. Biological control of weeds by means of plant pathogens: Significance for integrated weed management in modern agro-ecology. *Biol. Control* **2001**, *46*, 229–260.

27. Charudattan, R. Ecological, practical, and political inputs into selection of weed targets: What makes a good biological control target? *Biol. Control* **2005**, *35*, 183–196.

28. Glare, T.; Caradus, J.; Gelemter, W.; Jackson, T.; Keyhani, N.; Kohl, J.; Marrone, P.; Morin, L.; Stewart, A. Have biopesticides come of age? *Trends Biotechnol.* **2012**, *30*, 250–258.

29. Weaver, M.A.; Lyn, M.E.; Boyette, C.D.; Hoagland, R.E. Bioherbicides for weed control. In *Non-Chemical Weed Management*; Updhyaya, M.K., Blackshaw, R.E., Eds.; CABI International: Cambridge, MA, USA, 2007; pp. 93–110.

30. Duke S.O.; Scheffler, B.E.; Boyette, C.D.; Dayan, F.E. Biotechnology in weed control. In *Encyclopedia of Chemical Technology*; Kirk, O., Ed.; John Wiley & Sons, Inc.: New York, NY, USA, 2015; pp. 1–25.

31. De Gruyter, J. Contributions towards a monograph of *Phoma* (Coelomycetes) IX. Section Macrospora. *Persoonia* **2002**, *18*, 85–102.

32. Boyetchko, S.M.; Peng, G. Challenges and strategies for development of mycoherbicides. In *Fungal Biotechnology in Agricultural, Food, and Environment Applications*; Arora, D.K., Ed.; Marcel and Dekker: New York, NY, USA, 2004; pp. 111–121.

33. Tuite, J. *Plant Pathological Methods: Fungi and Bacteria*; Burgess Publication Co.: Minneapolis, MN, USA, 1969.

34. Steele, R.G.D.; Torrey, J.H.; Dickeys, D.A. Multiple Comparisons. In *Principles and Procedures of Statistics—A Biometrical Approach*; McGraw Hill: New York, NY, USA, 1997; p. 365.

35. Horsfall, J.G.; Barratt, R.W. An improved grading system for measuring diseases. *Phytopathology* **1945**, *35*, 655.

36. Zhao, S.; Shamoun, S.F. Effects of culture media, temperature, pH, and light on growth, sporulation, germination, and bioherbicidal efficacy of *Phoma exigua*, a potential biological control agent for salal (*Gaultheria shallon*). *Biocontrol Sci. Technol.* **2006**, *16*, 1043–1055.

37. Heiny, D.K.; Templeton, G.E. Effects of spore concentration, temperature, and dew period on disease of field bindweed caused by *Phoma proboscis. Phytopathology* **1991**, *81*, 905–909.

38. Auld, B.A.; Morin, L. Constraints in the development of bioherbicides. *Weed Technol.* **1995**, *9*, 638–652.

39. Heiny, D.K. *Phoma probocis* sp. nov. pathogenic on *Convolvoulus arvensis. Mycotaxon* **1990**, *36*, 457–471.

40. Heiny, D.K. Field survival of *Phoma proboscis* and synergism with herbicides for control of field bindweed. *Plant Dis.* **1994**, *78*, 1156–1164.

41. Gu, Z.M.; Ji, M.S.; Li, X.H.; Qi, Z.Q. Effects of environmental factors on effectiveness of *Phoma herbarum* strain SYAU-06 against *Commelina communis. Chin. J. Biol. Control* **2009**, *25*, 355–358.

42. Rajak, R.C.; Farkya, S.; Hasija, S.K.; Pandey, A.K. Fungi associated with congress weed (*Parthenium hysterophorus* L.). *Proc. Nat. Acad. Sci. India* **1990**, *60*, 165–168.

43. Pandey, A.K.; Luka, R.S.; Hasija, S.K.; Rajak, R.C. Pathogenicity of some fungi to *Parthenium* and obnoxious weed in Madhya Pardesh. *J. Biol. Control* **1991**, *5*, 113–115.

44. Pandey, S.; Pandey, A.K. Mycoherbicidal potential of some fungi against *Lantana camara* L.: A preliminary observation. *J. Trop. For.* **2000**, *16*, 28–32.

45. Pandey, A.K. Microorganism associated with weeds: Opportunities and challenges for their exploitation as herbicides. *Int. J. Mendel* **2000**, *17*, 59–62.

46. Deshmukh, P.; Rai, M.K.; Kövics, G.; Irinyi, L.M.; Karaffa, E.M. *Phomas*—Can these fungi be used as biocontrol agents and sources of secondary metabolites? In Proceedings of the 4th International Plant Protection Symposium at Debrecen University and 11th Trans-Tisza Plant Protection Forum, Debrecen, Hungary, 18–19 October 2006; pp. 224–232.

47. Quarles, W. New Biopesticides for IPM and Organic Production. *IPM Pract.* **2011**, *33*, 1–20.

48. Neumann, S.; Boland, G.J. Influence of host and pathogen variables on the efficacy of *Phoma herbarum*, a potential biological control agent of *Taraxacum officinale. Can. J. Bot.* **2002**, *80*, 425–429.

49. Stewart-Wade, S.M.; Boland, G.J. Selected cultural and environmental parameters influence disease severity of dandelion caused by the potential bioherbicidal fungi, *Phoma herbarum* and *Phoma exigua. Biocontrol Sci. Technol.* **2004**, *14*, 561–569.

50. Burgos, N.R.; Singh, V.; Tseng, T.-M.; Black, H.; Young, N.D.; Huang, Z.; Hyma, K.E.; Gealy, D.R.; Caicedo, A.L. The impact of herbicide-resistant rice technology on phenotypic diversity and population structure of United States weedy rice. *Plant Physiol.* **2014**, *166*, 1208–1220.

51. Scott, R.C.; Meins, K.B.; Smith, K.L. Tank-mix partners with Newpath herbicide for hemp sesbania control in a Clearfield rice-production system. *Ark. Agric. Res. Ser.* **2005**, *54*, 225–229.

52. Norsworthy, J.K.; Ward, S.M.; Shaw, D.R.; Llewellyn, R.S.; Nichols, R.L.; Webster, T.M.; Bradley, K.W.; Frisvold, G.; Powles, S.B.; Burgos, N.R.; *et al.* Reducing the risks of herbicide resistance: Best management practices and recommendations. *Weed Sci.* **2012**, *60*, 31–62.

53. Hoagland, R.E.; Boyette, C.D.; Vaughn, K.C. Interactions of quinclorac with a bioherbicidal strain of *Myrothecium verrucaria*. *Pest Technol.* **2011**, *5*, 88–96.

54. Boyette, C.D.; Hoagland, R.E.; Weaver, M.A.; Stetina, K.C. Interaction of the bioherbicide *Myrothecium verrucaria* and glyphosate for kudzu control. *Am. J. Plant Sci.* **2014**, *5*, 3943–3956.

55. Cartwright, K.; Boyette, C.D.; Roberts, M. Lockdown: Collego bioherbicide gets a second act. *Phytopathology* **2010**, *100*, doi:10.1080/09583157.2011.625398.

# 6

# Performance of Northwest Washington Heirloom Dry Bean Varieties in Organic Production

Carol Miles [1,†,*], Kelly Ann Atterberry [1,†] and Brook Brouwer [2,†]

[1] Department of Horticulture, Washington State University, Northwestern Washington Research and Extension Center, Mount Vernon, WA 98273, USA; E-Mail: kelly.atterberry@wsu.edu

[2] Department of Crop and Soil Sciences, Washington State University, Northwestern Washington Research and Extension Center, Mount Vernon, WA 98273, USA; E-Mail: brook.brouwer@wsu.edu

† These authors contributed equally to this work.

* Author to whom correspondence should be addressed; E-Mail: milesc@wsu.edu

Academic Editor: Herb Cutforth

**Abstract:** This two-year study compared nine northwest Washington dry bean (*Phaseolus vulgaris* L.) heirloom (H) varieties with 11 standard (S) commercial varieties in matching market classes using organic, non-irrigated production practices. Heirloom and standard varieties differed in days to harvest (DTH) (110 DTH and 113 DTH, respectively), while both days to harvest (113 DTH and 110 DTH) and yield (2268 kg·ha$^{-1}$ and 1625 kg·ha$^{-1}$) were greater in 2013 than in 2014. Varieties with the shortest DTH both years were "Bale" (H), "Coco" (H), "Decker" (H), "Ireland Creek Annie" (H and S), "Kring" (H) and "Rockwell" (H). Varieties that had the highest yield both years were "Eclipse" (S), "Lariat" (S) and "Youngquist Brown" (H). Only "Eclipse" (S) had the shortest cooking time both years, while "Rockwell" (H), "Silver Cloud" (S) and "Soldier" (S) had short cooking times in 2013, and "Orca" (S) and "Youngquist Brown" (H) had short cooking time in 2014. Varieties with the highest protein content both years were "Calypso" (S), "Coco" (S) and "Silver Cloud" (S). Further research should investigate yield of early maturing standard varieties, with a focus on color-patterned beans that are attractive for local markets.

**Keywords:** organic agriculture; crop rotation; *Phaseolus vulgaris*; heirloom varieties; niche market

# 1. Introduction

Common dry bean (*Phaseolus vulgaris* L.) is a sustainable crop, as it can be grown with less irrigation and fertilizer than many other crops, is a good rotation crop due to its nitrogen fixing capabilities (up to 44 kg·ha⁻¹) and can break some common disease cycles (such as Verticillium wilt caused by *Verticillium dahliae* Kleb.) [1–4]. Dry bean is also important in the human diet and is a good source of protein (18%–22%), dietary fiber, minerals (such as potassium, zinc and iron) and B vitamins (such as folic acid and B$_{12}$) [5]. Dry bean is one of a few pulse crops (other pulse crops include common dry pea (*Pisum sativum*), garbanzo bean or chickpea (*Cicer arietinum*) and lentil (*Lens culinaris*)) that can be grown successfully in relatively cool regions, such as northwest Washington.

In Washington, most dry beans are grown on a large scale in the Columbia Basin of eastern Washington, which accounted for 1.5% (9300 ha) of the dry beans produced in the U.S. in 2013 [6]. However, there is small-scale production of dry beans in every region of Washington, including northwest Washington, where dry beans have been grown organically on a small scale for over 100 years [7,8]. Regional heirloom dry bean varieties include a wide diversity of market classes, including navy, black, pinto and unique color-patterned types. Currently, in northwest Washington, organic heirloom and color-patterned varieties are sold at regional farmer's markets for $10–$30 per kg [9].

Grower experience suggests that local heirloom dry bean varieties are well adapted to the northwest Washington climate; however, available varieties have not been previously evaluated in systematic trials. Skagit County is the largest agricultural production region in northwest Washington and accounted for 19% of Washington State's vegetable production in 2012 [6]. The region has a cool, maritime climate with coastal influences from the Puget Sound (mean maximum of 23 °C in July and 24 °C in August at Mount Vernon, situated in the center of the Skagit Valley), with approximately 1300 growing degree days (GDD) (base 10 °C) during the dry bean growing period (mid-May–mid-September) [10,11]. In contrast, there are 1800–2800 GDD in the major dry bean growing regions in eastern Washington. The mean annual rainfall at Mount Vernon is 840 mm, with 110 mm occurring in June, 75 mm in July and 90 mm in August [11]. It is common practice for farmers in this region to not irrigate dry beans, as the presence of heavy dew, combined with precipitation and relatively low temperatures during the summer months, provides adequate moisture for crop production most years [12].

Heirloom dry bean varieties are an important crop genetic resource and are valued for local adaption, distinct colors and patterns, their use in traditional cooking and for providing a sense of self-sufficiency [13–16]. Dry beans were moved from their centers of origin, Central and South America, to North America through the migration of peoples up through Mexico, and they were moved to Europe, Africa and Asia by European explorers beginning in the 1500s [17–20]. Dry beans are a unique crop in that the bean is both the food and the seed crop, and they can be stored for several years with minimal processing. While dry beans remain an important component of sustainable farming systems worldwide, there is limited information regarding the performance of heirloom dry bean varieties, especially in modern organic production systems in the U.S. In Michigan, two heirloom varieties ("Michelite", a navy bean released in 1938, and "T-39", a 1970s "Black Turtle Soup" selection from California) and 30 newer varieties were compared in an organic field study [21]. The yields of "Michelite" (1514 kg·ha⁻¹) and "T-39" (yield not stated) were less than the overall mean (1909 kg·ha⁻¹), and the authors concluded that modern varieties appeared to be better suited to organic production due to improved disease resistance, plant architecture that keeps pods off the ground for better mechanical harvest and less pod mold and better allocation of biomass to the seed.

Irrigation is an important consideration for sustainable production systems. Dry bean yield and days to harvest (DTH) have been investigated in several regions of the U.S. under both irrigated and non-irrigated conditions. An irrigated field trial of 25 great northern, pink, pinto and red dry bean varieties from three locations in southern Idaho found a mean yield of 3465 kg·ha$^{-1}$, and the mean DTH was 93 [22]. In a comparison of irrigated and non-irrigated production systems in North Dakota, the mean yield of 20 dry bean varieties was 3069 kg·ha$^{-1}$ with irrigation and 2642 kg·ha$^{-1}$ without irrigation [23]. DTH was 87 under dryland and 100 under irrigation. In Michigan, two irrigated trials with 36 dry bean varieties were evaluated under organic and conventional production practices, and the mean yield was 2455 kg·ha$^{-1}$ for the organic system and 2700 kg·ha$^{-1}$ for the conventional system [24].

There is high demand for local organic crops throughout the U.S., and demand for organic and nutritious staple crops has opened a market opportunity for dry beans [25]. In western Washington, locally-grown dry beans are sold at farmers' markets, as well as to institutions (schools, hospitals, *etc.*) [9]. Cooking time is important to consumers who are purchasing beans in the dry form, especially for institutions, such as schools, that are preparing large quantities of beans. While several studies have investigated the influence of storage on cooking qualities, no studies have evaluated the cooking time of heirloom varieties [26,27]. The protein content of heirloom dry bean varieties is also generally unknown in comparison to newer varieties [28]. In a study in northwest Spain, an heirloom dry bean "Ganxet" had a higher protein level (28%) than the newer standard white kidney, navy and Tolosa black varieties (25% mean) [13].

In this study, dry bean varieties that have been grown in northwest Washington for 20–130 years were compared to standard varieties (seed grown outside the region) in the same market class, to determine which may be best suited for organic production in this region. Varieties were also measured for cooking time and protein content, as these are important considerations for consumers.

## 2. Materials and Methods

The experiment was carried out at Washington State University (WSU) Northwestern Washington Research and Extension Center (NWREC) at Mount Vernon. The field site was transitioned to certified organic during the study and was eligible for organic certification in fall 2014. The soil type is a silt loam, recently developed from alluvium and volcanic ash [29]. The field elevation was approximately 9 m above sea level and situated in the flood plains of the Skagit River, with a water table at 90–120 cm below the surface during the summer season.

Experimental design: The experiment utilized a randomized complete block design with four replications and was repeated 2 years (2013 and 2014). Plots were four rows wide and 3 m in length. Heirloom varieties were collected from growers in five counties (Clallam, Island, San Juan, Skagit and Whatcom) in northwest Washington, where they had been grown at each site as early as 1880 and as recently as 1990 [8]. A total of 13 heirloom (H) varieties were collected and planted in the first study year, each matched as closely as possible with a standard (S) variety in the same market class (Table 1); however, four heirloom varieties did not germinate well or had a climbing growth habit in the field plots and, so, were dropped from the study. A total of 20 dry bean varieties were included both years: 9 heirloom and 11 standard varieties.

**Table 1.** Heirloom dry bean varieties collected from counties in northwest Washington, USA, the date each heirloom variety was first grown in the region and standard varieties in the same market classes.

| Heirloom | County Collected | Date Introduction | Standard Variety | Market Class |
|---|---|---|---|---|
| Bale | Skagit | 1920s | Etna | Cranberry |
| Decker | Skagit | 1990 | Etna | Cranberry |
| Kring's Cranberry | San Juan | 1900s | Etna | Cranberry |
| Price's Soldier [1] | Skagit | 1920s | Soldier | Soldier |
| Wally's Soldier [1] | Clallam | 1940s | Soldier | Soldier |
| Rockwell | Island | 1880s | Soldier [2] | White with Partial Color |
| Hutterite | Skagit | 1990 | Hutterite | Yellow |
| Ireland Creek Annie | Skagit | 1990 | Ireland Creek Annie | Yellow |
| Youngquist Swedish Brown | Skagit | 1880s | Swedish Brown | Brown |
| Coco | Skagit | 1980 | Coco | Black |
| Skyriver Black | San Juan | 1990 | Eclipse | Black |
| Cannellinni [1] | San Juan | 1990 | Silver Cloud [3] | White Kidney |
| Calypso [1] | Skagit | 1990 | Calypso | Black and White |
|  |  |  | Orca [3] | White with Partial Color |
|  |  |  | Lariat [3] | Pinto |

[1] Heirloom variety was planted in 2013, but had low germination or a climbing growth habit and, so, was dropped from the study; standard variety in the same market class was grown both years; [2] color pattern of Rockwell is similar to Soldier, but bean shape is not; as no standard variety with a more similar appearance was found, Soldier was used as the standard variety for comparison; [3] new standard varieties grown in Washington, USA, included in this study to provide a contrast to western Washington heirloom varieties.

Planting and field maintenance: Beans were direct-seeded by hand 22 May 2013 and 15 May 2014. Both years, spacing in the row was 5 cm, and seeding depth was 5 cm. Rows were spaced 76 cm center-to-center in 2013 and 86 cm in 2014; spacing between rows was increased in the second year to provide better tractor access for cultivation to control weeds. Both years, the seed was inoculated with a granular *Rhizobium spp.* blend (Guard-N; Johnny's Select Seeds, Winslow, ME, USA). Plots were side dressed with organic fertilizer (Proganic 8N-0.9P-3.3K; Wilbur-Ellis, Mount Vernon, WA, USA) 21 days after planting (DAP) at a rate of 51 kg N·ha$^{-1}$ in 2013 and 60 kg N·ha$^{-1}$ in 2014. For weed control, tractor cultivation was done between rows with a 2-headed rototiller 6 and 21 DAP in 2013 and 14 and 36 DAP in 2014. In 2014, to reduce in-row weed pressure, plants were hilled (5–8 cm) 41 DAP with 25-cm discs mounted on a cultivation tractor (Allis-Chalmers G; Briggs and Stratton, Wauwatosa, WI, USA). Hand weeding was done within rows 42 and 75 DAP in 2013 and 53 DAP in 2014. The study was not irrigated either year, following common grower practices in the region.

Plant growth, flowering and yield: Data were collected from the center 1.5 m of the center 2 rows of each plot. Each year, emergence was measured twice weekly in each plot until 85% of seeds emerged. Fifty percent flowering was recorded for each plot when half of the plants had one flower (55 DAP in 2013 and 66 DAP in 2014); plant height was measured for 10 randomly-selected plants when all varieties had reached 50% flowering. Height was measured from the base of the plant to the top node by straightening the plant.

The center 1.5 m of the center 2 rows in each plot was harvested by hand when beans in the plot were dry, but before pod shatter. Plants from each plot were placed in a burlap sack and dried at 41 °C for 36 h or until the stems snapped and bean seed moisture content was approximately 12%. Whole dry plants were fed through a chipper-mulcher (Model 500; Roto-Hoe, Newbury, OH, USA) that had been converted to a small-scale standing thresher [30]. The large debris was removed, and beans were separated from small debris with a small-scale standing cleaner [31]. Beans were then sorted by hand, and moldy, split and immature beans were removed. The marketable yield and weight of 100 representative beans were recorded for each plot.

Bean cooking time, firmness and protein content: Six months after harvest in 2013 and 2014, bean cooking time was measured using a Mattson Bean Cooker (MBC) (Canadian Grain Commission, Vancouver, British Columbia, Canada). Bean varieties with a sufficient number of beans in at least three replicates were included in this measurement and included four heirloom and four standard varieties in the same market class, as well as one heirloom and seven standard varieties in different market classes (16 varieties total). For each replicate sample, 55 g of beans were placed in a separate 250-mL plastic bottle with 125 mL deionized (DI) water and soaked at room temperature for 12 h; the beans were then strained and patted dry. Beans that had absorbed water were selected and placed on the MBC cooking rack, with one "pin" placed on top of each bean. The MBC was placed into an electric cooking pot (Model DCP-6; Dazey Products Co., Industry Airport, KS, USA) filled within 5 cm from the top with boiling (205 °C) DI water; the temperature was immediately reduced to 150 °C (boiling point). For each sample, when 80% of the pins pierced the beans, this time was recorded as the cooking time for that sample [32]. The MBC was then removed immediately from the water, the beans were placed into a plastic weighing dish, which was inserted into a small plastic bag and placed in the freezer (0 °C) for approximately three minutes to prevent further cooking. Firmness of all 25 beans per sample was measured with a mechanical force gauge (L-500; Ametek, Hunter Spring Division, Hatfield, PA, USA) using a needle sharp tip (FG-M6PUNCTURE-ST; Shimpo, Kyoto, Japan) mounted on a drill press (Craftsmen Model 9-25921; Sears, Roebuck, and Co., Chicago, IL, USA). Firmness was recorded in units of gram force, averaged for each sample, and converted to Newtons (N) by multiplying by $9.807 \times 10^{-3}$.

Percent protein (g protein per 100 g uncooked dry beans) was measured for three replicates of the same 16 bean varieties as were evaluated for cooking time. Samples (175 g) of whole uncooked dry beans from each plot were analyzed by a commercial laboratory (Soiltest Farm Consultants, Inc., Moses Lake, WA, USA) eight months after harvest.

Statistical analysis: All data were analyzed using JMP (JMP v. 11.2.0, SAS Institute Inc., Cary, NC, USA) with $\alpha = 0.05$. Means were separated using the least significant difference test. When data violated the assumptions of ANOVA, a transformation was used for analysis following the range method outlined by Kirk [33].

## 3. Results

There were significant differences due to variety type (heirloom (H) and standard (S) varieties) for some parameters measured and due to variety and year for most parameters. Except for plant height, there were no significant interactions between type and year but there were many significant interactions between year and variety (Table 2).

**Table 2.** ANOVA tests and means for days after planting (DAP) to 50% flower, plant height (cm) at 50% flowering, days to harvest (DTH), yield (kg·ha⁻¹), cooking time with a Mattson bean cooker (MBC), bean firmness after cooking with MBC and protein content (%) of five heirloom and 11 standard dry bean varieties grown at Washington State University Northwestern Washington Research and Extension Center, Mount Vernon, WA, USA, in 2013 and 2014.

| Effect | DAP to 50% Flower [y] | Plant Height (cm) | DTH | Yield (kg ha⁻¹) | MBC Time (min) | MBC Firmness (N) | Protein (%) |
|---|---|---|---|---|---|---|---|
| | | | | *p*-value [z] | | | |
| Type | 0.0570 | 0.0398 | 0.0353 | 0.7049 | 0.4593 | 0.3510 | 0.0893 |
| Variety | <0.0001 | <0.0001 | <0.0001 | <0.0001 | 0.008 | 0.0005 | <0.0001 |
| Year | <0.0001 | 0.3889 | 0.0097 | <0.0001 | <0.0001 | <0.0001 | 0.0058 |
| Type × Year | 0.8375 | 0.0025 | 0.3341 | 0.5794 | 0.2146 | 0.7262 | 0.0973 |
| Year × Variety | 0.0002 | <0.0001 | 0.0968 | 0.0158 | <0.0001 | 0.0330 | <0.0001 |
| *Type* | | | | Mean [x] | | | |
| Heirloom | 51 | 35 | 110 [y] | 1852 | 25.9 | 8.0 | 20.3 |
| Standard | 54 | 36 | 113 | 1983 | 21.7 | 7.4 | 20.7 |
| *Year* | | | | Mean [x] | | | |
| 2013 | 54 | 36 | 113 [y] | 2268 | 19.7 | 8.6 | 21.2 |
| 2014 | 52 | 35 | 110 | 1625 | 26.3 | 6.6 | 20.0 |

[z] All data were analyzed using JMP (JMP v. 11.2.0, SAS Institute Inc., Cary, NC, USA) with $\alpha = 0.05$. When data violated the assumptions of ANOVA, a transformation was used for analysis following the range method outlined by Kirk [33]. [y] Planting date was 22 May 2013 and 15 May 2014. [x] All means are non-transformed.

Plant Emergence and Growth: In 2013, the mean emergence for all varieties at 12 DAP was 13% with a range of 0%–31%, and all varieties reached 85% emergence by 26 DAP. In 2014, almost all varieties reached 85% emergence by 12 DAP; Hutterite (H) reached only 50% emergence at 12 DAP and did not reach 85% emergence. Although soil temperature (5-cm depth) was above the recommended level for beans (13 °C) both years, soil temperature was lower at planting in 2013 (14 °C) than in 2014 (19 °C). There was no difference between heirloom and standard varieties in DAP to 50% flowering ($p = 0.06$), but standard varieties were slightly taller at 50% flowering than heirloom varieties ($p = 0.04$) (Table 2). There was a significant difference in mean days to 50% flower due to year (54 DAP and 36 cm, respectively; $p < 0.0001$), but not in plant height. Varieties differed in DAP to 50% flowering and in plant height ($p < 0.0001$, both), and there was a significant interaction between year and variety for both ($p = 0.0002$ and $p < 0.0001$, respectively). In 2013 the range in DAP to 50% flower was 50–62, and the range in plant height was 27–45 cm ($p < 0.0001$) (Table 3). In 2014, the range in DAP to 50% flower was 49–61, and the range in plant height was 22–57 cm ($p < 0.0001$) (Table 3). The overall mean air temperature was similar for May, June and July both years (16 °C in 2013 and 17 °C in 2014), and the growing degree days (GDD; base 10 °C) were slightly less for this time period in 2013 (822 GDD) as compared to 2014 (895 GDD).

**Table 3.** Mean days after planting (DAP) to 50% flower, plant height (cm) at 50% flowering and protein content (%) of nine heirloom and 11 standard dry bean varieties grown at Washington State University Northwestern Washington Research and Extension Center, Mount Vernon, WA, USA, in 2013 and 2014. H, heirloom varieties; S, standard varieties.

| Variety | 50% Flowering (DAP) [z,y] | | Height (cm) | | Protein (%) | |
|---|---|---|---|---|---|---|
| | 2013 | 2014 | 2013 | 2014 | 2013 | 2014 |
| Bale (H) | 50 e [x] | 49 c | 42.4 ab | 28.0 de | 19.47 efg | 20.43 de |
| Calypso (S) | 50 e | 49 c | 36.3 cdefg | 24.2 bc | 25.46 ab | 23.41 a |
| Coco (H) | 51 cde | 49 c | 34.3 defgh | 34.3 fgh | 24.57 abc | 22.71 ab |
| Coco (S) | 50 e | 49 c | 33.5 fgh | 33.7 fgh | -- [w] | -- |
| Decker (H) | 51 cde | 49 c | 36.4 bcdefgh | 21.6 a | -- | -- |
| Eclipse (S) | 62 a | 61 a | 26.5 i | 45.1 jk | 21.26 de | 17.24 h |
| Etna (S) | 59 b | 49 c | 37.3 bcdef | 26.7 cd | 18.24 gh | 19.23 f |
| Hutterite (H) | 51 cde | 49 c | 26.5 i | 30.5 def | -- | -- |
| Hutterite (S) | 55 bcd | 52 b | 36.0 defg | 43.2 ij | 18.89 fgh | 16.98 h |
| Ireland Creek Annie (H) | 50 e | 49 c | 38.6 abcdef | 30.5 def | -- | -- |
| Ireland Creek Annie (S) | 50 e | 50 bc | 39.8 bcde | 27.9 de | 19.42 efg | 18.78 fg |
| Kring (H) | 51 de | 49 c | 37.3 bcdef | 23.5 ab | 20.81 def | 21.31 cd |
| Lariat (S) | 57 bc | 61 a | 41.8 abc | 73.7 l | 17.11 h | 17.63 gh |
| Orca (S) | 62 a | 61 a | 28.3 hi | 34.3 fgh | 21.24 de | 21.88 bc |
| Rockwell (H) | 50 e | 49 c | 39.5 bcde | 29.8 def | 20.20 efg | 21.63 bc |
| Silver Cloud (S) | 51 de | 52 b | 31.5 ghi | 36.2 ghi | 26.36 a | 23.26 a |
| Skyriver (H) | 51 de | 49 c | 40.3 abcd | 38.1 hij | 22.83 cd | 21.41 cd |
| Soldier (S) | 51 de | 49 c | 34.3 efg | 36.2 ghi | 23.86 bc | 19.60 ef |
| Swedish Brown (S) | 53 bcd | 49 c | 30.8 ghi | 31.8 efg | 20.87 def | 17.74 gh |
| Youngquist Brown (H) | 62 a | 61 a | 45.4 a | 56.6 kl | 18.35 gh | 17.08 h |
| Mean of Response | 54 | 52 | 36 | 35.3 | 21.15 | 20.02 |
| *p* Value | <0.0001 | <0.0001 | <0.0001 | <0.0001 | <0.0001 | <0.0001 |
| Transformation | Rank | Rank | None | Reciprocal | None | None |

[z] All data were analyzed using JMP (JMP v. 11.2.0, SAS Institute Inc., Cary, NC, USA) with $\alpha = 0.05$. When data violated the assumptions of ANOVA, a transformation was used for analysis following the range method outlined by Kirk [33]. All means are non-transformed. [y] Planting date was 22 May 2013 and 15 May 2014. [x] Treatment means within a row followed by the same letter are not significantly different using Fisher's LSD test ($\alpha = 0.05$). [w] There were insufficient beans harvested from plots each year to complete protein analysis.

Days to harvest and yield: Overall for both years, heirloom varieties were harvested three days earlier than standard varieties (110 DTH and 113 DTH, respectively; $p = 0.04$) (Table 2). The average for all varieties was 113 DTH (13 September) in 2013 with a range of 104–127 DTH, as compared to 110 DTH (2 September) in 2014 with a range of 106–118 DTH ($p < 0.0001$ both) (Figure 1). While the overall mean air temperature for the growing season was similar both years (16 °C in 2013 and 17 °C in 2014), accumulated GDD for the growing season were slightly lower in 2013 (1490 GDD) as compared to 2014 (1524 GDD). There was no significant interaction for DTH between variety and year ($p = 0.10$; Table 2). Varieties with the shortest DTH both years were "Bale" (H), "Coco" (H), "Decker" (H), "Ireland Creek Annie" (H), "Ireland Creek Annie" (S), "Kring" (H) and "Rockwell" (H); while varieties with the longest DTH both years were "Eclipse" (S), "Hutterite" (S) "Skyriver" (H), "Youngquist Brown" (H) and "Orca" (S) (Figure 1).

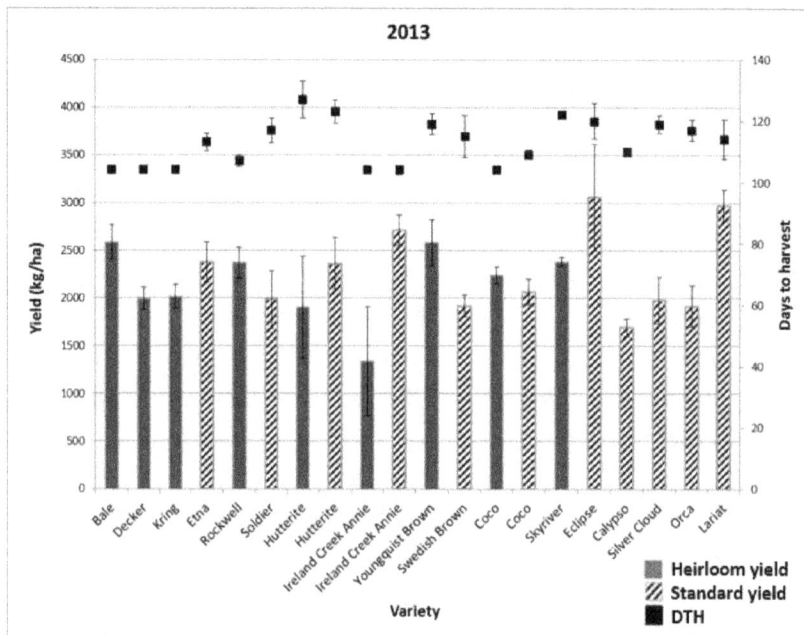

(a)

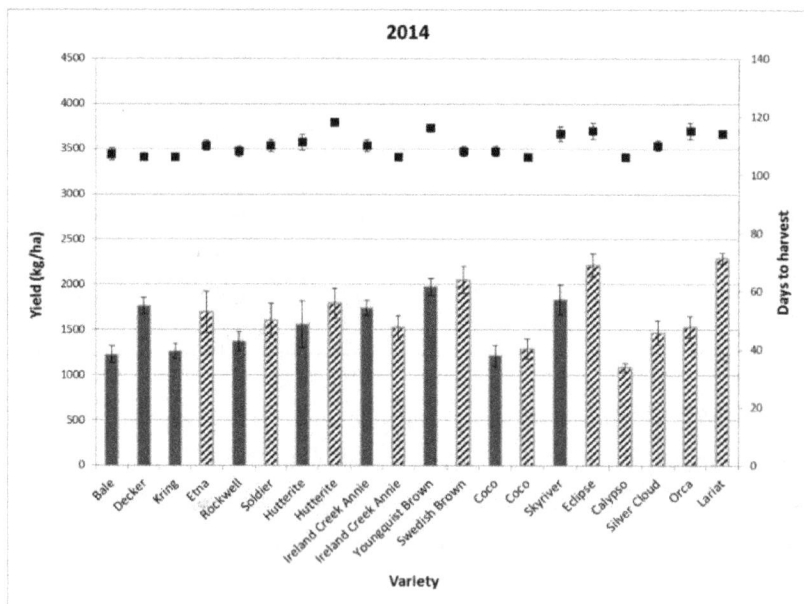

(b)

**Figure 1.** Days to harvest and mean yield of heirloom and standard dry bean varieties at Washington State University Northwestern Washington Research and Extension Center, Mount Vernon, WA, USA in 2013 (**a**) and 2014 (**b**); varieties differed significantly both years ($p < 0.0001$ for all). All data were analyzed using JMP (JMP v. 11.2.0, SAS Institute Inc., Cary, NC, USA) with $\alpha = 0.05$. When data violated the assumptions of ANOVA, a transformation was used for analysis following the range method outlined by Kirk [33]. All means are non-transformed. Error bars are based on the standard deviation.

Overall yield of heirloom varieties (1852 kg·ha$^{-1}$) did not differ significantly from standard varieties (1983 kg·ha$^{-1}$) ($p = 0.70$); there was no significant difference between type either year ($p = 0.75$ and $p = 0.20$, respectively); and there was no interaction between type and year ($p = 0.58$) (Table 2). Mean yield for all varieties was significantly greater in 2013 (2268 kg·ha$^{-1}$) than in 2014 (1625 kg·ha$^{-1}$) ($p < 0.0001$); yield differed significantly among varieties each year ($p = 0.002$ and $p < 0.0001$, respectively); and there was a significant interaction between variety and year ($p = 0.02$) (Table 2 and Figure 1). Varieties that were high yielding both years were "Eclipse" (S), "Lariat" (S) and "Youngquist Brown" (H); while only "Calypso" (S) was low yielding both years. While there were 11% more plants per hectare in 2013 than in 2014 (approximately 263,000 and 233,000 plants per hectare, respectively), due to different between-row spacing each year (76 cm and 86 cm, respectively), yield was 30% greater in 2013. There was little difference in mean temperature for the growing season both years (16 °C and 17 °C, respectively), and while precipitation was essentially equal both years (232 mm and 249 mm, respectively), there was no rainfall in July 2013, while there was 32 mm of rainfall in July 2014. Other studies have found that dry bean yield did not increase when plant populations increased from 168,000–240,000 plants per ha$^{-1}$ [34] or from 222,000–296,000 plants per ha$^{-1}$ [35]. The result from this study suggests that optimal planting density needs to be determined in northwest Washington; however, in organic production systems row spacing must be adequate to allow for mechanical weed control, and thus, plant population is likely to be determined by equipment needs and not yield optimization.

Bean cooking time, firmness and protein content. Cooking time measured with the MBC for heirloom varieties (26 min) did not differ significantly from standard varieties (22 min) ($p = 0.46$); there was no significant difference between type either year ($p = 0.13$ both years); and there was no significant interaction between type and year ($p = 0.21$) (Table 2). There was a significant difference in mean cooking time for all varieties ($p = 0.008$); mean cooking time was faster in 2013 (20 min) than 2014 (26 min) ($p < 0.0001$); and there was a significant interaction between variety and year ($p < 0.0001$) (Table 2). Only "Eclipse" (S) had the shortest cooking time both years, while "Rockwell" (H), "Silver Cloud" (S) and "Soldier" (S) had short cooking time in 2013, and "Orca" (S) and "Youngquist Brown" (H) had short cooking times in 2014 (Figure 2). "Coco" (S) had the longest cooking time both years, while "Etna" (S) and "Skyriver" (H) also had long cooking times in 2013. While cooking time measured with an MBC is shorter than with most home-cooking methods, it does provide relative values for comparison purposes.

Overall, firmness after cooking with the MBC also did not differ between heirloom varieties (8.0 N) and standard varieties (7.4 N) ($p = 0.35$) (Table 2). While there was no significant interaction between type and year ($p = 0.73$), there was a difference between years ($p < 0.0001$). In 2013, there was no significant difference between heirloom and standard varieties (8.9 N and 8.5 N, respectively; $p = 0.34$), while in 2014, the firmness of heirloom varieties (7.0 N) was greater than for standard varieties (6.4 N) ($p = 0.04$) (data not shown). There was a significant difference in bean firmness among varieties overall ($p = 0.0005$); there were significant differences among varieties each year ($p = 0.002$ in 2013, $p < 0.0001$ in 2014); and there was a significant interaction between year and variety ($p = 0.03$) (Table 2 and Figure 2). Mean bean firmness for all varieties was greater in 2013 (8.6 N) than in 2014 (6.6 N), and varieties with the lowest firmness both years were "Eclipse" (S) and "Hutterite" (S), while only "Soldier" (S) had the greatest firmness both years.

**(a)**

**(b)**

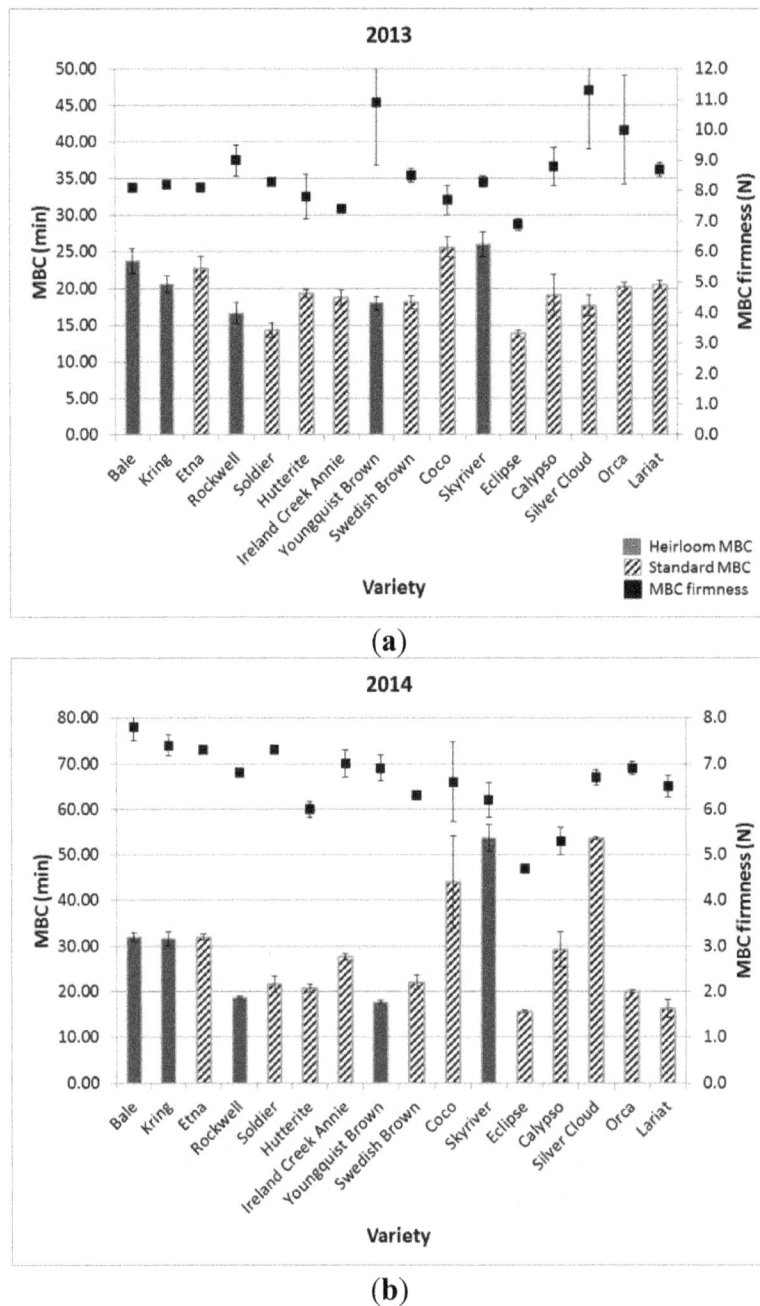

**Figure 2.** Mean cooking time (min) and firmness (N) using a Mattson bean cooker six months after harvest of 16 dry bean varieties grown at Washington State University Northwestern Washington Research and Extension Center, Mount Vernon, WA, USA, in 2013 (**a**) and 2014 (**b**); varieties differed significantly both years ($p < 0.0001$ and $p = 0.002$, respectively). All data were analyzed using JMP (JMP v. 11.2.0, SAS Institute Inc., Cary, NC, USA) with $\alpha = 0.05$. When data violated the assumptions of ANOVA, a transformation was used for analysis following the range method outlined by Kirk [33]. All means are non-transformed. Error bars based on standard deviation.

Protein content for heirloom varieties (20.3%) did not differ significantly from standard varieties (20.7%) ($p = 0.09$); there was no significant difference between type either year ($p = 0.17$ in 2013 and $p = 0.48$ in 2014); and there was no significant interaction between type and year ($p = 0.10$) (Table 2). There was a significant difference in protein content for varieties overall ($p < 0.0001$); protein content for varieties was greater in 2013 (21.2%) than 2014 (20.0%) ($p = 0.006$); and there was a significant interaction between variety and year ($p < 0.0001$) (Tables 2 and 3). Varieties with the highest protein both years were "Calypso" (S), "Coco" (S) and "Silver Cloud" (S), while "Hutterite" (S) and "Youngquist Brown" (H) had the lowest protein both years (Figure 2).

## 4. Discussion

The most important consideration when selecting a dry bean variety for a cool region, such as northwest Washington, is the number days to harvest. DTH with 1500 GDD on average for the two study years was 22 days longer than in other bean-growing regions [22,23] Northwest Washington heirloom varieties matured three days earlier than standard varieties, and six heirloom varieties had the shortest DTH both years. This difference could be significant as rainfall starting in mid to late September can prevent beans from reaching a sufficiently dry stage for harvest and storage. Studies of early maturing varieties are needed to determine if there are standard varieties that are well suited to this region. It is important to note that the average GDD during this two-year study (1500 GDD) was 15% greater than the previous 20-year average (1300 GDD).

Overall yield of dry beans in this study was 1950 kg·ha$^{-1}$, which was comparable to yield in one organic field study in Michigan (1909 kg·ha$^{-1}$, non-irrigated), but less than a different organic field study in Michigan (2455 kg·ha$^{-1}$, irrigated), as well as conventional field studies in Idaho (3465 kg·ha$^{-1}$, irrigated) and North Dakota (2642 kg·ha$^{-1}$, non-irrigated) [21–24]. Nevertheless, this study demonstrates that dry beans can be a productive crop in northwest Washington, even without irrigation, and, therefore, could be a suitable legume crop for crop rotation in the region. To better understand yield potential in the region, future studies should be carried out with irrigation.

Consumer demand for regionally-produced staple crops has opened a market opportunity for regionally-produced crops, such as dry beans [25]. While this study shows that there are many dry bean varieties that can be grown successfully in the region, consumers who shop at farmers' markets (as well as other direct market customers) may be more likely to purchase unique color-patterned beans or heirloom varieties [36]. These niche market varieties may be sold at premium prices ($13–$31 per kg) [9]. While not as profitable, schools could be secure and substantial customers for local crops, including dry beans [37,38]. The National School Lunch Program (NSLP) requires school cafeterias to serve 120 mL dry beans and/or dry peas per student per week [39]; in northwest Washington where 530,000 students are enrolled in NSLP, this equates to 51,000 kg of dry beans served in school cafeterias per week or 2000 tons per year (based on a school year of 36 weeks) [40]. For institutional buyers, it is likely that price will be the most important factor, so identifying the highest yielding cultivars regardless of appearance may be a priority for this market outlet.

Varieties that are faster cooking may be of particular interest to institutions where large quantities of beans are prepared. Cooking time measured with the MBC is shorter than can be obtained with most home cooking methods, and the MBC underestimates cooking time as beans are generally too firm for eating [32]. However, the MBC provides a comparative cooking time and may be used to identify beans that are fast or slow cooking. Of the beans with the shortest cooking time in this study, "Eclipse" (S) was the only variety that was among the softest after cooking. In contrast, "Silver Cloud" (S) and "Rockwell" (H) both had a shorter cooking time, but their beans were among the most firm after cooking. Protein content of dry bean varieties is also of interest to many consumers. The mean protein content for dry bean varieties in this study was 21% with a range of 17%–26%. One serving (120 mL or 96 g dry uncooked beans) of a high protein variety contains 24 g of protein, and one serving of a low protein variety contains 16 g of protein. In order to acquire the same amount of protein when eating a low protein bean variety, an additional 8 g of beans are needed, or the equivalent of approximately 10 dry uncooked beans, or approximately two spoonfuls. While this amount could be considered insignificant for a household, for institutions, such as schools in northwest Washington, this would amount to an additional 5300 kg of dry beans per week or 191 tons per year, and the selection of dry beans with high protein may be desirable.

## 5. Conclusion

This study shows that dry beans have the potential to be grown as a commercial organic crop in northwest Washington; however, other considerations that must first be addressed in the region are equipment needs for mechanical harvest, drying and cleaning [8]. Additionally, a cooking and packaging facility is likely needed for the school market, as soaking and cooking dry beans in school kitchens is currently perceived as a barrier by school food service staff [9]. Existing facilities, such as grain dryers, may be suitable for drying beans for long-term storage, and vegetable processing plants in the region may be suitable for cooking and canning the beans; however, these would need to be tested to ensure suitability for dry beans.

## Acknowledgments

We thank Ed Scheenstra for statistical analysis assistance and Ron Dralle, Jesse Wimer, Charlene Grahn and Carolyn Klismith for field work assistance. Funding for this project from American Pulse Association, Northwest Agriculture Research Foundation and Washington State University Center for Sustaining Agriculture and Natural Resources is gratefully acknowledged.

## Author Contributions

Carol Miles planned this study, wrote and managed the grants that funded this work, oversaw the field and laboratory experiments, carried out the final data analysis, and was the lead author for this manuscript. Kelly Atterberry carried out the field and laboratory experiments as part of her Master's in Science project, and Brook Brouwer assisted with field trial implementation and preliminary data analysis as part of his PhD project.

## Conflicts of Interest

The authors declare no conflict of interest.

## References

1.  Canevari, W.M.; Frate, C.A.; Godfrey, L.D.; Goodell, P.B.; Long, R.F.; Mickler, C.J.; Mueller, S.C.; Schmierer, J.L.; Temple, S.R. UC IPM Pest Management Guidelines: Dry Beans General Information. Available online: http://www.ipm.ucdavis.edu/PDF/PMG/pmgdrybeans.pdf (accessed on 18 June 2013).

2. Davis, J.G.; Brick, M.A. Colorado State University Extension. Fertilizing Dry Beans. Crop Series. Fact Sheet No. 0.539. Available online: http://www.ext.colostate.edu/pubs/crops/00539.html (accessed on 1 September 2014).

3. Kirkpatrick, A.; Browning, L.; Bauder, J.; Waskom, R.; Neibauer, M.; Cardon, G. A Practical Guide to Choosing Crops Well-Suited to Limited Irrigation. Available online: http://region8water.colostate.edu/PDFs/Irrigating%20with%20Limited%20Water%20Supplies.pdf (accessed on 18 September 2014).

4. Farr, D.; Bills, G.; Charmuris, G. *Fungi of Plants and Plant Products in the United States*; American Pathological Society: St. Paul, MN, USA, 1989.

5. Leterme, P. Recommendations by health organizations for pulse consumption. *Br. J. Nutr.* **2002**, *88*, 239–242.

6. USDA-NASS. Statistics by State: Washington, DC, USA, 2013. Available online: http://www.nass.usda.gov/Statistics_by_State/Idaho/Publications/Crops_Press_Releases/pdf/DB08_1.pdf (accessed on 10 January 2013).

7. Miles, C.; Sonde, M. Survey of Washington Dry Bean Growers. Available online: http://naldc.nal.usda.gov/download/IND43757215/PDF (accessed on 10 January 2014).

8. Brouwer, B.O. Plant Breeding for Regional Food Systems: Investigating Craft Malt, Disease Resistance and Production Potential of Barley and Dry Beans in Western Washington. Ph.D. Thesis, Washington State University, Pullman, WA, USA, 2015.

9. Atterberry, K.A. Increasing Knowledge, Preference, and Availability of Pulse Crops in K-12 Schools. Master's Thesis, Washington State University, Pullman, WA, USA, 2015.

10. AgWeatherNet. 2014. 24-h Data Report. Available online: http://weather.wsu.edu/awn.php (accessed on 9 November 2014).

11. NOAA. 1981–2010 Climate Normals. Available online: http://www.ncdc.noaa.gov/cdo-web/datatools/normals (accessed on 9 November 2014).

12. McMoran, D.W. 2011 Skagit County Agriculture Statistics. Available online: http://ext100.wsu.edu/skagit/wp-content/uploads/sites/5/2014/03/2011AgStats.pdf (accessed on 1 December 2013).

13. Casanas, F.; Bosch, L.; Pujola, M.; Sanchez, E.; Sorribas, X.; Baldi, M.; Nuez, F. Characteristics of common bean landrace (*Phaseolus vulgaris*) of great culinary value and selection of a commercial inbred line. *J. Sci. Food Agric.* **1999**, *79*, 693–698.

14. Gepts, P. *Phaseolus vulgaris* (Beans). Available online: http://www.plantsciences.ucdavis.edu/gepts/a1749.pdf (accessed on 10 October 2014).

15. Nabhan, G. *Conservation You Can Taste: Best Practices in Heritage Food Recovery and Success in Restoring Agricultural Biodiversity over the Last Quarter Century*; University of Arizona Southwest Center: Tucson, AZ, USA; Slow Food USA: Brooklyn, NY, USA, 2013.

16. Veteto, J. Seeds of persistence: Agrobiodoversity in the American mountain south. *Cult. Agric. Food Environ.* **2014**, *36*, 17–27.

17. Angioi, S.A.; Rau, D.; Attene, G.; Nanni, L.; Bellucci, E.; Logozzo, G.; Negri, V.; Zeuli, P.L.S.; Papa, R. Beans in Europe: Origin and structure of the European landraces of *Phaseolus vulgaris* L. *Theor. Appl. Genet.* **2010**, *121*, 829–843.

18. Asfaw, A.; Blair, M.W.; Almekinders, C. Genetic Diversity and Population Structure of Common Bean (*Phaseolus Vulgaris* L.) Landraces from the East African Highlands. *Theor. Appl. Genet.* **2009**, *120*, 1–12.

19. Gepts, P. Origin and evolution of common bean: Past events and recent trends. *HortScience* **1998**, *33*, 1124–1130.

20. Hart, J.P.; Scarry, C.M. The age of common beans (*Phaseolus vulgaris*) in the Northeastern United States. *Am. Antiq.* **1999**, *64*, 653–658.

21. Heilig, J.; Kelly, J. Performance of dry bean genotypes grown under organic and conventional production systems in Michigan. *Agron. J.* **2012**, *104*, 1485–1492.

22. Singh, S.H.T.; Lema, M.; Webster, D.; Strausbaugh, C.; Miklas, P.; Schwartz, H.; Brick, M. Seventy-five years of breeding dry bean of the western USA. *Crop Sci.* **2007**, *47*, 1–9.

23. Kandel, H.; Osorno, J.; VanderWal, J.; Kloberdanz, M. North Dakota Dry Bean Variety Trial Results for 2013 and Selection Guide. Available online: http://www.ag.ndsu.edu/pubs/plantsci/rowcrops/a654_13.pdf (accessed on 1 March 2015).

24. Kelly, J.D.; Wright, E.; Blakely, N.; Heilig, J. Dry Bean Yield Trials. Available online: http://www.varietytrials.msu.edu/wp-content/uploads/2012/12/2014_Dry_Bean_Report.pdf (accessed on 2 April 2015).

25. Martinez, S.; Hand, M.; Da Pra, M.; Pollack, S.; Ralston, K.; Smith, T.; Vogel, S.; Clark, S.; Lohr, L.; Low, S.; *et al.* Local Food Systems: Concepts, Impacts, and Issues. Available online: http://www.ers.usda.gov/media/122868/err97_1_.pdf (accessed on 16 December 2014).

26. Martin-Cabrejas, M.A.; Esteban, R.M. Hard-to-cook phenomenon in beans: Changes in antinutrient factors and nitrogenous compounds during storage. *J. Sci. Food Agric.* **1995**, *69*, 429–435.

27. Paredes-Lopez, O.; Carabez-Trejo, A.; Palma-Tirado, L.; Reyes-Moreno, C. Influence of hardening procedure and soaking solution on cooking quality of common beans. *Plants Foods Hum. Nutr.* **1990**, *41*, 155–164.

28. Tiwari, B.; Gowen, A.; McKenna, B. *Pulse Foods: Processing, Quality, and Nutraceutical Applications*; Elsevier Inc.: Atlanta, GA, USA, 2011.

29. Natural Resource Conservation Service. Web Soil Survey. Available online: http://websoilsurvey.sc.egov.usda.gov/ (accessed on 20 May 2014).

30. Miles, C. Washington State University Vegetable Research and Extension. WSU Seed Cleaner. Available online: http://vegetables.wsu.edu/NicheMarket/WSU-SeedCleaner.html (accessed on 1 March 2013).

31. Miles, C. Washington State University Vegetable Research and Extension. Small-Scale Thresher. Available online: http://vegetables.wsu.edu/NicheMarket/SmallScaleThresher.html (accessed on 1 March 2013).

32. Wang, N.; Daun, J.K. Determination of cooking times of pulses using an automated mattson cooker apparatus. *J. Sci. Food Agric.* **2005**, *85*, 1631–1635.

33. Kirk, R.E. *Experimental Design*, 2nd ed.; Wadsworth: Belmont, CA, USA, 1982.

34. Mehraj, K.; Brick, M.A.; Pearson, C.H.; Ogg, J.B. Effects of bed width, planting arrangement, and plant population on seed yield of pinto bean cultivars with different growth habits. *J. Prod. Ag.* **2013**, *9*, 79–82.

35. Schatz, B.G.; Zwinger, S.F.; Endres, G.J. Dry Edible Bean Performance as Influenced by Plant Density. Available online: https://www.ag.ndsu.edu/archive/carringt/agronomy/Research/ProdMgmt/Dry%20Bean%20Plant%20Density.pdf (accessed on 15 August 2015).

36. USDA-NAL. Heirloom Varieties. Available online: http://afsic.nal.usda.gov/alternative-crops-and-plants/specialty-heirloom-and-ethnic-fruits-and-vegetables/heirloom-varieties (accessed on 9 November 2014).

37. Berkenkamp, J.; Skaar, K.; Vanslooten, E. Using Regionally Grown Grains and Pulses in School Meals: Best Practices, Supply Chain Analysis, and Case Studies. Available online: http://www.iatp.org/files/2015_02_02_GrainsAndPulses_EMV.pdf (access on 13 June 2015).

38. Stone, M.; Brown, K.; Comnes, L.; Koulias, J. *Rethinking School Lunch: A Planning Framework from the Center of Ecoliteracy*, 2nd ed.; Center for Ecoliteracy: Berkeley, CA, USA, 2010.

39. USDA-FNS. National school lunch program. Available online: http://www.fns.usda.gov/nslp/national-school-lunch-program-nslp (accessed on 1 July 2014).

40. USDA-FNS. Annual Summary of Food and Nutrition Service Programs. Available online: http://www.fns.usda.gov/data-and-statistics (accessed on 15 May 2015).

# 7

# The Response of Sorghum, Groundnut, Sesame, and Cowpea to Seed Priming and Fertilizer Micro-Dosing in South Kordofan State, Sudan

**Elgailani A. Abdalla [1], Abdelrahman K. Osman [1], Mahmoud A. Maki [1], Fadlalmaola M. Nur [1], Salah B. Ali [1] and Jens B. Aune [2,\***

[1] Elobeid Research Station, Elobeid 611, Sudan; E-Mails: elgailani_ers@hotmail.com (E.A.A.); arkosman@hotmail.com (A.K.O.); mahmekki@yahoo.com (M.A.M.), fadelmoh75@yahoo.com (F.M.N.); salahbakor328@yahoo.com (S.B.A.)

[2] Department of International Environment and Development Studies, Noragric, Norwegian University of Life Sciences (NMBU) P.O. Box 5003, N-1432 Ås, Norway

\* Author to whom correspondence should be addressed; E-Mail: jens.aune@nmbu.no

Academic Editor: Yantai Gan

---

**Abstract:** This study was undertaken with the objective of evaluating micro-dosing of mineral fertilizer combined with seed priming in sorghum, groundnut, sesame, and cowpea. On-station and on-farm trials were conducted for two consecutive seasons (2009/2010 and 2010/2011) at Al-Tukma village (12°00′57.60″ N and 29°46′12.15″ E) in South Kordofan State, 15 km southeast of Dilling city. Heavy cracking clay soil is the dominant soil type in the region with low fertility. The experiments for each crop consisted of two priming levels (primed seeds *vs.* non-primed) and four micro-doses of NPK mineral fertilizer (0, 0.3, 0.6 and 0.9 g per planting pocket or hole). On-farm trials in 15 fields consisted of control, seed priming, and seed priming + micro fertilizer (0.3 g/planting hole). Data collected included plant vigor, stand count, plant height, grain and straw yield, seed weight, and other relevant agronomic traits. This study shows that it is possible to increase productivity of sorghum, sesame, groundnut, and cowpea in the semi-arid cracking clay of South Kordofan State at a low cost and with a moderate risk for farmers through seed priming and micro-dosing of fertilizers. Seed priming combined with micro-dosing NPK mineral fertilizer of 0.9 g was the best treatment for plant establishment, seedling vigor, grain yield, and hay yield in sorghum and groundnut, whereas the combination of seed priming and 0.3 g micro-doing of fertilizer was the best in sesame. Seed priming and micro-dosing of fertilizer of 0.6 g was the best combination for cowpea. On-farm trial results indicated that priming alone and priming combined with fertilizer application significantly increased the yields of sorghum, groundnut, and cowpea over the control ($P = 0.01$). Of the crops tested, groundnut responded most favorably to micro-dosing and seed priming, with a value to cost ratio (VCR) of 26.6, while the highest VCR for sorghum, sesame, and cowpea was 12.5, 8.0 and 4.4, respectively. For the best productivity and profitability, we recommend using seed priming in combination with the micro-dosing of 0.9 g/hole of 15:15:15 NPK fertilizer for sorghum and groundnut, of 0.3 g/hole for sesame, and of 0.6 g/hole for cowpea grown in the semiarid South Kordofan State of Sudan.

**Keywords:** seed treatment; fertilization; yield; gross margin; on-farm; intensificaton

## 1. Introduction

Traditional dry-land farming is the major production system and source of livelihood for more than 75% of the population in Western Sudan. The major food crops grown are millet and sorghum while groundnut and sesame are the major cash crops. Other crops grown are cowpea, maize, cotton, and okra. The productivity of the main crops are very low compared to other parts of the world [1]. This is due to a magnitude of natural and socio-economic constraints. Poor crop establishment and low soil fertility are particularly constraining for crop productivity.

The study site, South Kordofan State, falls in the semi-arid zone where heavy cracking clays constitute the dominant soil type. The region is characterized by seasonal variation in rainfall and low soil fertility. The maintenance of soil fertility is becoming one of the most important interventions required to increase crop productivity in the dry areas. Application of small amounts of mineral fertilizer in the planting hole is a more efficient way to apply mineral fertilizer compared to broadcasting. This method increases both yield and the efficiency of fertilizer application [2–6]. Another low-cost approach to increase yield under marginal dry land conditions is seed priming, a process of soaking seeds in water for a specific period prior to sowing [5,7,8]. Seed priming and fertilizer micro-dosing are recommended in the sandy soils of North Kordofan State as these methods have significantly improved crop establishment and increased the yield of rain-fed sorghum, pearl millet, groundnut, sesame, and cowpea [6,9,10]. The objective of the study was to evaluate the effect of placing small amounts of mineral fertilizer in the planting hole (micro-dosing) in combination with seed priming for sorghum, groundnut, sesame, and cowpea in South Kordofan State in Western Sudan, a location with heavy cracking clay soil and higher rainfall, compared to North Kordofan.

## 2. Materials and Methods

Four field experiments were conducted for two consecutive seasons (2009/2010 and 2010/2011) at Al-Tukma village in South Kordofan State (29°46'12.15" E and 12°00'57.60" N), 15 km southeast of Dilling city and 145 km north of Kadugli, the capital city of the state. This area is part of the central clay plain, where soils are dark, heavy cracking (vertisol), with high water-holding capacity and low nitrogen and phosphorus content.

Soil cores were taken to the depth of 0–10 cm prior to planting. The soil analysis showed a clay fraction of 74%, very low nitrogen content (0.02 ppm), moderate phosphorus content (30 ppm) and low potassium content (0.78 ppm). NPK mineral fertilizer (15:15:15) was used to compensate for low contents of the soil NPK.

The experiments were carried out under rain-fed conditions on sorghum (*Sorghum bicolor* (L.) Moench), groundnut (*Arachis hypogea* L.), sesame (*Sesamum indicum* L.), and cowpea (*Vigna unguiculata* (L.) WLAP). The experiments consisted two levels of priming (primed seeds *vs.* non-primed) and four levels of micro-dose of NPK (15:15:15) mineral fertilizer (0, 0.3, 0.6 and 0.9 grams per planting pocket or hole) giving eight ($2 \times 4$ factorial) treatment combinations. The treatments were laid out in a randomized complete block design (RCBD) with four replications. The experimental plots were five meters long and three meters wide with 60 cm between-row spacing and between-plant spacing of 40, 20, 40 and 30 cm for sorghum, groundnut, sesame, and cowpea, respectively. Total annual rainfall at the nearest main meteorological station at Dilling was 675 mm and 562 mm during 2009/2010 and 2010/2011 seasons, respectively. The varieties used were Yarwasha, Gubeish, Obeid-1, and Ainelgazal of sorghum, groundnut, sesame, and cowpea, respectively. The fertilizer type was NPK (15-15-15) (Yara). Seeds of the four crops were soaked in water for eight hours (overnight) and then surface-dried for less than an hour for planting the next day. Before sowing, the seeds were treated with Apron Star (20% Metalaxyl–m, 20% Thiamethoxam and 2% Difenoconazole) at a dose of 3 g/kg seed. Two seeds per hole were planted for groundnut and cowpea, while 4–5 seeds were planted for sorghum and sesame, then thinned to 2–3 plants per hole two weeks after planting. The planting hole was opened using a traditional hoe and the seed and fertilizer were placed together in the planting pit at a depth of 5 to 7 cm. The microdosing rates applied were 0.3, 0.6 and 0.9 g per pocket which corresponded to 6–7, 12–14, and 18–21 fertilizer granules per pocket, respectively.

The amount of fertilizer utilized per hectare for each crop differed according to the crop spacing (Table 1). Manual hand-weeding was conducted three times, the first before planting, the second, approximately two weeks after planting, and the third, three weeks after the second weeding. Data collected included:

1) Plant vigor: The plant vigor score was measured at two and four weeks after planting using a 1–4 rating scale (score): 1 = Low, 2 = Moderate, 3 = Vigorous, and 4 = Highly Vigorous.

2) Stand count: Number of plants per four central rows, two weeks after planting.

3) For sorghum: Plant height (average of five random plants, pre-harvesting), panicle length (cm), weight of straw, grain yield, and 1000 seed weight.

4) For groundnut: Number of pods per plant, shelling percentage, hay yield, pod yield, and 100 seed weight.

5) For sesame: Plant height, number of capsules per plant, hay yield, seed yield, and 1000 seed weight.

6) For cowpea: Pods per plant, seeds per pod, pod yield, hay yield, seed yield and 100 seed weight.

**Table 1.** Calculated fertilizer quantity in kg·ha$^{-1}$ corresponding to fertilizer micro-dose level per planting hole.

| Crop | Spacing (cm) | Micro dose (g/hole) | | | |
|---|---|---|---|---|---|
| | | 0 | 0.3 | 0.6 | 0.9 |
| | | Equivalent dose (kg/ha) | | | |
| Sorghum | 60 × 40 | 0 | 12.5 | 25.0 | 37.5 |
| Groundnut | 60 × 20 | 0 | 25.0 | 50.0 | 75.0 |
| Sesame | 60 × 40 | 0 | 12.5 | 25.0 | 37.5 |
| Cowpea | 60 × 30 | 0 | 16.7 | 33.4 | 50.1 |

On-farm trials were also conducted to study the effect of priming and micro-dosing. Each of the 15 selected farmers (men and women) was provided with 3 kg of 15-15-15 NPK fertilizer. The treatments in each field of the selected farmers consisted of:

1.    Control
2.    Seed priming
3.    Seed priming + micro-dosing (0.3 g/planting hole).

Plot size for each treatment in the on-farm experiment was 360 m$^2$ (15 × 24 m) with all cultural practices carried out by the farmers according to their preferences. Yields obtained from farmers' plots were analyzed according to the randomized complete block design, considering each farmer as a replicate [11]. The combined analysis was carried out over two seasons and means were separated using Duncan Multiple Range Test (DMRT) at levels of 0.01 and 0.05 [12]. The partial budgeting technique was used to assess and compare the economic returns and net benefits of the different treatments [13]. The average yield over seasons and replications in each treatment were used. The average field prices of the crops during 2011/2012 were taken from the markets in which the farmers sell their produce. Production cost, or the sum of all of the variable costs including labor and inputs costs (without fertilizer), was taken from the surveys, which were conducted annually by the Ministry of Agriculture, South Kordofan State, Sudan. In the micro-fertilizer treatment, the fertilizer cost was added based on the amount of fertilizer applied (Table 1. The fertilizer use efficiency (FUE) was computed according to the formula:

$$FUE_t = (Y_t - C_t)/F_t$$

where FUE is the fertilizer use efficiency for the treatment level, $Y_t$ is the grain yield for the treatment level, $C_t$ is the grain yield from the control, and $F_t$ is the fertilizer rate used in kg·ha$^{-1}$ for the treatment level.

The value cost ratio was calculated as:

$$VCR_t = (Y_t - Y_c) \times PG_t/CF_t$$

where $VCR_t$ denotes the value cost ratio for the treatment level, $Y_t - Y_c$ denotes the incremental grain yield resulting from fertilizer use in the treatment and the control, respectively, $PG_t$ denotes the grain price per kg and $CF_t$ denotes the cost of fertilizer per hectare of the treatment level.

The total production cost including labor and input costs, was calculated by adding the fertilizer cost for each treatment to the production cost for each crop. The production cost for sorghum, groundnut, sesame, and cowpea was 353, 428, 283 and 179 SDG/ha, respectively. The production costs data were obtained from North Kordofan State Ministry of Agriculture 2010–2011 annual survey reports. The prices in kg ha$^{-1}$ of sorghum, groundnut, sesame, and cowpea, according to ElObeid Auction Market 2010–2011, were 1.25, 4.7, 2.83 and 3 SDG, respectively. Hay prices from the local market were 0.41, 1.6, 0.2 and 1.0 SDG for the above crop order, respectively. The market price of 15-15-15 NPK fertilizer was 200 SDG per 50 kg sack (4 SDG/Kg).

## 3. Results

### 3.1. Sorghum

Results of the combined analysis over the two seasons (2009/2010 and 2010/2011) are shown in Table 2. Differences in stand count (plant population), plant vigor score (two and four weeks after planting), and plant height were highly and significantly affected by seed priming and micro-dose of mineral fertilizer ($p = 0.01$). Seed priming increased plant vigor, plant stand and plant height by 28%, 14%, and 3%, respectively ($p < 0.05$), when compared to the control.

**Table 2.** Effect of seed priming and fertilizer micro-dosing on some traits of sorghum (combined across two seasons).

| Treatments | Plant pop./ha | Vigor score 2 WAP | Vigor score 4 WAP | Plant height (cm) | No. of heads/plot | No. of seeds/head | 1000 seed weight (gram) |
|---|---|---|---|---|---|---|---|
| **Seed Priming** | | | | | | | |
| Non-primed | 67842 | 2.16 | 2.88 | 179 | 97 | 751 | 40.9 |
| Primed | 78446 | 3.00 | 3.35 | 185 | 113 | 930 | 42.7 |
| SE± | 2260 ** | 0.21 ** | 0.17 ** | 3.0 ** | 4.6 ** | 33 ** | 0.54 ** |
| **Fertilizing micro-dosing (gram/planting hole)** | | | | | | | |
| Control | 69252 | 2.19 | 2.94 | 175 | 97 | 699 | 41.2 |
| 0.3 g/hole | 68261 | 2.44 | 3.13 | 178 | 96 | 739 | 40.5 |
| 0.6 g/hole | 75572 | 2.75 | 3.25 | 186 | 110 | 791 | 42.1 |
| 0.9 g/hole | 79492 | 2.94 | 3.50 | 190 | 117 | 1134 | 42.5 |
| SE± | 1598 ** | 0.15 ** | 0.12 * | 2.1 ** | 3.2 ** | 47 ** | 0.38 ** |
| C.V% | 8.74 | 23.11 | 14.77 | 4.60 | 12.30 | 15.66 | 3.68 |

Ns = not significant; * significant at $p \leq 0.05$; ** significant at $p \leq 0.01$, WAP = weeks after planting.

Plant stand, plant vigor score, and plant height also increased with micro-dosing. The highest increments were recorded for the application of 0.9 g of fertilizer, which increased plant stand, plant vigor, and plant height by 13%, 25%, and 8%, respectively. A similar effect of seed priming and micro-dosing ($p < 0.01$) was also observed in heads per plot, and 1000 seed weight. Number of heads per plot increased due to seed priming and micro-dosing by 30% and 16% respectively and number of seeds per head by 24% and 27%, respectively ($p < 0.01$). Seed weight (g/1000 seeds) was significantly improved by seed priming and fertilization ($p < 0.01$).

The yield differences in sorghum were significantly ($p < 0.01$) affected by seed priming and micro-dosing (Table 3). Seed priming increased average grain yield from 619 kg/ha to 949 kg/ha. Compared to the control, micro-dosing of 0.3, 0.6, and 0.9 g/hole fertilizer increased grain yield by 12%, 42%, and 84%, respectively. Yield increased from 512 kg/ha in the control treatment to 1371 kg/ha in the "seed priming and 0.9 g fertilizer/pocket" treatment. This corresponds to a 167% yield increase. Seed priming and micro-fertilization also significantly increased straw yield by 21% and 23%, respectively (Table 3). The highest straw yield of 7700 kg/ha was recorded at the dose 0.9 g/hole, while the lowest straw yield of 4750 kg/ha was obtained from the control, without seed priming or fertilization. The interaction between priming and micro-fertilization was only significant for plant height.

**Table 3.** Effect of seed priming and fertilizer micro-dosing on sorghum grain and straw yields (kg/ha) combined across two seasons.

| Fertilizer dose/hole | Grain yield (kg/ha) | | | | Straw yield (kg/ha) | | | |
|---|---|---|---|---|---|---|---|---|
| | Non-primed | Primed | Mean | SE± | Non-primed | Primed | Mean | SE± |
| 0.0 g | 512 [g] | 652 [f] | 582 | | 4752 [h] | 5251 [f] | 5002 | |
| 0.3 g | 500 [g] | 799 [c] | 650 | 104 ** | 4822 [g] | 5513 [d] | 5168 | 323 ** |
| 0.6 g | 688 [e] | 974 [b] | 831 | | 5485 [e] | 7200 [b] | 6342 | |
| 0.9 g | 776 [d] | 1371 [a] | 1074 | | 6164 [c] | 7701 [a] | 6932 | |
| SE± | 73 ** | | | | 228 ** | | | |

Ns = not significant; * significant at $p \leq 0.05$; ** significant $p \leq 0.01$. Different letters signify statistically different.

## 3.2. Groundnut

Seed priming and micro-fertilization positively affected crop establishment in groundnut by increasing plant population and vigor score ($p < 0.01$, Table 4). Seed priming affected crop establishment in groundnut by increasing stand count by 16%, plant vigor score two weeks after planting by 25% and plant vigor score four weeks after planting by 15%. The number of pods per plant was not significantly affected by seed priming, but was by fertilization. Shelling out-turn and 100 seed weight were not significantly affected by neither seed priming nor fertilizer application ($p < 0.05$). Micro-fertilization significantly improved stand count (plant population) and vigor score from 4% to 24% and from 3% to 34%, respectively. The plant characteristics were improved with increased levels of fertilizer.

**Table 4.** Effect of seed priming and micro-dosing on some traits of groundnut (combined across two seasons).

| Treatments | Plant pop./ha | Vigor 2 WAP | Vigor 4 WAP | Shelling% | No. of pods/plant | 100 seed wt. |
|---|---|---|---|---|---|---|
| | | | Seed priming | | | |
| Non-primed | 102475 | 2.34 | 3.06 | 67.1 | 42.0 | 31.8 |
| Primed | 118984 | 2.94 | 3.59 | 66.8 | 42.8 | 32.2 |
| SE± | 7288.1 ** | 0.21 ** | 0.17 ** | 0.98 ns | 0.74 ns | 0.57 ns |
| | | | Fertilizer micro-dose (gram/planting hole) | | | |
| Control | 96468 | 2.38 | 3.00 | 66 | 39 | 32.2 |
| 0.3 g/hole | 100340 | 2.44 | 3.38 | 67 | 41 | 31.9 |
| 0.6 g/hole | 126361 | 2.56 | 3.38 | 67 | 43 | 31.9 |
| 0.9 g/hole | 119750 | 3.19 | 3.56 | 67 | 47 | 32.3 |
| SE± | 5153.4 ** | 0.15** | 0.12 * | 0.70 ns | 1.05 ** | 0.41 ns |
| C.V% | 18.62 | 22.9 | 14.31 | 4.16 | 7.02 | 5.07 |

Ns = not significant; * significant at $p \leq 0.05$; ** significant $p \leq 0.01$, WAP = weeks after planting.

Seed priming and micro-fertilization significantly increased pod and hay yields ($p < 0.01$, Table 5). Seed priming increased average pod yield from 1995 to 2404 kg·ha$^{-1}$, corresponding to a 20% yield increase, while micro-fertilization increased pod yield by 35% on average, from 1865 to 2629 kg·ha$^{-1}$. Groundnut pod yield increased from 1716 kg·ha$^{-1}$ in the control to 2955 kg in the treatment "priming and 0.9 g fertilizer per pocket", equivalent to a 72% increase in yield. Hay yield was also significantly increased by seed priming and micro-fertilization ($p < 0.01$). The best hay yield of 2637 kg·ha$^{-1}$ was obtained from a micro-dose of 0.9 gram fertilizer per hole. There was no significant interaction between priming and micro-fertilization for all traits, with the exception of shelling out-turn.

**Table 5.** Effect of seed priming and micro dosing on groundnut pod and hay yields (kg/ha) combined across two season.

| Fertilizer dose/hole | Pod yield (kg/ha) | | | | Hay yield (kg/ha) | | | |
|---|---|---|---|---|---|---|---|---|
| | Non-primed | Primed | Mean | SE± | Non-primed | Primed | Mean | SE± |
| 0.0 g | 1716 [f] | 2013 [d] | 1865 | | 1737 [h] | 2170 [f] | 1954 | |
| 0.3 g | 1931 [e] | 2291 [c] | 2111 | 95 ** | 1915 [g] | 2270 [c] | 2092 | 97 ** |
| 0.6 g | 2030 [d] | 2354 [b] | 2193 | | 2076 [f] | 2752 [b] | 2414 | |
| 0.9 g | 2303 [c] | 2955 [a] | 2629 | | 2212 [d] | 3061 [a] | 2637 | |
| SE± | 67 ** | | | | 69 ** | | | |

Ns = not significant; * significant at $p \leq 0.05$; ** significant $p \leq 0.01$. Different letters signify statistically different.

### 3.3. Sesame

Results showed a strong and significant ($p \leq 0.01$) effect of seed priming and micro fertilizing on plant stand, which increased by 56% with priming and 25% with micro-dosing, compared to the control (Table 6). Seed priming increased vigor score by 57% ($p \leq 0.01$), while there was no effect of micro-dosing on plant vigor score. Plant height increased by 3% and 8% due to seed priming and micro-dosing, respectively. The number of capsules per plant increased ($p < 0.05$) by 12% and 15%

with priming and micro-dosing, respectively. Seed priming showed no significant effect on sesame 1000 seed weight, while micro-dosing significantly increased 1000 seed weight by 8% ($p = 0.05$).

Sesame seed yield was significantly increased by priming and micro-dosing ($p < 0.01$). Seed yield increased from 276 to 383 kg·ha$^{-1}$ with priming and from 276 to 393 kg/ha with micro-fertilizer application (Table 7). Seed priming and micro fertilization improved sesame hay yield by 55% and 24%, respectively. The best seed and hay yields of 524 and 1942 kg·ha$^{-1}$, respectively, were obtained from the combination of seed priming and micro-dosing of 0.9 g per pocket. Significant interaction between priming and micro-fertilization was observed for plant height, 1000 seed weight, and seed yield ($p < 0.05$). Although the highest seed yield was recorded for a 0.9 g micro-dose in combination with seed priming, the difference between 0.3 g and 0.9 g micro-dosing rates in combination with seed priming was not significant for seed yield.

**Table 6.** Effect of seed priming and micro-dosing of fertilizers on sesame combined across two seasons.

| Treatments | Plant pop./ha | Vigour 2 WAP | Plant ht. (cm) | Av. no. of caps/plant | 1000 seed wt. (gm) |
|---|---|---|---|---|---|
| **Seed priming** | | | | | |
| Non-primed | 166683 | 1.97 | 132.6 | 61.2 | 2.37 |
| Primed | 259836 | 3.09 | 137.3 | 68.6 | 2.41 |
| SE± | 19264 ** | 0.22 ** | 3.3 * | 3.2 ** | 0.10 ns |
| **Fertilizer micro-dose (gram/planting hole)** | | | | | |
| Control | 180199 | 2.31 | 129 | 58 | 2.26 |
| 0.3 g/hole | 204993 | 2.56 | 135 | 63 | 2.43 |
| 0.6 g/hole | 246007 | 2.50 | 138 | 70 | 2.43 |
| 0.9 g/hole | 221838 | 2.75 | 139 | 68 | 2.44 |
| SE± | 13261 * | 0.15 ns | 2.3 * | 2.3 * | 0.03 * |
| C.V% | 25.55 | 24.48 | 6.88 | 14.15 | 6.58 |

Ns = not significant; * significant at $p \leq 0.05$; ** significant $p \leq 0.01$, WAP = weeks after planting. Polp./ha = population/ha, Plant ht (cm) = plant hight in cm, 1000 seed wt. (gm) = weight of 1000 seeds in gram.

**Table 7.** Effect of seed priming and micro-dosing on sesame seed and hay yields (kg/ha), combined across two seasons.

| Fertilizer Dose/hole | Seed yield (kg/ha) | | | | Hay yield (kg/ha) | | | |
|---|---|---|---|---|---|---|---|---|
| | Non-primed | Primed | Mean | SE± | Non-primed | Primed | Mean | SE± |
| 0.0 g | 276 [e] | 383 [c] | 330 | | 935 [h] | 1335 [d] | 1135 | |
| 0.3 g | 243 [f] | 507 [a] | 375 | | 970 [g] | 1801 [b] | 1385 | |
| 0.6 g | 348 [d] | 467 [b] | 408 | 19 ** | 1058 [f] | 1430 [c] | 1244 | 100 * |
| 0.9 g | 393 [c] | 524 [a] | 458 | | 1223 [e] | 1942 [a] | 1582 | |
| Mean | | | | | | | | |
| SE± | 13 ** | | | | 70 ** | | | |

Ns = not significant; * significant at $p \leq 0.05$; ** significant $p \leq 0.01$. Different letters signify statistically different.

## 3.4. Cowpea

Differences in plant stand (plant population) and 100 seed weight were highly affected by seed priming and micro-dose of mineral fertilizer ($p < 0.01$, Table 8). Seed priming improved plant stand by 23% compared to non-primed, while micro-dosing treatments increased plant stand (plant population) by up to 26%. Seed priming improved vigor score by 32% compared to non-primed, while no significant differences were observed among micro-dosing treatments for vigor score ($p < 0.05$). Seed priming did not affect the number of pods per plant, while micro-dosing increased pods number from 11 to 13 per plant. The number of seeds per pod was not significantly affected by either seed priming or fertilizer application ($p < 0.05$). Hundred seed weight was significantly increased due to seed priming and fertilizer micro-dose application.

Cowpea seed yield was significantly ($p < 0.05$) increased by 11%, 38% and 16% compared to the control with a micro-dose application of 0.3, 0.6 and 0.9 g, respectively, while seed yield increased by 36% with seed priming (Table 9). Hay yield highly and significantly ($P = 0.01$) increased with priming (115% increase) and fertilization (51% increase) compared to the control. There was no significant interaction between priming and micro fertilizing for grain and hay yields.

**Table 8.** Effect of seed priming and micro fertilizing on cowpea combined across two seasons.

| Treatments | Plant pop. | Vigour 2 WAP | No. pods/plant | No. seeds/pod | 100 seed wt. gram |
|---|---|---|---|---|---|
| **Seed priming** | | | | | |
| Non-primed | 37938 | 2.31 | 13 | 9.1 | 18.2 |
| Primed | 46656 | 3.06 | 13 | 9.0 | 18.4 |
| SE± | 3570 ** | 0.23 ** | 0.74 ns | 0.51 ns | 0.10 * |
| **Fertilizer micro-dose (gram/planting hole)** | | | | | |
| Control | 34438 | 2.44 | 11 | 9 | 18.1 |
| 0.3 g/hole | 39813 | 2.56 | 14 | 9 | 18.2 |
| 0.6 g/hole | 48563 | 2.69 | 13 | 9 | 18.4 |
| 0.9 g/hole | 46375 | 3.06 | 13 | 9 | 18.4 |
| SE± | 2524 ** | 0.17 ns | 0.52 ** | 0.36 ns | 0.10 ** |
| C.V% | 23.87 | 24.86 | 16.36 | 16.02 | 1.19 |

Ns = not significant; * significant at $p \leq 0.05$; ** significant $p \leq 0.01$, WAP = weeks after planting. 100 seed wt. gram=weight of 100 grains. Different letters signify statistically different.

**Table 9.** Effect of seed priming and micro-dosing on cowpea seed and hay yields (kg/ha), combined across two seasons.

| Fertilizer Dose/hole | Seed yield (kg/ha) | | | | Hay yield (kg/ha) | | | |
|---|---|---|---|---|---|---|---|---|
| | Non-primed | Primed | Mean | SE± | Non-primed | Primed | Mean | SE± |
| 0.0 g | 191 [e] | 240 [c] | 215 | | 219 [g] | 414 [d] | 316 | |
| 0.3 g | 203 [de] | 275 [b] | 239 | | 216 [g] | 488 [c] | 352 | |
| 0.6 g | 239 [c] | 356 [a] | 297 | 20 * | 283 [f] | 657 [b] | 470 | 41 ** |
| 0.9 g | 217 [d] | 283 [b] | 250 | | 395 [e] | 831 [a] | 613 | |
| Mean | 212 | 288 | 250 | | 278 | 597 | 438 | |
| SE± | | 14 ns | | | | 29 | | |

Ns = not significant; * significant at $p \leq 0.05$; ** significant $p \leq 0.01$. Different letters signify statistically different.

## 3.5. On-Farm Trials

Priming alone and priming combined with fertilizer application significantly increased the yield of sorghum, groundnut, and cowpea over the control in the two seasons ($p = 0.01$), while in sesame the increase was only significant in one season (Table 10).

**Table 10.** On-farm evaluation of seed priming and fertilizer micro-dosing at three villages (15 farmers from each village) in South Kordofan.

| Treatment | Sorghum | Groundnut | Cowpea | Sesame |
|---|---|---|---|---|
| 1st season: 2009/2010-(yield kg/ha) | | | | |
| Control | 686 [c] | 586 [c] | 128 [b] | 272 [b] |
| Primed seeds | 860 [b] | 731 [b] | 207 [a] | - |
| Primed+ fertilizer (0.3 gram) | 1022 [a] | 906 [a] | 233 [a] | 405 [a] |
| 2nd season: 2010/2011-(yield kg/ha) | | | | |
| Control | 607 [c] | 886 [b] | 253 [b] | 181 |
| Primed seeds | 837 [b] | 1237 [a] | 353 [a] | - |
| Primed + fertilizer (0.3 gram) | 986 [a] | 1438 [a] | 415 [a] | 198 |
| # of participating farmers | 15 | 15 | 15 | 15 |
| # of participating villages | 3 | 3 | 3 | 3 |

Ns = not significant; * significant at $p \leq 0.05$; ** significant $p \leq 0.01$. Different letters signify statistically different.

## 4. Economic Analysis

An economic analysis was undertaken to assess economic performance of the different treatments (Table 11). Generally, the highest returns were obtained in all crops when both seed priming and micro-fertilization were used. Micro-dosing without seed priming, even at low doses, negatively affected gross margin in sorghum, sesame and cowpea.

**Table 11.** Economic analysis of the different seed priming and micro-dosing treatments.

| Treatments | Net return (SDG/ha) | | | |
|---|---|---|---|---|
| | Sorghum | Groundnut | Sesame | Cowpea |
| Non priming no fertilizer (control) | 287 f | 7637 h | 498 g | 394 e |
| Non priming + 0.3 g per planting hole | 222 g | 8548 g | 355 h | 363 f |
| Non priming + 0.6 g per planting hole | 596 c | 8913 f | 602 f | 404 e |
| Non priming + 0.9 g per planting hole | 407 e | 10096 d | 719 e | 272 g |
| Priming no fertilizer | 462 d | 9033 e | 801 d | 541 c |
| Priming + 0.3 g per planting hole | 765 b | 10240 c | 1102 a | 580 b |
| Priming + 0.6 g per planting hole | 467 d | 10436 b | 939 c | 755 a |
| Priming + 0.9 g per planting hole | 1211 a | 13161 a | 1050 b | 470 d |
| SE± | 112 | 591 | 95 | 53 |

* 15-15-15 NPK fertilizer price is 200 SDG per 50 kg sack (4 SDG/Kg); ** Note: sorghum, groundnut, sesame and cowpea production costs are 353, 428, 283 and 179 SDG, respectively [14]; *** Note: sorghum, groundnut, sesame and cowpea price according to Obeid Auction Market, 2010–2011, are 1.25, 4.7, 2.83, and 3 SDG, respectively. Hay prices from the local market were 0.41, 1.6, 0.2 and 1.0 for the above crop order, respectively. Different letters signify statistically different.

Sorghum seed priming and application of 0.9 g fertilizer per hole increased the gross margin from 462 to 1211 SDG/ha, while the treatment of non-priming with application of 0.9 g fertilizer per hole increased the net revenue from 278 to 407 SDG/ha.

For groundnut, the net benefit increased from 9033 in the control to 13161 SDG/ha with the application of priming and 0.9 g fertilizer per hole. Priming with micro-dosing gave a higher return than non-priming with micro-dosing. However, unlike in other crops, groundnut yield increased with increasing micro-dosing rates up to 0.9 kg fertilizer per pocket. The revenue increase from micro-dosing and seed priming was much higher in groundnut compared to the other crops.

Sesame gross margin increased from 498 SDG/ha in the control to 1102 in the treatment "priming combined with application of 0.3 g fertilizer per hole". Priming alone with micro dosing also gave a good return in sesame.

For cowpea, the gross margin increased from 541 in the control to 755 SDG/ha with the application of 0.6 g fertilizer per hole combined with seed priming. Fertilizer use efficiency (FUE) and value cost ratio (VCR) for grain yield and the total biological yield were the highest at the treatment combinations with the highest yields for all crops (Tables 12 and 13). FUE and VCR were always the highest in treatments which combined micro-dosing with seed priming, and were higher in sorghum and groundnut compared to sesame and cowpea. Adding the hay value to the grain value increased the FUE and VCR. It is worth mentioning that hay yield is a valuable agricultural bi-product which is utilized for animal feed in the dry months of the year. Sorghum straw is also utilized as building material in rural areas.

**Table 12.** Value cost ratio (VCR) of NPK fertilizer micro-dose rates with and without seed priming (VCR1-grain yield, VCR2-grain + hay yield). a = non-primed seed, b = primed seeds.

| NPK rates | Sorghum | | Groundnut | | Sesame | | Cowpea | |
|-----------|---------|------|-----------|-------|--------|-------|--------|------|
| gram/hole | VCR1 | VRC2 | VCR1 | VCR2 | VCR1 | VCR2 | VCR1 | VCR2 |
| 0.3[a] | −0.3 | 0.27 | 10.11 | 12.96 | −1.70 | −1.56 | 0.54 | 0.41 |
| 0.6[a] | 2.2 | 5.21 | 7.38 | 10.09 | 2.04 | 2.29 | 1.08 | 2.44 |
| 0.9[a] | 2.2 | 6-06 | 9.20 | 11.70 | 2.21 | 2.69 | 0.39 | 3.02 |
| 0.3[b] | 3.68 | 5.78 | 13.07 | 14.67 | 7.02 | 8.02 | 1.57 | 2.68 |
| 0.6[b] | 4.03 | 11.83 | 8.01 | 19.65 | 2.38 | 2.60 | 2.60 | 4.42 |
| 0.9[b] | 5.99 | 12.52 | 14.76 | 26.64 | 2.66 | 3.26 | 0.64 | 2.72 |

**Table 13.** Fertilizer use efficiency (FUE kg·kg$^{-1}$) of NPK fertilizer micro-dose rates with and without seed priming (FUE1-grain yield, FUE2-grain + hay yield). a = non-primed seed, b = primed seeds.

| NPK rates | Sorghum | | Groundnut | | Sesame | | Cowpea | |
|-----------|---------|-------|-----------|-------|--------|-------|--------|-------|
| gram/hole | FUE1 | FUE2 | FUE1 | FUE2 | FUE1 | FUE2 | FUE1 | FUE2 |
| 0.3[a] | −0.96 | 4.64 | 8.60 | 22.84 | −2.64 | 0.16 | 0.72 | 0.54 |
| 0.6[a] | 7.04 | 36.36 | 6.28 | 13.06 | 2.88 | 7.80 | 1.44 | 3.36 |
| 0.9[a] | 7.04 | 44.69 | 7.83 | 14.16 | 3.12 | 10.80 | 0.52 | 4.03 |
| 0.3[b] | 11.76 | 32.72 | 11.12 | 15.12 | 09.92 | 29.92 | 2.10 | 06.53 |
| 0.6[b] | 12.88 | 90.84 | 06.82 | 18.46 | 03.36 | 7.72 | 3.47 | 10.75 |
| 0.9[b] | 19.17 | 84.50 | 12.56 | 24.44 | 3.76 | 15.68 | 0.86 | 09.18 |

The economic analysis of the on-farm experiment confirmed the positive impact of seed priming and micro-dosing (Table 14). There was an increase for all crops, with the highest increase observed in groundnut. The increase in gross margin from the control to the treatment combining seed priming and micro-dosing was 2178, 444, 372, and 469 for groundnut, sorghum, cowpea, and sesame, respectively.

**Table 14.** The effect of seed priming and micro-dosing on gross margin in on-farm farmer managed plots.

| Treatment | Combined (30 farmers) | | | |
|-----------|-----------|---------|--------|--------|
|           | **Groundnut** | **Sorghum** | **cowpea** | **Sesame** |
| Control   | 3265      | 513     | 387    | 13     |
| Priming   | 4662      | 824     | 690    |        |
| P + M     | 5443      | 957     | 759    | 482    |
| SE        | 283 **    | 40 **   | 54 **  | 74 **  |
| CV        | 35        | 28      | 48     | 163    |

Ns = not significant; ** significant $p \leq 0.01$.

## 5. Discussion

Seed priming and micro-dosing represent low-cost approaches to increase yields of small-holders under marginal dry land conditions [5,8,9].

Seed priming is a simple strategy to improve plant establishment and alleviate the negative effects associated with stress exposure. Seed priming has been shown to reduce germination time, improve plant stand, increase vigor, shorten the growing cycle, and increase crop yield [5,7,8,15]. It is a technology that is particularly suited to adverse environmental conditions. The results of the present study have shown that seed priming significantly improved crop establishment and seedling vigor of rain-fed sorghum, groundnut, sesame, and cowpea grown on clay soil in South Kordofan State. This is important as seed priming and micro-dosing can reduce the need for re-sowing. Similar results were obtained with these crops under rain-fed condition on sandy soils in North Kordofan State [10].

Previous studies have shown that seed priming and micro-dosing can increase yield at a low cost [2–5,9,10,16]. Micro-dosing was also reported to improve the fertilizer use efficiency compared to broadcasting of fertilizer [17]. The results from this study indicate that seed priming combined with micro-dosing is not only an approach for sandy soils, but also works well on the cracking clay soil of South Kordofan State. Most previous studies on micro-dosing have been with cereal crops, but this study shows that micro-dosing and seed priming is also an appropriate technology in groundnut, sesame, and cowpea. The yield increase observed for the best treatment compared to the control was 85%, 41%, 84% and 48% in sorghum, groundnut, sesame, and cowpea, respectively. The best treatment with regard to net return in sorghum and groundnut was found to be seed priming in combination with 0.9 g of NPK fertilizer, while 0.3 g and 0.6 g in combination with priming were the doses with highest net returns in sesame and cowpea, respectively. These combinations were significantly better than all other combinations including the control. Higher micro-dose response was found in groundnut in the clay soils of South Kordofan State compared to the sandy soils of North Kordofan State [10]. The highest fertilizer use efficiency (FUE) and value cost ratio (VCR) corresponded with the combination which gave the highest total grain and biological yield, except for sorghum. The results of this study show that the agronomic and economic benefits of micro-dosing can be increased if it is combined with seed priming, as the VCR were generally higher when seed priming was combined with micro-dosing, compared to when micro-dosing was used alone. Seed priming therefore makes micro-dosing a safer investment. The on-farm study confirms that seed priming and micro-dosing can greatly increase gross margin. This study shows that the type of crop onto which the fertilizer is applied is of importance. If the objective is to have the highest possible economic return, it is far better to apply the fertilizer in groundnut as the VCR for the best treatment in groundnut was 26.6, whereas the best VCR for sorghum, sesame, and cowpea were 12.5, 8.0, and 4.4, respectively. This illustrates that applying micro-dosing combined with seed priming is a very safe investment as the VCR should be above 2 and preferably above 4 under dry land conditions where the risk is high [18]. This point can also be illustrated by looking at the increase in gross return from the control to the best treatment. For groundnut, this increase was 5524 SDG/ha whereas it was 924, 607, and 285 for sorghum, sesame, and cowpea, respectively. The on-farm experiments also confirmed that the highest return is found in groundnut. The main reason the return on fertilizer is higher in groundnut compared to other crops is the high price of this crop compared to the other crops.

## 6. Conclusions

This study showed that it is possible to increase the productivity of sorghum, sesame, groundnut, and cowpea in the semi-arid cracking clay of South Kordofan State at a low cost and at a very moderate risk for farmers through seed priming and fertilizer micro-dose application. Seed priming with micro-doses of 0.9 g improved crop establishment and seedling vigor of sorghum and groundnut, while the combination of seed priming and 0.3 g micro-fertilization improved crop establishment, seedling vigor and grain yield in sesame. Seed priming and micro-fertilization of 0.6 g improved crop establishment and grain yield in cowpea. The highest economic return, FUE, and VCR corresponded to the treatment which produced the highest grain and total biological yield in each crop. Seed priming and micro-dosing can be considered as a safe option for farmers, particularly if used in groundnut and sorghum.

## Acknowledgments

Sincere thanks are due to the Sudan and Norway Drylands Coordination Group (DCG) for their assistance and support that made this work possible. Our thanks are also extended to the Dry Lands Research Center (ARC) and ADRA (Sudan) for their cooperation and facilitation. Thanks are due to Mutaz M. Elsadig, Sudan, DCG coordinator, for his keen follow up and for facilitating the implementation of the project. Special thanks to Edaw Mohamed Idris, technician at El-Obeid Research Station, for his help in trial conduction and data collection. Sincere thanks are also extended to Khalid Ali Issa and Faiz Ali Ahmed, the drivers who made the journeys to and from trial sites easy and enjoyable. We are very grateful to the members of the communities who participated in the implementation of the activities.

## Author Contributions

The authors contributed to the design, implementation, analyzing and write-up of the paper.

## Conflicts of Interest

The authors declare no conflict of interest.

## References

1.  Osman, A.K.; Ali, M.K. Crop production under traditional rain-fed agriculture. In Proceedings of the National Symposium on: Sustainable Rain-Fed Agriculture, Khartoum, Sudan, 17–18 November 2009; pp. 113–131.
2.  Hayashi, K.; Abdoulaye, T.; Gerard, B.; Bationo, A. Evaluation of application timing in fertilizer micro-dosing technology on millet production in Niger, West Africa. *Nutr. Cycl. Agroecosyst.* **2008**, *80*, 257–265.
3.  Klaij, M.C.; Genard, C.; Reddy, K.C. Low input technology options for millet based cropping systems in the Sahel. *Exp. Agric.* **1994**, *30*, 77–82.
4.  Aune, J.; Doumbia, M.; Berthe, A. Microfertilization sorghum and pearl millet in Mali. *Outlook Agric.* **2007**, *36*, 199–203.
5.  Aune, J.; Bationo, A. Agricultural intensification in the Sahel: The Ladder Approach. *Agric. Syst.* **2008**, *98*, 119–125.
6.  Aune, J.B.; Ousman, A. Effect of seed priming and micro dosing of fertilizer on sorghum and pearl millet in western Sudan. *Exp. Agric.* **2011**, *47*, 419–430.
7.  Harris, D.; Pathan, A.K.; Gothkar, P.; Joshi, A.; Chivasa, W.; Nyamudeza, P. On-farm seed priming: Using participatory methods to revive and refine a key technology. *Agric. Syst.* **2001**, *69*, 151–164.
8.  Harris, D. Development and testing of "on-farm" seed priming. *Adv. Agron.* **2006**, *90*, 129–178.
9.  Osman, A.K.; Abdalla, E.A.; Mekki, M.A.; Elhag, F.M.A.; Aune, J. Effect of seed priming and fertilizer micro dosing on Traditional Rain-Fed Crops of North Kordofan. In Proceedings of the 49th National Crop Husbandry Committee Meeting, Wad Medani, Sudan, 21 December 2010.
10. Osman, A.K.; Aune, J.B. Effect of seed priming and micro dosing of fertilizer on groundnut, sesame and cowpea in western Sudan. *Exp. Agric.* **2011**, *47*, 431–443.
11. Gomez, K.A.; Gomez, A.A. *Statistical Procedures for Agricultural Research*; Wiley: New York, NY, USA, 1984; p. 680.
12. MSTAT-C. *User's Guide to MSTAT-C*; MSTAT Development Team, Michigan State University: East Lansing, MI, USA, 1983.
13. CIMMYT. *From Agronomic Data to Farmer Recommendations: An Economics Training Manual*; CIMMYT: Batan, Mexico, 1998
14. MOA. *Ministry of Agriculture, North Kordofan State, Department of Planning and Agricultural Statistics, Agricultural Season Evaluation Report*; MOA: Auckland, New Zealand, 2011.
15. Ashraf, C.M.; Abu-Shakra, S. Wheat seed germination under low temperature and moisture stress. *Agron. J.* **1978**, *70*, 135–139.

16.  Buerkert, A.; Bationo, A.; Piepho, H.P. Efficient phosphorus application strategies for increased crop production in sub-Saharan West Africa. *Field Crop. Res.* **2001**, *72*, 1–15.

17.  Tabo, R.; Bationo, A.; Hassane, O.; Amadou, B.; Fosu, M.; Sawadogo-Kabore, S.; Ndjeunga, J.; Fatondji, D.; Korodjouma, O.; Abdou, A.; *et al.* Fertilizer Micro Dosing for the Prosperity of the Resource Poor Farmers: A success story. In Proceedings of Increasing the Productivity and Sustainability of Rainfed Cropping Systems of Poor Smallholder Farmers, Tamale, Ghana, 22–25 September 2008.

18.  Koning, N.; Heerink, N.; Kauffman, S. *Integrated Soil Improvement and Agricultural Development in West Africa: Why Current Policy Approaches Fail*; Wageningen Agricultural University: Wageningen, The Netherlands, 1998.

# 8

# Differential Evolution's Application to Estimation of Soil Water Retention Parameters

**Zhonghui Ou**

School of Mathematics and Computer Science, Fujian Normal University, 350117 Fuzhou, China;
E-Mail: zhou@fjnu.edu.cn

Academic Editors: Ole Wendroth and Peter Langridge

**Abstract:** A Differential Evolution (DE) is introduced to predict the parameters of the soil water retention curve (SWRC) and it is configured for reliability and efficiency with the Unsaturated Soil Hydraulic Property Database (UNSODA). The main investigated dataset is 235 samples from lab_drying_h-t table and the testing shows that the data resource is reliable and steady. Some specific statistical computations are designed to investigate the convergence speed and the fitness precision of DE, different measurements of hydraulic data, and parametric characteristics of textural groups. The statistical results on UNSODA show that DE has higher performance in parameter fitness and time saving than some previous optimization methods and the statistical values of soil water retention parameters (SWRP) can be directly applied in the agricultural research and practice.

**Keywords:** van Genuchten model; Differential Evolution; soil water retention parameter; UNSODA

## 1. Introduction

The Van Genuchten model (vG model) describes the soil-water content-pressure head curve and the closed-form relative hydraulic conductivity expression in unsaturated soils derived from the predictive conductivity models of Burdine or Mualem [1–5]. At the beginning, Genuchten numerically obtained the residual soil-water content $\theta_r$ (cm$^3$cm$^{-3}$) , scaling parameter $\alpha$ (>0, in cm$^{-1}$) inversely proportional to the air entry pressure, and the pore-size distribution index $n$ (>1) by the semi-analytical and semi-graphical method and the nonlinear least-squares curve-fitting method while the saturated soil-water content $\theta_s$ and the saturated hydraulic conductivity $K_s$ were measured experimentally [3]. The unsaturated hydraulic conductivity $K_r$ was predicted well in the cases of Hygiene sandstone, Touchet Silt Loam G.E. 3, Silt Loam G.E.3 and Guelph Loam, but Beit Netofa Clay was not well predicted [3].

It is difficult to estimate SWRPs because of the spatial and temporal variability of the soil hydraulic properties in the field. In addition, pedotransfer functions (PTF) help to convert the directly-measured data from soil survey (e.g., field morphology, soil texture, structure, pH, *etc.*) into estimates of soil hydraulic parameters [6–9]. Five hierarchical neural network PTFs fulfilled this conversion in vG model from different levels of input data in Unsaturated Soil Hydraulic Property Database (UNSODA), and the neural network analyses combining with the bootstrap method generated uncertainty estimates of the predicted hydraulic properties and statistically appraised the reliability of the predictions [10–15].

Some quasi-physical models have estimated the SWRC according to the shape similarity between SWRC and the cumulative particle-size distribution [16–18].

The fractal geometry model can mathematically and physically describe porous media and build functional relationships between the water content and the matric potential. It has three SWRC types based on the fractal organization of soil structure: fractal mass, fractal surface and fractal pore-size distribution [19–22]. Ghanbarian *et al.* [23] used a relationship between the pore size distribution index of the Brooks and Corey model (BC model) and the fractal dimension of SWRC to evaluate two approaches for estimating parameter $m$ in vG model [3] and the statistical parameters showed that the approach proposed by Lenhard *et al.* [24] provided better estimates of $m$.

The predictions of parameters in SWRC are severely restricted by the experiment method and data. The accurate identification of SWRP ($\theta_r$, $\alpha$ and $n$) and hydraulic conductivity made demand on cumulative outflow as input data in the one-step pressure outflow experiments and the initial parameter evaluation to be reasonably close to their true values [25,26]. An integral method with Richards' equation and the closed-form equations of soil hydraulic properties was used to estimate $\alpha$ and $n$ if both infiltration and wetting front with time in a horizontal absorption experiment were recorded and it then provided a transient water flow approach to estimate SWRC instead of the usual equilibrium method [27,28].

UNSODA is a database with unsaturated soil hydraulic properties and other soil information [14]. UNSODA version 1 consists of 791 and version 2 of 790 soil materials with water retention, saturated and unsaturated hydraulic conductivity data measured in the field or laboratory, as well as particle size distribution and bulk density data. Each soil material has an identifier code, the minimum is 1010, the maximum is 4960, the soil material with same identifier code in one table is defined as a sample, and the data size of a sample is the number of data records with the same identifier code in one table. The data of each soil material are classified to different tables as a consecutive series of records with the same identifier code according to three hierarchic levels: measurement methodology (filed or lab); hydraulic drainage curve (drying or wetting); data relationship (preshead-conductivity, preshead-$\theta$, $\theta$-diffusivity or $\theta$-conductivity). For instance, there are 700 soil samples with available preshead-$\theta$ paired data and 90 samples missing data in the lab_drying_h-t table. Kosugi developed a general conductivity model for soils with lognormal pore-size distribution based on the Mualem-Dagan pore-scale model and two predictive methods reducing the average prediction error more than 77% compared with the Burdine and Mualem predictive models with use of 200 soil samples in UNSODA [29,30]. UNSODA was the database of Neural network analysis, bootstrap method and ROSETTA model implementing five

PTFs for hierarchical estimation of the soil water retention and the saturated or unsaturated hydraulic conductivity [10–12].

DE invented by Kenneth Price and Rainer Storn is a very simple population based, stochastic function minimizer for continuous function optimization and optimizes a problem by iteratively trying to improve a candidate solution with regard to a given measure of quality [31–33]. The crucial idea behind DE is a scheme for generating trial parameter vectors. Basically, DE adds the weighted difference between two population vectors to a third vector. DE optimizes a problem by maintaining a population of candidate solutions and creating new candidate solutions by combining existing ones according to its simple formulae, and then keeping whichever candidate solution has the best score or fitness on the optimization problem at hand. It has become a popular optimization method widely used in Informatics, Thermodynamics, dynamic systems, etc., and it is already integrated in Mathematica and MATLAB. At present, we mainly focus on the compatibility and adaptation of DE and UNSODA.

In this paper, we will show how to generate different datasets from UNSODA, estimate the SWRPs in vG model on each dataset, and obtain more reliable statistical results with DE. First we will correctly configure and test DE with UNSODA; then we will design statistical computations to estimate SWRPs and compare with previous algorithms and results; finally we will present the SWRP tables which can be referred directly by agricultural research and practice.

## 2. Differential Evolution

DE is a parallel direct search method which utilizes NP D-dimensional parameter vectors $\mathbf{x}_{i,G}$ ($i = 1, 2, \cdots, \text{NP}$) as a population for each generation G and it consists of 4 basic steps: initialization, mutation, crossover and selection [31,33]. $N_{\mathbf{p}}$ is the number of parameters to be optimized.

### 2.1. Initialization

The initial vector population is randomly selected and should cover the entire parameter space assuming a uniform probability distribution for all random decisions. If a preliminary solution is ready, the initial population might be generated by adding normally distributed random deviations to the nominal solution,

$$\mathbf{x}_{i,G} = \mathbf{x}_{min} + \text{rand(D)}(\mathbf{x}_{max} - \mathbf{x}_{min}), \quad G = 0 \tag{1}$$

where D is the dimension of parameter vector, and $\mathbf{x}_{max}$ and $\mathbf{x}_{min}$ are the lower and upper bounds of the parameter vectors $\mathbf{x}_{i,G}$.

### 2.2. Mutation

DE adds the weighted difference between two population vectors to a third one in order to generate new parameter vectors. For each target vector $\mathbf{x}_{i,G}$ ($i = 1, 2, \cdots, \text{NP}$), a mutant vector is generated by

$$\mathbf{v}_{i,G+1} = \mathbf{x}_{r_1,G} + F(\mathbf{x}_{r_2,G} - \mathbf{x}_{r_3,G}) \tag{2}$$

where the mutually different random indexes $r_1$, $r_2$, $r_3 \in \{i = 1, 2, \cdots, NP\}$ are different from the running index $i$, and $F \in [0, 2]$ is a constant factor which controls the amplification of the differential variation $(\mathbf{x}_{r_2,G} - \mathbf{x}_{r_3,G})$.

## 2.3. Crossover

Crossover is to increase the diversity of the perturbed parameter vectors. The mutated vector's parameters are mixed with the parameters of another predetermined vector, *i.e.*, the target vector, to yield a trial vector

$$\mathbf{u}_{i,G+1} = (\mathbf{u}_{1i,G+1}, \mathbf{u}_{2i,G+1}, \cdots, \mathbf{u}_{Di,G+1}) \tag{3}$$

where

$$\mathbf{u}_{ji,G+1} = \begin{cases} \mathbf{v}_{ji,G+1} & \text{if } (\text{randb}(j) \leq \text{CR}) \text{ or } j = \text{rnbr}(i) \\ \mathbf{x}_{ji,G} & \text{if } (\text{randb}(j) > \text{CR}) \text{ and } j \neq \text{rnbr}(i) \end{cases} \tag{4}$$

$j = 1, 2, \cdots, D$. In Equation (4), randb$(j)$ is the $j$th evaluation of a uniform random number generator within the range $[0, 1]$, CR is the crossover constant in $[0, 1]$ determined by the investigated problem, and rnbr$(i)$ is the randomly chosen index in $[1, 2, \cdots, D]$ ensuring that $\mathbf{u}_{i,G+1}$ gets at least one parameter from $\mathbf{v}_{i,G+1}$.

## 2.4. Selection

To decide whether or not it should become a member in generation $G + 1$, the trial vector $\mathbf{u}_{i,G+1}$ is compared to the target vector $\mathbf{x}_{i,G}$ using the greedy criterion. If $\mathbf{u}_{i,G+1}$ yields a smaller cost function value than $\mathbf{x}_{i,G}$, $\mathbf{x}_{i,G+1}$ is set to $\mathbf{u}_{i,G+1}$ in the next generation $G + 1$; otherwise, the old value $\mathbf{x}_{i,G}$ is retained to $\mathbf{x}_{i,G+1}$.

Moreover, the iteration termination conditions of the main algorithm in DE are relevant to the fitness effect and the convergence precision, and we adopt some usual methods: Maximum Evolution Generation (MEG), Maximum Number of Iterations (MIT) or some given objective value (GOV) for RMSE$_w$.

During the test and selection of algorithms, we found that DE can get higher precision and faster speed mainly depends on such characteristics: random initialization without predetermined initial parameter values, vector-based computation, global optimization and evolution strategy. Some algorithms, e.g., neural network, support vector regression, genetic algorithms, do not include all these characteristics or are not easily implemented in pSWRP estimation [6–9,19–22,35–37].

## 3. Fitting Soil Water Retention Parameters to Hydraulic Data

The soil water retention function in vG model is given by

$$\theta(h) = \theta_r + \frac{\theta_s - \theta_r}{[1 + (\alpha h)^n]^m} \tag{5}$$

where $m = 1 - 1/n$, and $\theta(h)$ is the volumetric water content ($cm^3cm^{-3}$) at the pressure head $h$ (cm, taken positive). The dimensionless effective water content is

$$S_e = \frac{\theta(h) - \theta_r}{\theta_s - \theta_r} \qquad (6)$$

The objective function that is minimized for fitting SWRC, *i.e.*, Equation (5) to the prehead-$\theta$ data in UNSODA and then gets the optimized SWRPs is

$$O_w(\mathbf{p}) = \sum_{i=1}^{N_w} (\theta_i - \theta_i')^2 \qquad (7)$$

where $\theta_i$ and $\theta_i'$ are the measured and the estimated water contents respectively, $N_w$ is the number of measured water retention points for each sample and is taken as the data size of a sample in computation, and $\mathbf{p}$ is the parameter vector $(\theta_r, \theta_s, \alpha, n)$. $\theta_i$ is from UNSODA, and $\theta_s$ and $\theta_i'$ will be calculated by DE. The goodness of fit of Equation (5) is quantified with the root mean square error:

$$RMSE_w(\mathbf{p}) = \sqrt{\frac{O_w(\mathbf{p})}{N_w - N_{\mathbf{p}}}} \qquad (8)$$

$N_{\mathbf{p}}$ is the number of parameters to be optimized. Results of Equation (8) will be presented as averages for each textural group or for the investigated dataset and also be taked as an iteration termination condition in programming [12].

Before we apply DE to estimate SWRPs, it is necessary to configure DE correctly: choose an appropriate dataset from UNSODA for specific task, balance the convergency and time consumption, and evaluate the control variables.

- The choice of dataset. There are four tables about preshead-$\theta$ data in UNSODA: field_drying_h-t, field_wetting_h-t, lab_drying_h-t and lab_wetting_h-t. There are 127 samples in field_drying_h-t table, 0 samples in field_wetting_h-t table, 700 samples in lab_drying_h-t table, and 28 samples in lab_wetting_h-t table with available data and this dataset is called dataset 1. The data in dataset 1 are diverse and heterogeneous (field or lab, wetting or drying), and appropriate for looking into the universal characteristics of UNSODA and the feasibility and robustness of DE. Schaap and Leij [12] and Schaap *et al.* [13] chose 235 codes in lab_drying_h-t table as a dataset based on the criteria: quality of the data; presence of sufficient texture data; the data from the laboratory drying (drainage) branches; eliminating samples with low bulk density values ($<0.5$ g/cm$^3$), and it is called dataset 2 here. Moreover, the designated 235 identifier codes with available data also partially appear in other two h-t tables: 26 samples in field_drying_h-t table and 1 sample in lab_wetting_h-t, the sizes of which are not enough for statistics. Tables 1 and 2 show: The $RMSE_w$ mean of dataset 1 becomes numerically stable at MEG = 400 and dataset 2 at MEG = 200; the $RMSE_w$ maximum of dataset 1 is 14 times of the mean, and only 3 times for dataset 2; the discrepancy between $RMSE_w$ maximum and minimum of dataset 1 is 3 times of dataset 2; the $RMSE_w$ mean of dataset 2 is 20% higher than dataset 1; the means of loop times are different only at MEG $\geq$ 10,000; the minimum loop times of dataset 2 is 4~5 times of dataset 1. The above differences of $RMSE_w$ and loop times between datasets 1 and 2 lie in: Because the dataset 2 is completely from lab_drying_h-t table and the data size of any sample in dataset 2 is not less than 6, the data quality will be higher and the statistical characteristics are more homogeneous.

- The iteration termination. DE is globally convergent, the convergence speed will become slower, $RMSE_w$ will not decrease for long time after reaching certain relatively steady value, and this value can be taken as a numerical convergence value (NCV). MEG or MIT shall satisfy that every sample in dataset can reach a NCV. The loop times that a sample takes to reach its own NCV is called the actual loop times and it should be less than MEG. The NCVs of $RMSE_w$ mean in Tables 1 and 2 can be respectively taken as 0.0106~0.0107 and 0.0121~0.0122 while means of loop times remarkably increase with MEG. In Table 2, $RMSE_w$ becomes stable at MEG = 200, but the corresponding mean loop times 197 is close to MEG and it can not ensure that some samples already get NCV when MEG and mean loop times are close. Therefore it is ideal to set MEG = 1000 or above as an iteration termination condition according to the first and the sixth columns in Table 2. If a sample reaches GOV, the iteration will terminate. Hence we utilize both MEG and GOV as the termination conditions in DE programming.

- The relationship between the loop times and the data size of a soil sample. The data size is the number of data records of a soil sample, *i.e.*, $N_w$ in Equation (3). In Table 3, MEG = 10,000; in the minimum case of soil samples, the loop times is the minimum for all samples, and the maximum likewise; number is the count of the minimum or maximum cases; but the average data sizes, 13 and 12.23 of the minimum and the maximum cases are very close and it means that loop times are unrelated to data size; the maximum cases have much lower $RMSE_w$ mean and more loop times will get better fitness. It implies that DE time consumption depends on the convergency speed rather than the data size.

- The range of control variables. The rule of thumb values for the control variables in DE is: $F \in [0.5, 1.0]$, $Cr \in [0.8, 1.0]$ and $Np = 10 \cdot D$ [33]. The designated 235 identifier codes have only 26 samples with available data in the field_drying_h-t table, which composes dataset 3. We compare dataset 2 with 235 samples and dataset 3 with 26 samples by different values of Cr, 0.3 and 0.8 in Table 4 as MEG = 10,000: In dataset 2, different Crs do not generate evidently different estimates of water retention parameters; compared with dataset 2, the estimation differences in dataset 3 can not be ignored except $\theta_r$ and $RMSE_w$. It means the value of Cr is not crucial to the parametric estimation in UNSODA if the dataset is big enough and this conclusion can also be safely applied to other control variables and even the initial values of parameters.

After DE configuration and implement, we now apply it to predict the SWRPs of different soil texture groups and investigate its computation speed and goodness of fit. The composition of dataset 2 with 235 samples is: 112 sands, 37 loams, 55 silts and 31 clays. This composition is different from that used by Schaap and Leij [12], and Schaap *et al.* [13]. We think that UNSODA version 2 has been updated from version 1. We will examine this viewpoint by comparing the results in Table 5 with Table 1 in Schaap and Leij [12]: Most estimates of parameters $\theta_r$, $\theta_s$, lg($\alpha$), lg($n$) and even their standard deviations in the first row of two tables for two datasets are close and the highest relative difference is only 3%; however, $\theta_s$ of sands, lg($\alpha$) of loams, $\theta_r$ and lg($n$) of loams, silts and clays in both tables appear different and their relative differences are more than 12%. According to this comparison, we can confirm that the dataset 2 and the dataset used by Schaap and Leij [12] have selected the samples with the same 235 identifier codes, but the compositions are different: some soil materials have been reclassified to another textural groups in UNSODA version 2. $RMSE_w$s in the first 4 rows of Table 5 are almost 0.001 and is one order of magnitude lower than the values in Table 1 [12], which indicates SWRP estimates by DE are improved after grouping soil texture. The higher-precision prediction and fit of DE are also found in the comparison with nonlinear least-squares curve-fitting method or damper least square method [12]. Plant stresses (drought stress, flood stress) are related to $\theta_r$ and $\theta_s$, and tensile and shear modulus and the constitutive variables of soil to $\lg(\alpha)$ and $\lg(n)$. Table 5 indicates that soil texture should be taken into the agriculture engineering and the precaution of geological disasters.

The first hierarchic level in UNSODA is the measurement methodology, field or lab. Dataset 3 has only 26 of 235 designated identifier codes with available data in field_drying_h-t table. The samples with the same 26 identifier codes in lab_drying_h-t table are called dataset 4. We can compare the field and lab measurements by dataset 3 and dataset 4 in Table 6: $RMSE_w$ of field is 10 times of the lab; the field and lab values of $\theta_r$ and $\lg(n)$ are evidently different, but $\theta_s$ and $\lg(\alpha)$ otherwise; the standard deviation of each entry in the lab row are much smaller than the field one. Theses results verify again that the data gotten from lab measurement are more steady and accurate.

**Table 1.** $RMSE_w$ and loop times of dataset 1.

| MEG | $RMSE_w$ | | | | Loop Times | | | Time Consumption |
| | Mean | Maximum | Minimum | Std. Deviation | Mean | Maximum | Minimum | (h:m:s) |
| --- | --- | --- | --- | --- | --- | --- | --- | --- |
| 50 | 0.018 | 0.193 | 0.001 | 0.014 | 46 | 49 | 13 | 00:02:38 |
| 100 | 0.012 | 0.182 | 0.001 | 0.011 | 97 | 99 | 58 | 00:04:29 |
| 200 | 0.011 | 0.141 | 0.001 | 0.009 | 196 | 199 | 55 | 00:08:16 |
| 300 | 0.011 | 0.141 | 0.001 | 0.009 | 286 | 299 | 55 | 00:11:56 |
| 400 | 0.011 | 0.141 | 0.001 | 0.009 | 356 | 399 | 52 | 00:15:28 |
| 500 | 0.011 | 0.141 | 0.001 | 0.009 | 426 | 499 | 51 | 00:19:08 |
| 1000 | 0.011 | 0.141 | 0.001 | 0.009 | 748 | 999 | 50 | 00:37:11 |
| 10,000 | 0.011 | 0.141 | 0.001 | 0.009 | 3322 | 9999 | 49 | 07:51:09 |
| 40,000 | 0.011 | 0.141 | 0.001 | 0.009 | 10,247 | 39,999 | 57 | 27:15:55 |
| Std. deviation | $2.53 \times 10^{-3}$ | $2.07 \times 10^{-2}$ | $1.52 \times 10^{-4}$ | $1.44 \times 10^{-3}$ | | | | |

**Table 2.** $RMSE_w$ and loop times of dataset 2.

| MEG | $RMSE_w$ | | | | Loop Times | | |
| | Mean | Maximum | Minimum | Std. Deviation | Mean | Maximum | Minimum |
| --- | --- | --- | --- | --- | --- | --- | --- |
| 50 | 0.018 | 0.065 | 0.002 | 0.010 | 47 | 49 | 25 |
| 100 | 0.013 | 0.039 | 0.001 | 0.007 | 97 | 99 | 83 |
| 200 | 0.012 | 0.039 | 0.001 | 0.007 | 197 | 199 | 183 |
| 300 | 0.012 | 0.039 | 0.001 | 0.007 | 287 | 299 | 218 |
| 400 | 0.012 | 0.039 | 0.001 | 0.007 | 353 | 399 | 221 |
| 500 | 0.012 | 0.039 | 0.001 | 0.007 | 424 | 499 | 221 |
| 1000 | 0.012 | 0.039 | 0.001 | 0.007 | 722 | 999 | 254 |
| 10,000 | 0.012 | 0.039 | 0.001 | 0.007 | 2963 | 9999 | 240 |
| 40,000 | 0.012 | 0.039 | 0.001 | 0.007 | 7838 | 39,999 | 265 |
| Std. deviation | $2.067 \times 10^{-3}$ | $9.24 \times 10^{-3}$ | $3.26 \times 10^{-4}$ | $9.84 \times 10^{-4}$ | | | |

**Table 3.** Relationship between loop times and the data size of soil sample.

| Case | Loop Times | Number | Data Size | RMSE$_w$ Mean |
|---|---|---|---|---|
| minimum | 240 | 1 | 13 | 0.017 |
| maximum | 9999 | 31 | 12.23 | 0.009 |

**Table 4.** CR range.

| | Dataset 2 | | Dataset 3 | |
|---|---|---|---|---|
| | Cr = 0.3 | Cr = 0.8 | Cr = 0.3 | Cr = 0.8 |
| $\theta_r$ | 0.054 | 0.057 | 0.166 | 0.152 |
| $\theta_s$ | 0.463 | 0.466 | 0.442 | 0.476 |
| lg($\alpha$) | −1.625 | −1.607 | −1.581 | −1.495 |
| lg($n$) | 0.214 | 0.208 | 0.604 | 0.536 |
| RMSE$_w$ | 0.001 | 0.001 | 0.015 | 0.015 |

**Table 5.** Average hydraulic parameters for each soil textural group with standard deviations in parentheses.

| | N | $\theta_r$ (cm$^3$cm$^{-3}$) | $\theta_s$ | lg($\alpha$) (cm$^{-1}$) | lg($n$) | RMSE$_w$ (cm$^3$cm$^{-3}$) |
|---|---|---|---|---|---|---|
| All | 235 | 0.057 (0.081) | 0.467 (0.139) | −1.606 (0.550) | 0.204 (0.197) | 0.001 (0.002) |
| Sands | 112 | 0.050 (0.044) | 0.442 (0.154) | −1.534 (0.421) | 0.321 (0.321) | 0.001 (0.002) |
| Loams | 37 | 0.096 (0.137) | 0.525 (0.139) | −1.239 (0.488) | 0.106 (0.109) | 0.001 (0.002) |
| Silts | 55 | 0.035 (0.064) | 0.436 (0.097) | −1.890 (0.594) | 0.106 (0.057) | 0.002 (0.003) |
| Clays | 31 | 0.071 (0.104) | 0.542 (0.099) | −1.799 (0.644) | 0.075 (0.051) | 0.051 (0.001) |

Sands: sand, loamy sand, sandy loam, sandy clay loam; Loams: loam, clay loam; Silts: silt loam, silt; Clays: clay, sandy clay, silty clay, silty clay loam.

**Table 6.** Comparison between field and lab measurements.

| Measurement | $\theta_r$ | $\theta_s$ | lg($\alpha$) | lg($n$) | RMSE$_w$ |
|---|---|---|---|---|---|
| field | 0.155 (0.102) | 0.454 (0.196) | −1.562 (0.557 ) | 0.533 (0.354) | 0.015 (0.064) |
| lab | 0.075 (0.051) | 0.494 (0.131) | −1.568 (0.311) | 0.288 (0.186) | 0.001 (0.003) |

## 4. Results and Discussion

In Section 3, DE has already applied to estimate SWRPs after the dataset filtering, DE and UNSODA configuring and the statistical calculation compiling, and we have gotten some significant results from the calculation and Tables 1–6: The dataset composed of the designated 235 identifier codes in lab_drying_h-t table is more appropriate for statistical research and the results based on this dataset are credible; the data size of a soil sample has no direct relationship with the actual loop times in DE main iteration; the different values of control variables will not produce evident change on the average estimates of SWRPs if the dataset is big enough; the average RMSE$_w$s of dataset 2, sands, loams and silts groups are almost one order of magnitude lower than some previous optimization methods, and DE has higher precision and better fitness; the estimates of $\theta_s$ and lg($\alpha$) from field and lab measurements are close, and $\theta_r$ and lg($n$) otherwise; parameter estimates on the whole dataset 2 are close to the dataset used by Schaap and Leij [12], but there appear differences on the estimates of $\theta_r$ and lg($n$) of loam, silt and clay groups; $\theta_r$ has showed different change trend from other parameters in statistical process.

The convergence speed is a fascinating issue when using database. It is already shown that there is no evident relationship between the loop times and the data size of a soil sample in Table 3. In Table 7: a and b represent two program runnings with the same MEG so that we can estimate the effect of stochastic factors in DE; the first column is actual loop times and the second is $RMSE_w$ under same MEG. Loop times change irregularly while $RMSE_w$s maintain steady in a and b runnings under the same MEG; the sample with identifier code = 2763 has 13 pairs of data and gets the minimum loop times 240, and the sample with identifier code = 2660 has 12 pairs of data and gets the maximum loop times 19,999 when MEG = 20,000 a, b; in the row of code = 2763, $RMSE_w = 0.017$, but the actual loop times are extraordinarily higher when MEG = 5000 b and MEG = 20,000 a; in the row of code = 2660, the actual loop times fluctuate abruptly while $RMSE_w$s swing in narrow range $(0.012, 0.014)$, and the three smallest loop times 1117, 1982 and 1578 contrarily get the smallest $RMSE_w = 0.012790$ under MEG = 3000 a, 3000 b and 5000 b. These statistical results in Table 7 seem to contradict to the common sense: Similar $RMSE_w$s or similar data sizes of a sample should take similar loop times; smaller $RMSE_w$ should take more loop times. We guess that the stochastic factors (e.g., random seed, mutation step, crossover step, etc.) in DE might have considerable impact on the computation speed and the optimization efficacy.

**Table 7.** Convergence speed.

| Code | MEG | | | | | | | | | | | | | | |
|------|-----|---|-----|---|-----|---|-----|---|-----|---|-----|---|-----|---|-----|
| | 3000 a | | 3000 b | | 5000 a | | 5000 b | | 10,000 a | | 10,000 b | | 20,000 a | | 20,000 b | |
| 2763 | 328 | 0.017 | 293 | 0.017 | 399 | 0.017 | 504 | 0.017 | 240 | 0.017 | 310 | 0.017 | 849 | 0.017 | 372 | 0.017 |
| 2660 | 1117 | 0.013 | 1982 | 0.013 | 4999 | 0.013 | 1578 | 0.013 | 9999 | 0.013 | 9998 | 0.013 | 19,999 | 0.014 | 19,999 | 0.013 |

## 5. Conclusions

In this paper we first introduced UNSODA and DE, and then applied DE to estimate SWRPs in vG model based on UNSODA. We have configured DE and tested specific datasets in order to derive statistically tenable results. We have discussed the relationship between the data size of a soil sample and the actual loop times in the DE main iteration for reaching a NCV. Some specific statistical computations have been designed to select appropriate datasets and compare the parameter estimates between different values of control variables, different measurements or different textual groups, etc., and the calculation results shows that DE is capable to derive the values of soil water retention parameters from UNSODA, UNSODA is a reliable database for soil hydraulic property indices and statistical Tables 1–7 can be applied directly in agricultural research and practice.

We have also statistically analysed the factors on convergence or fitness as the fundamental issues of DE. The methods above are illuminative for correctly using UNSODA and DE, and the conclusions are valuable for soil hydraulic parameters. In the future, We will predict SWRC and unsaturated hydraulic conductivity expression together with basic or advanced forms of DE, explore some fundamental issues in DE in order to explain some basic problems in the determination of soil hydraulic parameters and then propose new algorithmic, survey or experimental methods.

## Acknowledgments

This work was financially supported by the Science and Research Startup Fund of Fujian Normal University, Nonlinear Analysis and Its Applications IRTL1206, and the Scientific Research Foundation for the Returned Overseas Chinese Scholars from State Education Ministry DB-179.

## Conflicts of Interest

The authors declare no conflict of interest.

## References

1.  Burdine, N.T. Relative permeability calculations from pore-size distribution data. *J. Pet. Technol.* **1953**, *5*, 71–78.
2.  Mualem, Y. A new model predicting the hydraulic conductivity of unsaturated porous media. *Water Resour. Res.* **1976**, *12*, 513–522.
3.  Van Genuchten, M.T. A closed-form equation for predicting the hydraulic conductivity of unsaturated soils. *Soil Sci. Soc. Am. J.* **1980**, *44*, 892–898.
4.  Van Genuchten, M.T.; Hopmans, J.W. A Decade of Multidisciplinary Research. *Vadose Zone J.* **2013**, *12*, doi:10.2136/vzj2013.08.0150.
5.  Dettmann, U.; Bechtold, M.; Frahm, E.; Tiemeyer, B. On the applicability of unimodal and bimodal van Genuchten—Mualem based models to peat and other organic soils under evaporation conditions. *J. Hydrol.* **2014**, *515*, 103–115.
6.  Rawls, W.J.; Pachepsky, Y.; Shen, M.H. Testing soil water retention estimation with the MUUF pedotransfer model using data from the southern United States. *J. Hydrol.* **2001**, *251*, 177–185.
7.  Pachepsky, Y.A.; van Genuchten, M.T. Pedotransfer functions. *Encycl. Agrophysics* **2011**, 556–561.
8.  Pan, F.; Pachepsky, Y.; Jacques, D.; Guber, A.; Hill, R.L. Data assimilation with soil water content sensors and pedotransfer functions in soil water flow modeling. *Soil Sci. Soc. Am. J.* **2012**, *76*, 829–844.
9.  Ramos, T.B.; Gonçalves, M.C.; Brito, D.; Martins, J.C.; Pereira, L.S. Development of class pedotransfer functions for integrating water retention properties into Portuguese soil maps. *Soil Res.* **2013**, *51*, 262–277.
10. Schaap, M.G.; Leij, F.J.; van Genuchten, M.T. Neural Network Analysis for Hierarchical Prediction of Soil Hydraulic Properties. *Soil Sci. Soc. Am. J.* **1998**, *62*, 847–855.
11. Schaap, M.G.; Feike, J.L. Using neural networks to predict soil water retention and soil hydraulic conductivity. *Soil Tillage Res.* **1998**, *47*, 37–42.
12. Schaap, M.G.; Leij, F.J. Improved Prediction of unsaturated hydraulic conductivity with the Mualem-van Genuchten Model. *Soil Sci. Soc. Am. J.* **2000**, *64*, 843–851.
13. Schaap, M.G.; Leij, F.J.; van Genuchten, M.T. ROSETTA: A computer program for estimating soil hydraulic parameters with hierarchical pedotransfer functions. *J. Hydrol.* **2001**, *251*, 163–176.
14. Nemes, A.; Schaap, M.G.; Leij, F.J.; Wösten, J.H.M. Description of the unsaturated soil hydraulic database UNSODA version 2.0. *J. Hydrol.* **2001**, *251*, 151–162.

15. Pachepsky, Y.; Pan, F.; Martinez, G. Sensor Network Data Assimilation in Soil Water Flow Modeling. In *Application of Soil Physics in Environmental Analyses*; Springer International Publishing: Geneva, Switzerland: 2014; pp. 239–260.

16. Fredlund, M.D.; Wilson, G.W.; Fredlund, D.G. Use of the grain-size distribution for estimation of the soil-water characteristic curve. *Can. Geotech. J.* **2002**, *39*, 1103–1117.

17. Haverkamp, R.; Leij, F. J.; Fuentes, C.; Sciortino, A.; Ross, P. J. Soil water retention: I. Introduction of a shape index. *Soil Sci. Soc. Am. J.* **2005**, *69*, 1881–1890.

18. Leij, F.J.; Haverkamp, R.; Fuentes, C.; Zatarain, F.; Ross, P.J. Soil water retention: II. Derivation and application of shape index. *Soil Sci. Soc. Am. J.* **2005**, *69*, 1891–1901.

19. Bird, N.R.A.; Perrier, E.; Rieu, M. The water retention function for a model of soil structure with pore and solid fractal distributions. *Eur. J. Soil Sci.* **2000**, *51*, 55–63.

20. Perfect, E. Modeling the primary drainage curve of prefractal porous media. *Vadose Zone J.* **2005**, *4*, 959–966.

21. Huang, G.; Zhang, R. Evaluation of soil water retention curve with the pore-solid fractal model. *Geoderma* **2005**, *127*, 52–61.

22. Ghanbarian-Alavijeh, B. Modeling Physical and Hydraulic Properties of Disordered Porous Media: Applications from Percolation Theory and Fractal Geometry. Ph.D. Thesis, Wright State University, Dayton, OH, USA, 2014.

23. Ghanbarian-Alavijeh, B.; Liaghat, A.; Huang, G.H.; van Genuchten, M.T. Estimation of the van Genuchten soil water retention properties from soil textural data. *Pedosphere* **2010**, *20*, 456–465.

24. Lenhard, R.J.; Parker, J.C.; Mishra, S. On the correspondence between Brooks-Corey and van Genuchten models. *J. Irrig. Drain. Eng.* **1989**, *115*, 744–751.

25. Kool, J.B.; Parker, J.C.; van Genuchten, M.T. Determining soil hydraulic properties from one-step outflow experiments by parameter estimation: I. Theory and numerical studies. *Soil Sci. Soc. Am. J.* **1985**, *49*, 1348–1354.

26. Londra, P.A.; Valiantzas, J.D. Soil water diffusivity determination using a new two-point outflow method. *Soil Sci. Soc. Am. J.* **2011**, *75*, 1343–1346.

27. Shao, M.; Horton, R. Integral method for estimating soil hydraulic properties. *Soil Sci. Soc. Am. J.* **1998**, *62*, 585–592.

28. Peng, H.; Horton, R.; Lei, T.; Dai, Z.; Wang, X. A modified method for estimating fine and coarse fractal dimensions of soil particle size distributions based on laser diffraction analysis. *J. Soils Sediments* **2015**, *15*, 937ï£¡C-948.

29. Kosugi, K. Three-parameter lognormal distribution model for soil water retention. *Water Resour. Res.* **1994**, *30*, 891–901.

30. Kosugi, K. General model for unsaturated hydraulic conductivity for soils with lognormal pore-size distribution. *Soil Sci. Soc. Am. J.* **1999**, *63*, 270–277.

31. Storn, R.; Price, K. Minimizing the Real Functions of the ICEC'96 Contest by Differential Evolution. In Proceedings of the 1996 IEEE Conference on Evolutionary Computation, Nagoya, Japan, 20–22 May 1996; pp. 842–844.

32. César Trejo Zúñiga, E.; López Cruz, I.L.; García, A.R. Parameter estimation for crop growth model using evolutionary and bio-inspired algorithms. *Appl. Soft Comput.* **2014**, *23*, 474–482.

33. Storn, R. Differential Evolution Research—Trends and Open Questions. In *Advances in Differential Evolution*; Chakraborty, U.K., Ed.; Springer-Verlag: Heidelberg, Germany, 2008; pp. 1–32.

34. Van Genuchten, M.T.; Leij, F.J.; Yates, S.R. *The RETC Code for Quantifying the Hydraulic Functions of Unsaturated Soils*; EPA Report 600/2-91/065; US Salinity Laboratory, USDA, ARS: Riverside, CA, USA, 1991.

35. Garg, A.; Vijayaraghavan, V.; Wong, C.H.; Tai, K.; Sumithra, K.; Gao, L.; Singru, P.M. Combined ci-md approach in formulation of engineering moduli of single layer graphene sheet. *Simul. Model. Pract. Theory* **2014**, *48*, 93–111.

36. Garg, A.; Tai, K. Stepwise approach for the evolution of generalized genetic programming model in prediction of surface finish of the turning process. *Adv. Eng. Softw.* **2014**, *78*, 16–27.

37. Garg, A.; Garg, A.; Tai, K. A multi-gene genetic programming model for estimating stress-dependent soil water retention curves. *Comput. Geosci.* **2014**, *18*, 45–56.

# 9

# Inter-Taxa Differences in Iodine Uptake by Plants: Implications for Food Quality and Contamination

**Eleni Siasou [†] and Neil Willey [†,*]**

Centre for Research in Bioscience, Faculty of Health and Applied Sciences,
University of the West of England, Coldharbour Lane, Frenchay, Bristol BS16 1QY, UK;
E-Mail: eleni.siasou@uwe.ac.uk

[†]  These authors contributed equally to this work.

[*]  Author to whom correspondence should be addressed; E-Mail: neil.willey@uwe.ac.uk

Academic Editor: Gareth J. Norton

**Abstract:** Although iodine is not essential for plants, they take it up readily and, in foodchains, are significant sources of iodine for organisms with an essential requirement for it. During several nuclear accidents radioiodine has been an important component of releases of radioactivity and has caused serious contamination of foodchains. Differences in iodine uptake by different plant taxa are, therefore, important to nutritional and radioecological studies. Using techniques we have developed for a range of other elements, we analyzed inter-taxa differences in radioiodine uptake by 103 plant species and between varieties of two species, and analyzed them using a recent, phylogenetically-informed, taxonomy. The results show that there are significant differences in uptake above and below the species level. There are significant differences between Monocots and Eudicots in iodine uptake, and, in particular, hierarchical ANOVA revealed significant differences between Genera within Families. These analyses of the taxonomic origin of differences in plant uptake of iodine can help the prediction of crop contamination with radioiodine and the management of stable iodine in crops for nutritional purposes.

**Keywords:** iodine; radioiodine; radioecology; angiosperm phylogeny; soil-to-plant transfer; foodchains

# 1. Introduction

Iodine (I) is readily taken up by plants if available [1], which is important to both agronomy and radioecology because, although I is not an essential element for plants, food crops are a major conduit for the entry of I to human foodchains. Stable I ($^{127}$I) is an essential trace element for humans whilst the radioisotopes $^{131}$I and $^{129}$I can be significant radioactive contaminants of the environment [2]. About two hundred million people worldwide suffer from I deficiency disorders (IDD), with an at risk population potentially in excess of one billion [3]. It has been estimated that around 44% of children and adults in Europe have a mild iodine deficiency [4]. Low I intake from food crops is partly responsible for IDD and thus, as for many other trace elements [3], to redress deficiencies it is potentially useful to understand the agronomy of $^{127}$I in the soil-crop system. The iodized-salt enrichment diet has reduced I deficiency in some areas but there are still areas with a significant level of IDD.

The toxicologically important radioisotopes $^{131}$I and $^{129}$I, which are both fission products, can be a significant component of releases of radioactivity to the environment, and, as isotopes of an essential element, tend to accumulate in animals if they are present in food. $^{131}$I has a short-half life (eight days), is quite a high energy β/γ emitter and is primarily of concern as a food contaminant in the immediate aftermath of releases from accidents or from fall-out from above-ground nuclear weapons detonation. $^{129}$I has a long half-life ($15.7 \times 10^6$ years) and has been of importance in accidental releases, but it is a major, and potentially mobile, constituent of high and medium level nuclear waste [2], and is released into the marine environment from nuclear-fuel reprocessing plants [5]. $^{129}$I has the potential to be drawn upwards through soil profiles from repositories [6] and to be transferred from sea to land [5], provoking interest in its transfer characteristics from soil-to-plants during assessments of nuclear waste repositories and marine releases.

The transfer of I from soils-to-plants is possible because its isotopes can be both available in soil and taken up by plants. In fact, in comparison to many other nutrients and radionuclides, I isotopes are highly available in many soils with, for example, compilations of soil-solution distribution coefficients ($K_d$) for radionuclides suggesting that $^{129}$I is amongst the least strongly adsorbed isotopes in a range of soils [7]. There is little sorption of $^{129}$I on clay minerals and any sorption is primarily to organic matter [6]. I is more labile under anoxic than oxic soil conditions. For example, the flooding of paddy soils has long been known to produce the "Akagare" phenomenon in rice, which results from I toxicity caused by large increases in availability brought on by anoxia [8]. It is also clear that there can be significant changes in I mobility between the water-table and the vadose zone in soils [6]. In many soils I$^-$ and IO$_3^-$ are the most common ionic forms, with I$^-$ most likely to be taken up by plants [1] because they have substantial capacity for the uptake of the chemically similar Cl$^-$ [9]. Overall, although soil-to-plant transfer factors can be quite low from, for example, Andosols with high anion exchange capacities [10], hydroponic experiments show that plants can take up large quantities of I if it is available to them [1] and most soils produce transfers to crops that can contribute significantly to food I content and to radiocontamination if $^{127}$I or $^{131/129}$I are available in the soil.

It has been suggested, based on a limited number of species, that inter-species differences exist in the plant uptake of I under comparative conditions (e.g., [10]) and concentrations of almost all elements across different plant species do not simply reflect soil availability, *i.e.*, there are significant inter-taxa differences in uptake under the same conditions of availability [11]. It seems likely, therefore, that there might be inter-taxa differences in the concentration to which plants take up I, and that these might be useful to understanding the agronomy of I. There are, however, few data on this phenomenon and no studies that have attempted to link these differences to recent phylogenies of angiosperms (flowering plants), nor to compare them to inter-varietal differences. The understanding of the phylogeny (evolutionary relationships) of angiosperms has been transformed in recent years by molecular and computer methodologies, resulting in new phylogenies for angiosperms (e.g., [12]). Given that many phenotypes can be affected by phylogeny, angiosperm phylogenies specifically for use in comparative biological experiments have been published [13]. These have now been used to analyze inter-species differences in the concentrations to which plants concentrate numerous elements [11,14–18], and to establish that there is a significant influence of angiosperm phylogeny on plant mineralogy, including that of crop plants. Such analyses require quite large databases of inter-species comparisons, often produced by collating data from a variety of sources through, for example, Residual Maximum Likelihood (REML) analysis. Here we utilize techniques successfully used to investigate inter-species differences of other elements to construct a database of relative I concentrations following root exposure in 103 angiosperm species, analyze their differences using a recently published phylogenetic hierarchy for the angiosperms, compare them to inter-varietal differences in two species, and assess their influence on I concentrations in food crops. The usefulness of the results to predicting the transfer of I isotopes from soils to plants in agricultural and radioecological contexts is then discussed.

## 2. Results and Discussion

### 2.1. Inter-Species Differences in I Concentration

REML-estimated relative mean I concentrations in 103 species of plants are shown in Table 1, which is the most taxonomically wide-ranging comparison of relative I concentrations in plants yet published. Given that raw data is $\log_e$-transformed prior to REML analysis, the large range in REML transformed values ($-6.83$ to $3.72$) indicates that there are substantial inter-species differences in I uptake after the exposures used to generate data contributing to the database. This was confirmed using species grown for this work, in which replicate values for individual species grown under the same conditions allowed inter-species differences to be analyzed statistically (Figure 1). The International Atomic Energy Agency (IAEA) has recommended, based on data compilations across a range of different soil and crop types, mean soil-to-plant concentration ratios (CRs) for radioelements including radioiodine [19]. Iodine CR is calculated by dividing the concentration in the plants by the concentration in the soil. To enable comparison we estimated an overall CR for radioiodine of 0.075 based on values for cereal stems and leaves, leafy vegetables and non-leafy vegetables (Table 17.1 in [19]). $^{125}$I activity values measured in the experiments for Figure 1 were therefore transformed to have a geomean CR of 0.075. In this selection of species, inter-species differences in transformed CR following acute 4 h exposure varied from 0.0005 (*Bergenia cordifolia*) to 0.127 (*Salvia splendens*), a variability of recommended CR values that if it had been produced, for example, by different soil types would be regarded as very significant.

**Table 1.** Residual Maximum Likelihood I concentrations in 103 species of angiosperm listed according to the Angiosperm Phylogeny Group III system. (Study 1: [20]; Studies 2–14: [21]; Study 15: [22]; Studies 16–19: [23]; Studies 20–23: [24]; Studies 24–26: [25]; Study 27: [26]; Study 28: [27]; Studies 29–35 *Experiments for this study*; Study 36: [10]; Study 37: [28]).

| "Class" | "Subclass" | "Superorder" | Order | Family | Genus + Species | Relative Mean (I) | Studies |
|---|---|---|---|---|---|---|---|
| MONOCOTS | Lilianae | Commelinids | Commelinales | Commelinaceae | Commelina coelestis | 2.646 | 30 |
| | | | Zingiberales | Cannanaceae | Canna indica | 2.333 | 31 |
| | | | | Zingiberaceae | Zingiber officinale | 0.016 | 31 |
| | | | Poales | Cyperaceae | Carex nigra | 0.842 | 30 |
| | | | | Juncaceae | Juncus effusus | 0.393 | 31 |
| | | | | Poaceae | Agrostis tenuis | 1.871 | 1, 16 |
| | | | | | Agrostis alba | 2.021 | 16 |
| | | | | | Arrenatherum elatius | 1.887 | 16 |
| | | | | | Cynosuarus cristatus | 2.244 | 16 |
| | | | | | Dactylis glomerata | 1.804 | 1, 16 |
| | | | | | Festuca arundinacea | 2.138 | 16 |
| | | | | | Festuca rubra | 1.733 | 16 |
| | | | | | Festuca pratensis | 2.021 | 16 |
| | | | | | Holcus lanatus | 0.512 | 1,19 |
| | | | | | Hordeum vulgare | 2.802 | 2, 3, 4, 5, 6, 7, 8, 9, 10, 11, 12, 13, 14, 15 |
| | | | | | Lolium perenne | 2.31 | 1, 16, 20, 21, 22, 23 |
| | | | | | Lolium multiflorum | 2.299 | 16 |
| | | | | | Lolium hybridum | 1.702 | 1 |
| | | | | | Phleum pratense | 2.296 | 16, 20, 21, 22, 23 |
| | | | | | Poa annua | 0.666 | 1 |
| | | | | | Poa trivialis | 1.59 | 1, 16 |

**Table 1.** *Cont.*

| "Class" | "Subclass" | "Superorder" | Order | Family | Genus + Species | Relative Mean (I) | Studies |
|---|---|---|---|---|---|---|---|
| | | | | | *Poa pratense* | 2.021 | 16 |
| | | | | | *Triticum aesivum* | 2.852 | 24, 25, 26 |
| | | | | | *Zea mays* | 0.108 | 15, 29 |
| | | Non-Commelenids | Asparagales | Iridaceae | *Sisyrinchium striatum* | 1.484 | 32 |
| | | | | Amaryllidaceae | *Allium cepa* | 0.622 | 27, 30, 36, 37 |
| | | | | | *Allium sativum* | 2.321 | 27 |
| | | | | | *Allium porum* | 2.565 | 27 |
| EUDICOTS | Ranunculanae | ranunculiids | Ranunculales | Papaveraceae | *Papaver commutatum* | −0.004 | 33 |
| | Rosanae | fabids | Cucurbitales | Cucurbitaceae | *Cucumis sativa* | −2.131 | 27 |
| | | | | | *Cucumis melo* | 1.872 | 27 |
| | | | | | *Cucurbita maxima* | 0.611 | 27 |
| | | | | | *Potentilla anserina* | −0.367 | 18 |
| | | | Fabales | Fabaceae | *Cicer arietinum* | 2.719 | 29 |
| | | | | | *Faba vulgaris* | −1.117 | 27 |
| | | | | | *Lupinus angustifolius* | 0.412 | 29 |
| | | | | | *Medicago sativa* | 2.099 | 31 |
| | | | | | *Medicago lupulina* | 1.844 | 17 |
| | | | | | *Phaseolus vulgaris* | 1.292 | 15, 27 |
| | | | | | *Pisum sativum* | 2.505 | 2, 3, 4, 5, 6, 7, 8, 9, 10, 11, 12, 13, 27 |
| | | | | | *Trifolium repens* | 1.355 | 1, 16, 17, 19, 20, 21, 22, 23 |
| | | | | | *Trifolium subterraneum* | 1.611 | 1 |
| | | | | | *Trifolium pratense* | 0.923 | 1,17,18,20,21,22,23 |
| | | | Malpighiales | Euphorbiaceae | *Euphorbia cyparasissus* | −0.148 | 32 |
| | | | | | *Euphorbia myrsinites* | −0.232 | 31,32 |
| | | | | Linaceae | *Linum lewisii* | 0.328 | 31,33 |
| | | | | | *Linum usitatissimum* | 0.47 | 31 |

**Table 1.** *Cont.*

| "Class" | "Subclass" | "Superorder" | Order | Family | Genus + Species | Relative Mean (I) | Studies |
|---|---|---|---|---|---|---|---|
| | | | | Violaceae | *Viola wittrockiana* | -4.134 | 33 |
| | | | Rosales | Cannabaceae | *Humulus japonica* | 2.701 | 35 |
| | | | | Rosaceae | *Fragaria vesca* | 2.583 | 32 |
| | | malvids | Brassicales | Brassicaceae | *Aubretia x cultorum* | 0.184 | 33 |
| | | | | | *Brassica napus* | 3.258 | 27 |
| | | | | | *Brassica oleracea* | 1.072 | 27, 36 |
| | | | | | *Brassica chinensis* | 0.857 | 37 |
| | | | | | *Eruca vesicaria* | 1.461 | 29 |
| | | | | | *Lepidium sativum* | 0.138 | 15 |
| | | | | | *Raphanus sativus* | 2.642 | 15, 27, 36 |
| | | | Geraniales | Geraniaceae | *Geranium pratense* | 1.155 | 31 |
| | | | Saxifrgales | Saxifrgaceae | *Bergenia cordifolia* | -0.497 | 30 |
| | | | Malvales | Cistaceae | *Helianthemum nummularium* | -0.838 | 33 |
| | Caryophyllanae | caryophylids | Caryophyllales | Amaranthaceae | *Amaranthus paniculatus* | 1.912 | 34 |
| | | | | | *Beta rapa* | 3.718 | 27 |
| | | | | | *Beta cycla* | 1.025 | 27 |
| | | | | | *Beta vulgaris* | 2.749 | 15, 29 |
| | | | | | *Celosia cristata* | 1.184 | 30 |
| | | | | | *Spinacia oleracea* | 1.072 | 15, 27, 33, 37 |
| | | | | Caryophyllaceae | *Cerastium holosteiodes* | 2.443 | 17, 18 |
| | | | | Polygonaceae | *Dianthus caryophyllus* | 1.4 | 30 |
| | | | | | *Rheum tataricum* | 0.986 | 30 |
| | | | | | *Rumex acetosa* | 1.801 | 17, 18, 19, 27 |
| | | | | | *Rumex obtusifolius* | 2.485 | 17, 18 |
| | Asteranae | lamiids | Ericales | Polemoniaceae | *Gilia tricolor* | 1.755 | 31 |
| | | | Gentianales | Rubiaceae | *Cinchona pubescens* | -5.533 | 35 |
| | | | | | *Galium vernum* | -0.299 | 31 |
| | | | Lamiales | Lamiaceae | *Glechoma hederacea* | 2.488 | 18 |

**Table 1.** *Cont.*

| "Class" | "Subclass" | "Superorder" | Order | Family | Genus + Species | Relative Mean (I) | Studies |
|---|---|---|---|---|---|---|---|
| | | | | | *Mentha spicata* | 1.933 | 30 |
| | | | | | *Plectranthus blumei* | 2.22 | 34 |
| | | | | | *Salvia Blaze of Fire'* | 2.199 | 33 |
| | | | | Oleacae | *Fraxinus excelsior* | −1.835 | 31 |
| | | | | Plantaginaceae | *Plantago major* | 1.763 | 17, 18, 19 |
| | | | | | *Plantago lanceolata* | 1.911 | 17, 18, 19 |
| | | | | | *Antirrhinum major* | 1.624 | 31 |
| | | | | | *Digitalis purpurea* | 1.58 | 31 |
| | | | | | *Veronica spicata* | 2.366 | 32 |
| | | | | | *Veronica chamaedrys* | 2.488 | 18 |
| | | | Solanales | Solanceae | *Lycopersicon esculentum* | 3.615 | 27, 32 |
| | | | | | *Solanum tuberosum* | −6.828 | 27 |
| | | | | | *Solanum melongena* | 0.08 | 27, 36 |
| | | | | | *Solanum macrocarpon* | 0.587 | 30 |
| | | | | Convolulaceae | *Ipomea aquatica* | 2.385 | 37 |
| | | campanulids | Apiales | Apiaceae | *Apium graveolens* | −0.928 | 30, 37 |
| | | | | | *Daucus carota* | 2.265 | 27, 29, 37 |
| | | | | | *Heracleum sphondylium* | −6.828 | 18 |
| | | | | | *Petroselinum crispum* | 2.719 | 27 |
| | | | | | *Scandix cerifolium* | 1.813 | 27 |
| | | | Asterales | Asteraceae | *Aster x frikartii* | 1.153 | 30 |
| | | | | | *Lactuca sativa* | 0.159 | 15, 24, 25, 26, 27, 29, 36 |
| | | | | | *Helianthus annuus* | 2.862 | 29 |
| | | | | | *Taraxicum officinales* | −6.828 | 17, 18 |
| | | | | | *Cichorum intybus* | −6.828 | 27 |
| | | | | | *Cichorum angustifolium* | 1.684 | 27 |
| | | | Dipsacales | Dipsacaceae | *Dipsacus fullonium* | 1.096 | 34 |
| | | | | | *Lamium album* | 3.076 | 18 |

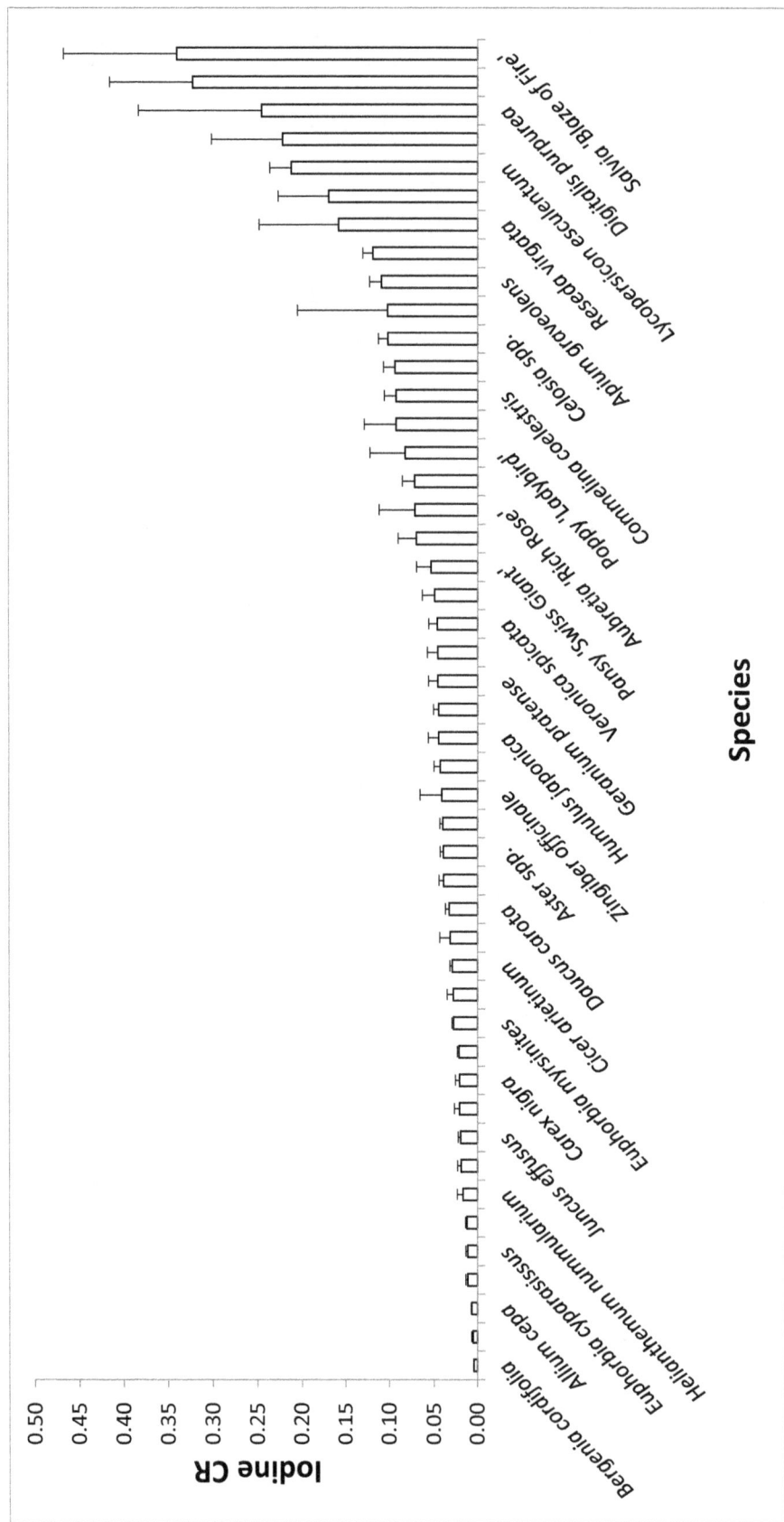

**Figure 1.** Mean I concentration ratios for 47 different species of plants all grown under the same conditions. Plants were exposed to $^{125}$I for 4 h and above ground green shoots harvested and analyzed for $^{125}$I activity. Activities were then transformed to give a mean value of 0.075—the mean soil-to-plant concentration ratio recommended by the IAEA (2014) ($n = 5$, $\pm 1 \times$ SE).

## 2.2. Taxonomic Influence on Relative Mean I Concentration

Although Q-Q normality plots showed that the REML-modeled mean iodine concentrations in Table 1 approached a normal distribution (Figure 2), they failed the Shapiro-Wilkes test primarily because of a few low values, mostly from study 27 (*Heracleum sphondylium*, *Taraxacum officinale*, and *Cichorium intybus*). Given that ANOVA is relatively robust to the assumption of normality and that there were other values from study 27 in the database, the whole dataset of 103 REML-modeled mean species values was used to analyze for taxonomic effects. Nested ANOVA coded with the APG III angiosperm phylogeny showed that REML-estimated relative mean I concentrations were significantly different between genera within families (Table 2) and between "Superorder". Using unbalanced nested factors in ANOVA can produce artifacts so we used the same unbalanced taxonomic hierarchy to analyze 103 random numbers with a range of −6.83 to 3.71 (Table 3). This confirmed that, in the database we compiled, the significant nested differences we found were real, indicating that there are significant differences in iodine concentrations in plants associated with taxonomic categories above the species level. A T-test on REML-estimated data between the Monocots and Eudicots ("Classes") indicated a highly significant difference between them ($t = 2.88$, df $= 100$, $p = 0.005$). Transformation of these values into CRs with a geometric mean of recommended IAEA values predicts significant differences between Monocots and Eudicots in CRs (Figure 3a). One-way ANOVA of REML-modeled values showed that there were significant differences between the Lilianae, Rosanae, Caryophyllanae and Asteranae ($F = 2.95$, $p = 0.037$) and produced predictions of significantly different mean CRs for these "Subclasses" (Figure 3b). At the ordinal level, when Orders with 3 or fewer species were excluded, there were significant differences between Orders ($F = 2.05$, $p = 0.045$). The Poales (incl. cereals and relatives), Asparagales (incl. onions and relatives), Caryophyllales (incl. beets and their relatives) and Lamiales (incl. mints and their relatives) had I concentrations significantly higher than the Malphigiales (flax and its relatives) and the Apiales (incl. carrot and its relatives) ($F = 2.05$, $p = 0.045$), again allowing us to predict significant differences in CR for these orders (Figure 3c). For *Beta vulgaris* one-way ANOVA showed that there were significant differences between varieties ($p = 0.002$, $F = 6.544$), Holm–Sidak tests indicating that Italian Chard had significantly higher concentrations than Mangel–Wurzel, Cheltenham Green and Perpetual Spinach (Figure 4). There were no inter-varietal differences in the concentrations of $^{125}$I in the *C. arietinum* varieties (Figure 4). Overall, it seems clear that the differences between taxa in I concentrations are not all associated with species and that not only is there a significant amount of variance associated with categories above the species level that can be used to make general predictions of CR but also that there can be differences in I uptake between varieties.

**Table 2.** Results of nested ANOVA of Residual Maximum Likelihood (REML)-modeled concentrations of iodine. Concentrations of 103 species of plants were analyzed using nested taxonomic units as factors—"Subclass" was nested within "Class", then "Superorder" within "Subclass" within "Class" and so on. A general linear model that excluded the intercept was used.

| Factor | df | SS | %SS | Cumulative SS | MS | $F$ | $p$-value |
|--------|----|----|----|----|----|----|----|
| "Class" | 2 | 38.28 | 4.80 | 4.02 | 19.14 | 6.26 | 0.053 |
| "Subclass" | 3 | 7.37 | 0.92 | 5.72 | 2.45 | 2.53 | 0.136 |
| "Superorder" | 3 | 1.96 | 0.24 | 5.97 | 0.65 | 0.15 | 0.928 |
| Order | 12 | 48.21 | 6.05 | 12.02 | 4.01 | 1.03 | 0.494 |
| Family | 12 | 50.37 | 6.32 | 18.34 | 4.19 | 0.82 | 0.623 |
| Genus | 45 | 239.06 | 30.0 | 48.34 | 5.31 | 2.78 | 0.004 |
| Species | 24 | 45.79 | 5.74 | 54.09 | 1.90 | 0.35 | 0.895 |
| Residual | 98 | 365.78 | 45.90 | 100 | | | |

**Table 3.** Results of nested ANOVA of random numbers. Random numbers between −6.83 and 3.70 for 103 species of plants were analyzed using the same taxonomy as for Table 2.

| Factor | df | SS | %SS | Cumulative SS | MS | $F$ | $p$-value |
|--------|----|----|----|----|----|----|----|
| "Class" | 2 | 67.16 | 3.32 | 3.32 | 33.58 | 2.85 | 0.211 |
| "Subclass" | 3 | 35.57 | 1.76 | 4.08 | 11.86 | 1.39 | 0.358 |
| "Superorder" | 3 | 24.71 | 1.22 | 6.30 | 8.23 | 0.64 | 0.603 |
| Order | 12 | 157.34 | 7.78 | 14.08 | 13.11 | 1.19 | 0.394 |
| Family | 12 | 129.00 | 6.38 | 20.45 | 10.75 | 1.28 | 0.254 |
| Genus | 45 | 355.43 | 17.57 | 38.02 | 7.89 | 0.78 | 0.774 |
| Species | 24 | 244.65 | 12.09 | 50.12 | 10.19 | 14.62 | 0.204 |
| Residual | 98 | 1009.18 | 49.88 | 100 | | | |

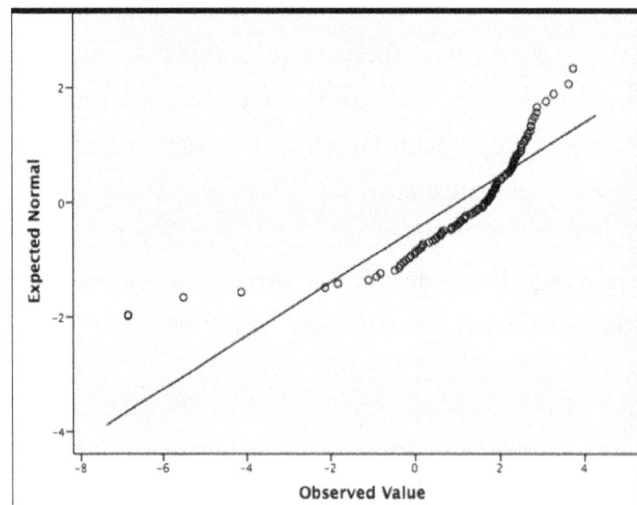

**Figure 2.** Normal Q-Q (Quantile-Quantile) plot of REML-modeled relative iodine concentrations in 103 species of plants. The data approach a normal distribution but fail the Shapiro–Wilkes test of normality primarily because of a few low values.

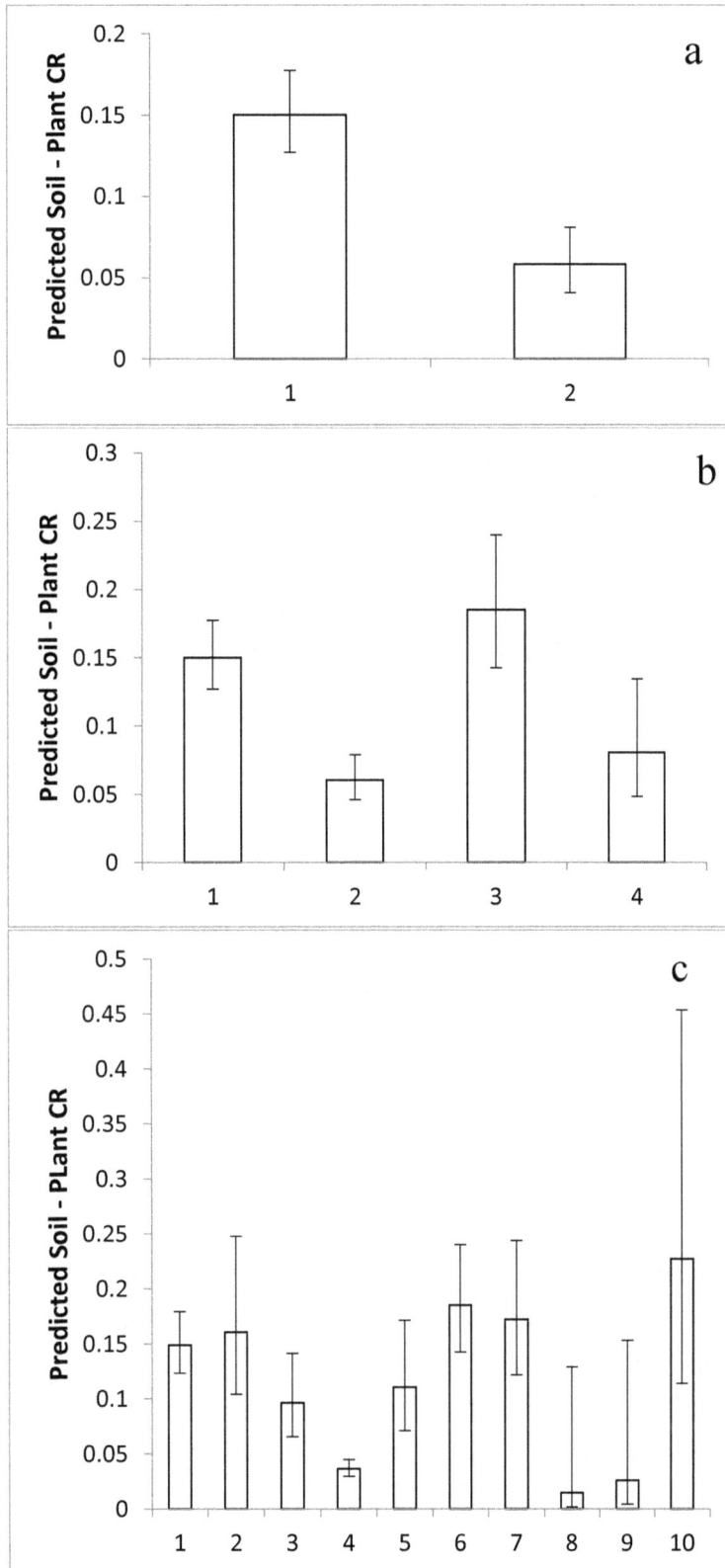

**Figure 3.** The mean Concentration Ratios predicted in angiosperm taxa according to the phylogeny of APG III (2009). (**a**) "Class", 1 = Monocots (*n* = 28), 2 = Eudicots (73); (**b**) "Subclass", 1 = Lilianae (*n* = 28), 2 = Rosanae (*n* = 30), 3 = Caryophyllanae (*n* = 11), 4 = Asteranae (*n* = 31); (**c**) Order. 1 = Poales (21), 2 = Asparagales (4), 3 = Fabales (*n* = 10), 4 = Malphigiales (*n* = 5), 5 = Brassicales (*n* = 7), 6 = Caryophyllales (*n* = 11), 7 = Lamiales (*n* = 12), 8 = Solanales (*n* = 4), 9 = Apiales (*n* = 5), 10 = Asterales (*n* = 6).

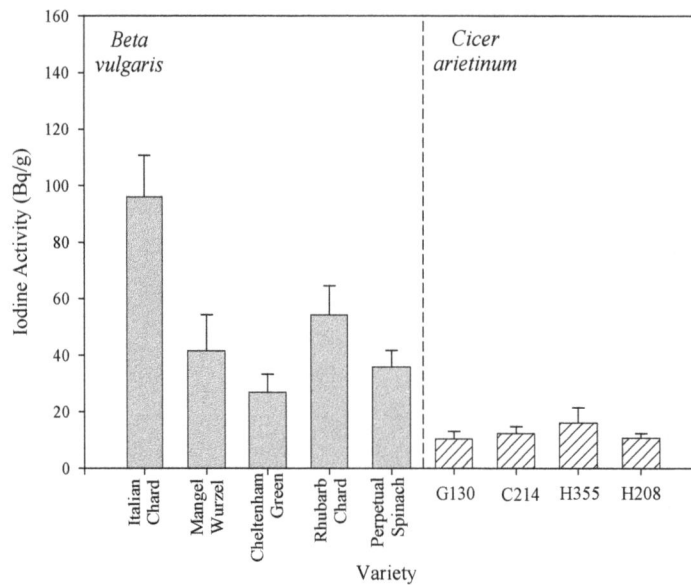

**Figure 4.** Mean I activities of shoots in varieties of seven-week-old *Beta vulgaris* and *Cicer arietenum* exposed to an acute dose of $^{125}$I. ($n = 5$, SEs).

## 2.3. Comparison of Taxonomic Effects

In total about 12% of the variance in I concentrations was accounted for by ranks of Order (Table 2). This is greater than the influence on N (3.3%) and P (6.8%), [16], but less than that for Cs (23.3%) [17], Pb (20%), Cr (24%), Cu (24%), [11], Na (23%) [16], Cd (27%) [11], Zn (44%), Ni (46%) [11], K (49%) [16], and Ca (63%) [15]. Of the elements for which phylogenetic influences have been detected, I is most closely related to Cl. In the only published analysis of relative mean concentration for Cl [18] there are 24 species that are also in Table 1. There was no correlation between relative mean I and Cl values in these species.

## 3. Discussion

Table 1 and Figure 1 indicate that there is a wide range of I concentrations between plant species. They confirm the existence of inter-species differences in I concentrations of sufficient magnitude to support previous suggestions that crop species might be as important as soil type in determining I concentrations in crop plants [21]. For elements in which there are very small phylogenetic influences above the species level there is very little variance attributable to taxonomic levels above the species. Table 2 shows that there is a phylogenetic influence on inter-species differences in iodine concentration, especially for Genera within Families. Figure 4 indicates that inter-varietal differences in I concentrations do occur, strengthening the assertion that there is nothing especially significant about the species as a taxonomic unit to describe inter-taxa differences in I concentrations in plants. These observations suggest, for the first time, (a) that angiosperm phylogeny influences the I concentrations of plants; (b) that the species is not an independent sampling unit for I concentrations in plants; and (c) that it might be possible to make general predictions of relative I concentrations in food crops based on phylogeny. If correct, such insights are potentially useful for understanding the agricultural chemistry and food toxicology of I.

The validity of the general insights above, however, depends on a number of assumptions about the relative mean concentrations reported in Table 1. First, a high proportion of the data in Table 1 are from experiments with [125]I. There is no evidence of discrimination between I isotopes during plant uptake and [125]I has previously been used as a proxy for other I isotopes in uptake experiments [22], so we assume that the data for [125]I are very likely to represent the behavior of I isotopes of more agricultural and toxicological significance. Second, it is likely that the acute exposures to I used to generate much of the data in Table 1 will not produce exactly the same relative mean concentrations in plants as chronic exposures. However, much nutrient uptake takes place during the exponential phase of growth when our plants were exposed so we assume that our observations will approximate inter-species differences that might be found following chronic exposures. Nevertheless the data in Table 1 may be more directly relevant to acute exposures to [129]I (which can be radioecologically significant, for example, during pulsed movement up through soil profiles [6]) than to long-term uptake of [127]I. And third, it is important to acknowledge that the relative mean concentrations between plant species reported in Table 1 might not be the same under all environmental conditions, *i.e.*, there might be an interaction between environment and inter-species differences. Despite these assumptions, it is notable that similar observations to those we make above for I have been reported for numerous other elements using a variety of isotopes, exposure times and environmental regimes [11,14–18]. Thus, given that it is the most taxonomically wide-ranging database yet reported for inter-species differences in plant I concentrations and that it is compatible with results for other elements, Table 1 provides a basis for initiating assessments of the influence of phylogeny on I concentrations in plants. There have been detailed studies of the translocation of I in plants and its partitioning between plant parts [22], which clearly affect I concentrations in food stuffs. As the data in Table 1 focuses on green shoots *in toto*, phylogenetic influences identified might provide background concentrations upon which internal partitioning is imposed.

If there is no effect of phylogeny on inter-species differences in uptake then, as is approximately the case for N and P [16], there will be no variance associated with taxonomic levels above the species. This is not the case for I and we conclude that Tables 1 and 2 show that there is an influence of phylogeny on differences in I concentrations between plant species. This gives further support to the assertion that particular taxa of plants have characteristic mineralogies and that a phylogenetic perspective on plant contribution to the transfer of elements in the soil-crop system might be useful [16]. A phylogenetic influence also means that plant species are not independent sampling units for I, *i.e.*, great care must be taken in statistical analysis of I transfer in the soil-crop system as many techniques, such as regression, make the assumption that samples from different species are independent. In contrast to the frequency distribution of relative concentrations of some elements [15], I concentrations in plants are not normally distributed. This indicates that the parametric statistics often used in soil-crop transfer analysis must be used with care in analyses with numerous species. Normal distributions of phenotypes are often characteristic of polygenic, "quantitative", traits. Quantitative techniques, such as the Quantitative Trait Loci analysis used to locate genes impacting on the concentration of other elements in plants [29,30] might have to be used with care for analysis of the genetic factors affecting the I chemistry of crops.

Detailed analyses of the effects of phylogeny necessitate concentration values for more than 103 species but some patterns do emerge from the analysis carried out here. It seems clear that, as is the case with some other elements [16,17], there is a significant difference in I concentration between Monocot and Eudicot plants. Of Orders with significant numbers of food crops the analysis reported here indicates that plants in the Poales (cereals and relatives), Asparagales (onions and relatives), and Caryophyllales (beets, amaranths, buckwheat and relatives) might have higher than average I concentrations. These Orders might worth further investigations if explanations for dietary loadings of I are being sought, particularly as some of these Orders are represented by few species in Table 1. Further investigations might, for example, test the suggestion that at a given soil availability of I, cereal grains such as amaranths and buckwheat might provide higher I concentrations than grains such as wheat or rice. These effects might be used to expand the reported general pattern of I concentrations in foodstuffs of legumes > vegetables > fruit [31] because they suggest that there are groups of plants with significantly higher I concentrations than legumes.

Figure 4 suggests that although there might be some inter-varietal differences in I concentrations in some crops, they might be small compared to inter-specific differences. This supports conclusions of previous studies with numerous varieties of clover, grasses and other herbage crops [23,25,32]. Nevertheless, further analyses might very usefully compare the amount of variation above and below the species level in order to determine the extent to which I concentrations in plant biomass can be altered by choosing different varieties or different species. The phylogenetic effects described above have some similarity to those we have reported for Cl [18], especially the higher than average values in the Caryophyllales. However, we found no direct correlation in relative mean values for 24 species that occur in both data sets. It might be interesting to investigate, using a dataset with more species, if this lack of correlation reflects real differences in the behavior of I and Cl.

In the database compiled here, $log_e$-transformed values subject to REML-modeling are approximately normally distributed. Using the IAEA recommended value and back transforming modeled values to CRs, confirms the $log_e$-normal distribution of I concentrations in plants and enables us to predict geometric mean CRs for different plant groups and 95% confidence intervals (Figure 2). These suggest that significantly improved predictions of CR for radioiodine can be made by taking taxonomic group into account, with splitting the recommended CR into two, one for Monocots and one for Eudicots, bringing about a significant improvement in predicted CR very simply (Figure 2a). Such overall predictions for groups of many species are very useful in the case of a contamination event in which many Monocots and Eudicots might be contaminated simultaneously. In different ecosystems that have different proportions of Monocots and Eudicots, the predicted CRs in Figure 2 could significantly improve predictions of overall radioiodine transfer from soils to plants.

Overall, in both agriculture and radioecology inter-taxa differences in I uptake by plants are important—in addition to iodized salt and irrigation water, it has been suggested that crop selection and/or breeding might help to provide increases in I concentration in food [31] and predictions of radioiodine movement into foodchains use CRs for soil-to-plant transfer. The data reported here improve the understanding of inter-taxa differences in I concentrations, and by initiating investigations of the phylogenetic distribution of the diversity in I uptake, might help to identify those groups of plants with particular I concentrations thus benefitting both agricultural supply of iodine and predictions of radioiodine transfer to food, flora and fauna.

## 4. Experimental Section

### 4.1. Plant Growth

For each of the 47 species, 5 replicates, in 12 cm pots, each filled with 250 g of Levington's F2S compost (Fisons, Ipswich, UK) (Table 1), were grown for approximately 7 weeks, *i.e.*, to the exponential phase of growth and before anthesis, in a greenhouse with 16 h day and 8 h night at *ca.* 24 °C and 16 °C, respectively. The 47 species were radiolabeled in 7 experimental sets ("studies") (Table 1). Forty-seven species were selected to provide a spread across the angiosperm phylogeny. In addition, 5 replicates of each of 5 additional varieties of *Beta vulgaris* and 4 additional varieties of *Cicer arietinum* were also radiolabeled—which gave 54 taxa in total. Five replicates of *Carex nigra*, *Canna indica*, *Geranium pratense*, *Euphorbia myrsinites* and *Linum lewisii* were labeled in two experimental blocks to provide linking species between blocks. Seeds were supplied by Chiltern Seeds (Cumbria, UK), Kings Seeds (Essex, UK) and the Institute for Crop Research in the Semi-Arid Tropics (Patencheru, Telangana, India).

### 4.2. Radiolabeling with $^{125}I$

Following trial experiments to establish appropriate labeling volumes, carriers, activities and exposures, 50 mL of 50 μM KI radiolabeled with 74 kBq of $^{125}I$ were added to each pot. During radiolabeling, replicate pots of each species were arranged in a randomized block design in an arena in the laboratory with artificial lighting at *ca.* 350 E$\mu^{-2}$s$^{-1}$. Pots were not watered for 24 h prior to radiolabeling and were placed in the arena with the lights on at least 1 h before the addition of radioactivity. Entire green shoots were harvested 1 cm above soil level 4 h after the radiolabel was added, dried for at least 48 h at 80 °C and then ground up. Ground plant samples were analyzed for $^{125}I$ γ emissions, with appropriate calibrations and blanks, on an LKB Wallac Compugamma "1282" (NaI (Tl) detector).

### 4.3. Residual Maximum Likelihood Analysis (REML)

As used in previous studies [11,14–18] REML modeling was used to estimate relative mean I concentrations across the seven experimental blocks reported here plus the literature data of similar inter-species comparisons of I uptake. There were 30 "studies" identified from 10 literature sources [10,20–28] that reported I concentration values after similar experiments, *i.e.*, I concentrations in green shoots after simultaneous root exposures in two or more plant species. There were concentration values for 56 species in these literature sources of which 6 also occurred in the experiments carried out for this analysis. Overall, therefore, the data for analysis included 103 species (47 + 65 − 9) from 37 "studies" (7 + 30) (Table 1). In the REML modeling of I concentrations there were 440 concentration values in 37 "blocks" (studies) from 103 "treatments" (species). The REML modeling accounts for the effect of block (study) on concentration to estimate relative mean concentrations in the treatments (species) across the whole dataset. Differences in absolute concentrations arising from different substrates and exposure conditions are, therefore, accounted for statistically. Raw data were log$_e$-transformed before REML analysis. REML analysis can produce negative as well as positive values [33] and was run on the statistical package SPSS.

*4.4. Analysis of Taxonomic Effects*

Following REML analysis, the mean I concentrations in species were analyzed using an unbalanced hierarchical Analysis of Variance (ANOVA) coded with the angiosperm phylogeny group III [34]. The nominal designations of "Class", "Subclass" and "Superorder" were used for categories above the Order, although their application to recent phylogenies is unresolved. Tests for normality of I concentrations using the Shapiro–Wilkes test, and all ANOVAs were carried out on SPSS v 22.0 for Mac (SPSS, Armonk, New York, USA).

## Acknowledgments

We would like to thank Judy Brown, Roy Bennet and Janine Wilkins of the University of the West of England (UK) for radioanalytical support, the Natural Environment Research Council, the Environment Agency and the Nuclear Decommissioning Authority (all UK) who provided financial support through the RATE programme and ICRISAT, Telengana, India for chickpea varieties.

## Author Contributions

Eleni Siasou compiled the database. Eleni Siasou and Neil Willey analyzed the data and wrote the manuscript.

## Conflicts of Interest

The authors declare no conflicts of interest.

## References

1.  Zhu, Y.-G.; Huang, Y.-Z.; Hu, Y.; Liu, Y.-X. Iodine uptake by spinach (*Spinacia oleracea* L.) plants grown in solution culture: Effects of iodine species and solution concentrations. *Environ. Int.* **2003**, *29*, 33–37.
2.  Sheppard, S.C. Interpolation of solid/liquid partition coefficients, K$_d$, for iodine in soils. *J. Environ. Radioact.* **2003**, *70*, 21–27.
3.  Welch, R.M.; Graham, R.D. A new paradigm for world agriculture: Meeting human needs—Productive, sustainable, nutritious. *Field Crops Res.* **1999**, *60*, 1–10.
4.  Zimmermann, M.B.; Andersson, M. Prevalence of iodine deficiency in Europe in 2010. *Ann. Endocrinol.* **2011**, *72*, 164–166.
5.  Fréchou, C.; Calmet, D. $^{129}$I in the environment of the La Hague nuclear fuel reprocessing plant—From sea to land. *J. Environ. Radioact.* **2003**, *70*, 43–59.
6.  Ashworth, D.J.; Shaw, G.; Butler, A.P.; Cociani, L. Soil transport and plant uptake of radio-iodine from near-surface groundwater. *J. Environ. Radiact.* **2003**, *70*, 99–114.

7.  Sheppard, M.; Thibault, D.H. Default soil solid/liquid partition coefficients, $K_d$s, for four major soil types: A compendium. *Health Phys.* **1990**, *59*, 471–482.

8.  Sheppard, S.C.; Motycka, M. Is the Akgare phenomenon important to iodine uptake by wild rice (*Zizania aquatica*)? *J. Environ. Radioact.* **1997**, *37*, 339–353.

9.  White, P.J.; Broadley, M.R. Chloride in soils and its uptake and movement within the plant. *Ann. Bot.* **2001**, *88*, 967–988.

10. Ban-Nai, T.; Muramatsu, Y. Transfer factors of radioiodine from volcanic-ash soil (Andosol) to crops. *J. Radiat. Res.* **2003**, *44*, 23–30.

11. Broadley, M.R.; Willey, N.J.; Mead, A. A method to assess taxonomic variation in shoot caesium concentration among flowering plants. *Environ. Pollut.* **1999**, *106*, 341–349.

12. APG (Angiosperm Phylogeny Group) II. An update of the Angiosperm Phylogeny Group classification for the orders and families of flowering plants: APG II. *Bot. J. Linnean Soc.* **2003**, *141*, 399–436.

13. Soltis, P.S.; Soltis, D.E.; Chase, M.W. Angiosperm phylogeny inferred from multiple genes as a research tool for comparative biology. *Nature* **1999**, *402*, 402–404.

14. Broadley, M.R.; Willey, N.J.; Wilkins, J.; Baker, A.J.M.; Mead, A.; White, P.J. Phylogenetic variation in heavy metal accumulation in angiosperms. *New Phytol.* **2001**, *152*, 9–27.

15. Broadley, M.R.; Bowen, H.C.; Cotterill, H.L.; Hammond, J.P.; Meacham, M.C.; Mead, A.; White, P.J. Variation in the shoot calcium content of angiosperms. *J. Exp. Bot.* **2003**, *54*, 1–16.

16. Broadley, M.R.; Bowen, H.C.; Cotterill, H.L.; Hammond, J.P.; Meacham, M.C.; Mead, A.; White, P.J. Phylogenetic variation in the shoot mineral concentration of angiosperms. *J. Exp. Bot.* **2004**, *55*, 321–336.

17. Willey, N.J.; Tang, S.; Watt, N. Predicting inter-taxa differences in plant uptake of [134/137]Cs. *J. Environ. Qual.* **2005**, *34*, 1478–1489.

18. Willey, N.J.; Fawcett, K. Species selection for phytoremediation of [36]Cl/[35]Cl using angiosperm phylogeny and inter-taxa differences in uptake. *Int. J. Phytoremediation* **2005**, *7*, 295–306.

19. International Atomic Energy Agency. *Handbook of Parameter Values for the Prediction of Radionuclide Transfer in Terrestrial and Freshwater Environments*; Technical Report Series; IAEA: Vienna, Austria, 2010; p. 472.

20. Butler, G.W.; Johnson, J.M. Factors influencing the iodine content of pasture herbage. *Nature* **1957**, *179*, 216–217.

21. Moiseyev, I.T.; Tikhomirov, F.A.; Perevezentsev, V.M.; Rerikh, L.A. Role of soil properties, inter-specific plant differences, and other factors affecting the accumulation of radioactive iodine in crops. *Sov. Soil Sci.* **1984**, *16*, 60–66.

22. Cline, J.F.; Klepper, B. Iodine-125 accumulation in plant parts: Influence of water use rate and stable iodine content of soil. *Health Phys.* **1975**, *28*, 801–804.

23. Hartmans, J. Factors affecting the herbage iodine content. *Neth. J. Agric. Sci.* **1974**, *22*, 195–206.

24. Sheppard, S.C.; Evenden, W.G.; Amiro, B.D. Investigation of the soil-to-plant pathway for I, Br, Cl and F. *J. Environ. Radioact.* **1993**, *21*, 9–32.

25. Whitehead, D.C. Uptake and distribution of iodine in grass and clover plants grown in solution culture. *J. Sci. Food Agric.* **1973**, *24*, 43–50.

26. Kashparov, V.; Colle, C.; Zvarich, S.; Yoshenko, V.; Levchuk, S.; Lundin, S. Soil-to-plant halogen transfer studies. *J. Environ. Radioact.* **2005**, *79*, 187–204.

27. Bourchet, M.P. Sur l'absorption de l'iode par les végétaux. *Acad. Sci. Paris* **1899**, *129*, 768–770.

28. Dai, J.L.; Zhu, Y.G.; Zhang, M.; Huang, Y.Z. Selecting Iodine-Enriched Vegetables and the residual effect of iodate application to soil. *Biol. Trace Elem. Res.* **2004**, *101*, 265–276.

29. White, P.J.; Bowen, H.C.; Willey, N.J.; Broadley, M.R. Selecting plants to minimise radiocaesium contamination of food chains. *Plant Soil* **2003**, *249*, 177–186.

30. Payne, K.C.; Bowen, H.C.; Hammond, J.P.; Hampton, C.R.; Lynn, J.R.; Mead, A.; Swarup, K.; Bennett, M.J.; White, P.J.; Broadley, M.R. Natural egentic variation in caesium (Cs) accumulation by *Arabidopsis thaliana*. *New Phytol.* **2004**, *162*, 535–548.

31. Johnson, C.C.; Fordyce, F.M.; Stewart, A.G. *Environmental Controls in Iodine Deficiency Disorders*; Project Summary Report, CR/03/058N; British Geological Survey: Keyworth, UK, 2003.

32. Crush, J.R.; Caradus, J.R. Cyanogenesis potential and iodine concentration in white clover (*Trifolium repens* L.) cultivars. *N. Z. J. Agric. Res.* **1995**, *38*, 309–316.

33. Thompson, R.; Welham, S.J. REML analysis of mixed models. In *The Guide to Genstat, Part 2—Statistics*; Payne, R.W., Ed.; VSN International: Oxford, UK, 2001; pp. 413–503.

34. Chase, M.W.; Reveal, J.L. A phylogenetic classification of the land plants to accompany APG III. *Bot. J. Linnean Soc.* **2009**, *161*, 122–127.

# Polyethylene Glycol (PEG)-Treated Hydroponic Culture Reduces Length and Diameter of Root Hairs of Wheat Varieties

**Arif Hasan Khan Robin \*, Md. Jasim Uddin and Khandaker Nafiz Bayazid**

Department of Genetics and Plant Breeding, Bangladesh Agricultural University, Mymensingh 02202, Bangladesh; E-Mails: jasimpstu998@gmail.com (M.J.U.); nafizbayazid@gmail.com (K.N.B.)

\* Author to whom correspondence should be addressed; E-Mail: gpb21bau@bau.edu.bd

Academic Editor: Leslie A. Weston

**Abstract:** Wheat is an important cereal crop worldwide that often suffers from moisture deficits at the reproductive stage. Polyethylene glycol (PEG)-treated hydroponic conditions create negative osmotic potential which is compared with moisture deficit stress. An experiment was conducted in a growth chamber to study the effects of PEG on root hair morphology and associated traits of wheat varieties. Plants of 13 wheat varieties were grown hydroponically and three different doses of PEG 6000 (*w/v*): 0% (control), 0.3% and 0.6% (less than −1 bar) were imposed on 60 days after sowing for 20 days' duration. A low PEG concentration was imposed to observe how initial low moisture stress might affect root hair development. PEG-treated hydroponic culture significantly decreased root hair diameter and length. Estimated surface area reduction of root hairs at the main axes of wheat plants was around nine times at the 0.6% PEG level compared to the control plants. Decrease in root hair diameter and length under PEG-induced culture decreased "potential" root surface area per unit length of main root axis. A negative association between panicle traits, length and dry weight and the main axis length of young roots indicated competition for carbon during their development. Data provides insight into how a low PEG level might alter root hair development.

**Keywords:** wheat; root hairs; root morphology; phytomer; PEG

# 1. Introduction

Wheat is grown on over 216 million hectares of land, yielding 634 million tons worldwide by 30 million farmers [1]. In Bangladesh, wheat is the second most important cereal crop after rice. It grows in 11% of the cropped area in *rabi* season and yields around 1.0 million tons [1,2]. To fulfill domestic demand of cereals, Bangladesh requires importing 3.0 million tons of wheat every year to meet the yield gap [3]. Unfavorable environmental conditions including heat stress, drought stress, and salinity are the main reasons for over 52% of the difference between current and recorded wheat production [1] and a 55% difference between average yields and yield potential [3]. To ensure future food security, the wheat production area needs expansion. Efforts are, therefore, required to overcome various limiting factors that shrink wheat production areas including abiotic stresses. In Bangladesh, drought stress which alone decreased cereal grain production by 3.5 million metric tons in the year 1994–1995 [4] is considered a more serious threat to crop production than floods and cyclones as its frequency has increased over the years [5]. The best option to meet yield gap through yield improvement and yield stability under drought stress conditions is to develop drought tolerant crop varieties. One important avenue for follow up is development of wheat varieties with an efficient root system which can exploit residual soil moisture in the dry season to overcome low to moderate water deficits in soil.

A better understanding of various plant developmental and physiological processes that ultimately influence yield determining factors is essential to breeding for specific, suboptimal environments [6,7]. Under a low to moderate level of soil water deficit, osmotic adjustment enables plants to maintain uptake activity in many species, and is ultimately the reason for maintenance of growth and optimum yield under drought [8,9]. Extensive research efforts over the last five decades revealed physiological mechanisms relevant to osmotic stress response of crop plants [8–13]. Awareness related to root morphology of annual crop plants, such as wheat, barley, or perennial forage grasses under limited water supply, has developed within the last 20 years [14–23]. However, little awareness has been created as to how root hairs get modified, and are possibly associated with crop yield under different levels of osmotic potential. In dry environments, large root systems of wheat capture more water and nitrogen to facilitate access to additional water for grain filling [23]. Osmotic stress results in significant genotypic variation for a number of morphological traits including root length, dry weight and root-shoot ratio [24]. In another study, osmotic stress is reported to suppress discrimination among genotypes for root length along with seed vigour index, germination percentage, shoot length, coleoptile length and osmotic membrane stability [25]. A thorough understanding of root systems and root hair morphology and their responses towards various degrees of drought is required for future wheat breeding programs. This study therefore planned to explore how Polyethylene glycol (PEG) treated hydroponic culture affects root hair development and associated traits. This study involved establishing experimental plants of wheat varieties hydroponically, imposing PEG-treatment and measuring root hairs, associated root morphology and panicle traits. PEG is a widely used chemical compound and maintains lower osmotic potential at a comparatively lower temperature under hydroponic culture [26,27]. Plants were grown at an osmotic potential of less than 1 bar for a longer duration of 20 days.

# 2. Results

## 2.1. Treatment Effect

Number of adventitious roots per tiller recorded the highest under 0.6% level, which was significantly different from control ($p = 0.035$, Table 1). Root hair length and diameter of the root hairs gradually decreased with increasing levels of PEG ($p < 0.001$, Table 1). Length of root hairs greatly reduced at 0.6% PEG level compared to control and 0.3% PEG level ($p < 0.001$, Table 1). Even though root hair density was slightly increased under osmotic stress, estimated overall surface area provided solely by the root hairs per $mm^2$ main axis was about nine times less at 0.6% PEG level compared to control, due to a severe decrease in diameter and length of root hairs ($p = 0.003$, Table 1). Second principal component (PC2) explained 17.5% data variation, which indicated a contrast between the three treatments for root hair density with other root hair traits like: diameter, length and surface area (Table 2, $p < 0.001$). PC2 clearly revealed that plants under 0.6% PEG level yielded much reduced surface area with simultaneous decrease in root hair diameter and root hair length (Table 2). PC scores indicated that 0.6% PEG level had average negative PC scores where the other two treatments had average positive PC scores (Table 2).

**Table 1.** Variation in morphological characters of 13 wheat varieties 20 days after treatment due to three different PEG treatments. Data represents average of 39 replicates under each treatment. SEM, standard error of mean; p, probability of statistical significance.

| Treatments (% PEG) | Number of Adventitious Roots per Tiller | Root Hair Density (no. $mm^{-2}$) | Root Hair Diameter ($\mu$) | Root Hair Length ($\mu$) | Root Hair Surface Area per $mm^2$ Main Axis ($mm^2$) |
|---|---|---|---|---|---|
| 0 | 10.2 | 70.8 | 7.75 | 367 | 0.725 |
| 0.3 | 9.3 | 96.2 | 5.90 | 341 | 0.592 |
| 0.6 | 11.1 | 106 | 4.04 | 59 | 0.085 |
| SEM | 0.37 | 7.9 | 0.24 | 30.9 | 0.071 |
| p value | 0.035 | 0.12 | <0.001 | <0.001 | <0.001 |

**Table 2.** Principal component analysis (PCA) of selected morphological traits of wheat varieties under PEG-treated hydroponic culture. PC, principal component; $p$, statistical significance; Pr, root bearing phytomer.

| Traits | PC1 | PC2 | PC3 |
|---|---|---|---|
| Live leaves (no.) | 0.396 | −0.117 | −0.096 |
| Total number of roots per tiller | 0.394 | −0.175 | −0.021 |
| Main axis length at Pr1 (cm) | 0.109 | 0.192 | −0.358 |
| Main axis length at Pr2 (cm) | 0.164 | 0.131 | −0.502 |
| Diameter of root hair ($\mu$) | 0.115 | 0.531 | 0.085 |
| Length of root hair ($\mu$) | 0.161 | 0.543 | 0.109 |
| Root hair density (no. $mm^{-2}$) | 0.291 | −0.254 | 0.008 |
| Root hair surface area $mm^{-2}$ axis | 0.231 | 0.473 | 0.137 |
| Panicle length (cm) | 0.297 | −0.090 | 0.470 |
| Panicle dry weight (mg) | 0.232 | −0.050 | 0.485 |

**Table 2.** *Cont.*

| Traits | PC1 | PC2 | PC3 |
|---|---|---|---|
| Root dry weight per tiller (mg) | 0.393 | −0.159 | −0.109 |
| Shoot dry weight per tiller (mg) | 0.202 | 0.005 | −0.304 |
| % variation explained | 31 | 17.5 | 13.1 |
| $p$ (treatment) | 0.27 | <0.001 | 0.292 |
| $p$ (variety) | <0.001 | 0.222 | <0.001 |
| $p$ (treatment × variety) | 0.169 | 0.982 | <0.001 |
| Mean PC scores for treatment | | | |
| 0.0% PEG | 0.23 ± 0.22 | 0.77 ± 0.25 | 0.13 ± 0.14 |
| 0.3% PEG | 0.02 ± 0.22 | 0.28 ± 0.25 | 0.09 ± 0.15 |
| 0.6% PEG | −0.26 ± 0.22 | −1.02 ± 0.25 | −0.17 ± 0.14 |
| Mean PC scores for variety | | | |
| Shotabdi | 0.91 ± 0.51 | −0.09 ± 0.58 | 0.60 ± 0.34 |
| Shourov | −0.61 ± 0.48 | −0.49 ± 0.55 | 1.55 ± 0.32 |
| Sufi | −1.92 ± 0.48 | 0.36 ± 0.55 | −0.55 ± 0.32 |
| Prodeep | 3.81 ± 0.44 | −1.51 ± 0.51 | 0.59 ± 0.29 |
| Bijoy | −0.87 ± 0.44 | 0.88 ± 0.51 | −0.07 ± 0.29 |
| BARI gom25 | 1.87 ± 0.44 | −0.22 ± 0.51 | −0.30 ± 0.29 |
| BARI gom26 | −0.23 ± 0.44 | 0.11 ± 0.51 | −0.29 ± 0.29 |
| BARI gom27 | −1.01 ± 0.44 | −0.38 ± 0.51 | 0.45 ± 0.29 |
| BARI gom28 | −0.99 ± 0.44 | 0.21 ± 0.51 | −0.07 ± 0.29 |
| Kheri | 0.43 ± 0.44 | 0.49 ± 0.51 | −0.78 ± 0.29 |
| Sonalika | −0.90 ± 0.44 | 0.14 ± 0.51 | −0.24 ± 0.29 |
| Kanchan | −0.37 ± 0.44 | 0.27 ± 0.51 | −0.42 ± 0.29 |
| Akbar | −0.15 ± 0.48 | 0.36 ± 0.55 | −0.26 ± 0.32 |

## 2.2. Varietal Difference

A remarkable number of morphological traits were significant among the 13 wheat varieties including: number of live leaves, number of root bearing phytomers, number of adventitious roots, average number of roots per phytomer, number of seminal roots per tiller, main axes length at Pr1 and Pr2; main axes diameter, root hair diameter and root dry weight per tiller (Table 3). First principal component (PC1) explaining 30% data variation indicated a significant contrast among varieties for number of live leaves, number of adventitious roots, root hair diameter, surface area of root hairs, panicle length and root and shoot dry weights per tiller (Table 3). PC1can be regarded as "size PC". The variety *Prodeep* which yielded the largest biomass along with the highest number of leaves and adventitious roots obtained the largest PC score compared to *Sufi* and *BARI gom27* which had the lowest number of live leaves and root dry weight, respectively (Table 3).

**Table 3.** Variation in different morphological characters of 13 wheat varieties on 20 days after PEG- treatment (DAT). Data represents average of nine replicates on 12 DAT from three treatments. Pr, root bearing phytomer; SEM, standard error of mean; *p*, statistical significance.

| Variety | Live Leaves (No.) | No. of Root Bearing Phytomer | No. of Adventitious Roots per Tiller | Average no. of Roots per Phytomer | No. of Seminal Roots | Main Axis Length at Pr1 (cm) | Main Axis Length at Pr2 (cm) | Main Axis Diameter (mm) | Root Hair Density (No. mm$^{-2}$) | Root Dry Weight (mg) |
|---|---|---|---|---|---|---|---|---|---|---|
| Shotabdi | 5.3 | 5.77 | 11.7 | 2.01 | 3.78 | 10.2 | 16.9 | 0.57 | 34.7 | 65 |
| Shourov | 4.0 | 3.89 | 7.3 | 1.89 | 3.00 | 12.8 | 12.4 | 0.55 | 21.9 | 43 |
| Sufi | 3.0 | 5.22 | 9.2 | 1.77 | 2.56 | 12.1 | 17.6 | 0.53 | 14.5 | 38 |
| Prodeep | 7.8 | 7.00 | 17.8 | 2.52 | 4.56 | 11.5 | 20.0 | 0.71 | 76.7 | 134 |
| Bijoy | 3.4 | 4.89 | 9.4 | 1.95 | 2.44 | 13.2 | 17.2 | 0.60 | 28.5 | 53 |
| BARI gom25 | 6.2 | 6.44 | 13.8 | 2.12 | 4.56 | 17.4 | 23.6 | 0.66 | 31.6 | 81 |
| BARI gom26 | 4.1 | 4.89 | 10.0 | 2.03 | 3.22 | 13.2 | 19.9 | 0.51 | 26.3 | 54 |
| BARI gom27 | 4.1 | 4.33 | 9.2 | 2.14 | 3.00 | 8.9 | 17.3 | 0.49 | 17.1 | 37 |
| BARI gom28 | 3.9 | 4.33 | 7.7 | 1.78 | 3.11 | 15.5 | 16.3 | 0.59 | 10.6 | 67 |
| Kheri | 5.7 | 5.11 | 9.7 | 1.88 | 3.00 | 28.2 | 21.0 | 0.57 | 34.5 | 62 |
| Sonalika | 3.6 | 4.78 | 8.4 | 1.78 | 2.78 | 19.0 | 16.2 | 0.52 | 33.2 | 46 |
| Kanchan | 4.7 | 5.00 | 9.0 | 1.79 | 2.78 | 17.5 | 19.1 | 0.47 | 26.7 | 41 |
| Akbar | 4.2 | 5.22 | 9.3 | 1.81 | 3.44 | 14.9 | 24.2 | 0.51 | 43.2 | 44 |
| SEM | 0.16 | 0.12 | 0.37 | 0.035 | 0.11 | 0.96 | 0.75 | 0.012 | 3.12 | 3.84 |
| *p* value | <0.001 | <0.001 | <0.001 | <0.001 | 0.001 | 0.03 | 0.02 | 0.003 | 0.004 | <0.001 |

## 2.3. Treatment–Variety Interaction

Third principal component (PC3), which explains 13.1% data variation, indicated a significant contrast for varieties and variety–treatment interaction. The panicle length and panicle dry weight had positive coefficients and main axes length at Pr1 & Pr2 and shoot dry weight had negative coefficients in PC3 (Table 2). Analysis of variance indicated that panicle length (Figure 1, $p < 0.001$), panicle dry weight (Figure 2, $p < 0.001$) and main axis length at Pr2 ($p < 0.001$) were significant for variety–treatment. PC3 indicated a "competition between roots and panicle of the varieties for dry matter partitioning" (Table 2). *Shourov, Shotabdi, Prodeep, BARI gom27* which are high yielding inbred varieties of wheat had the positive PC scores compared to low yielding varieties and landrace, *Kheri*, which obtained the lowest PC score (Table 2).

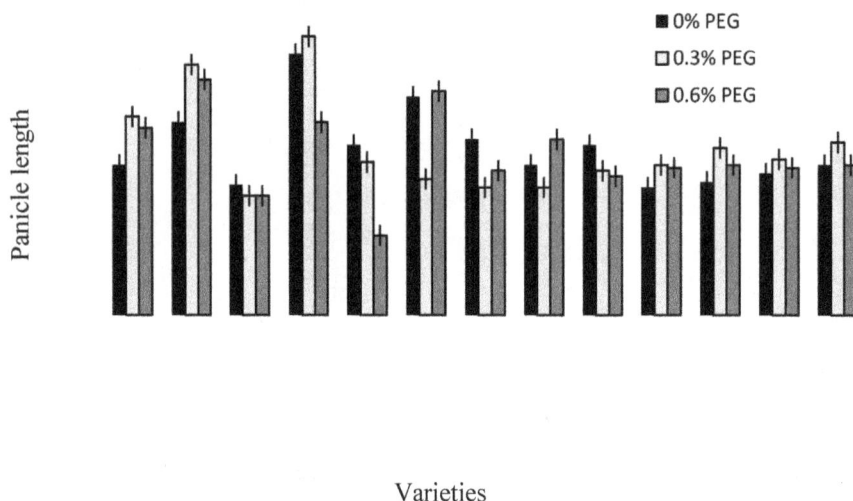

**Figure 1.** Panicle length (cm) of 13 wheat varieties with three different PEG treatments. PEG treatment was imposed for 60 days after sowing for 20 days.

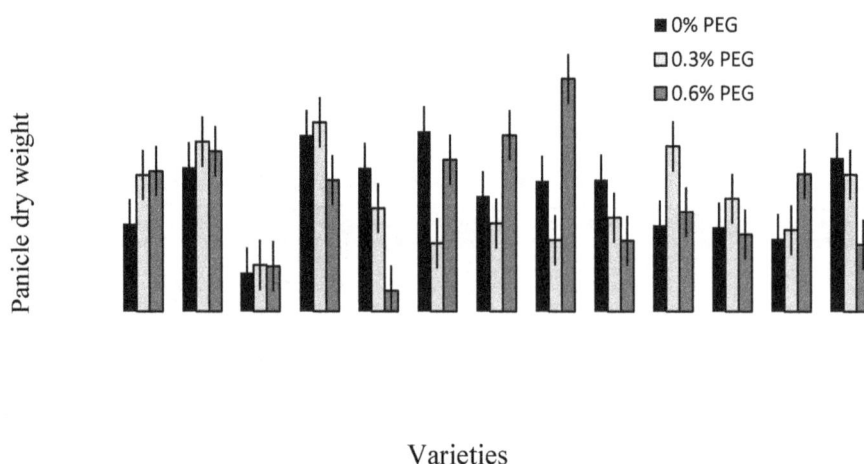

**Figure 2.** Panicle dry weight (mg) of 13 wheat varieties with three different PEG treatments. PEG treatment was imposed for 60 days after sowing for 20 days.

## 2.4. Discussion

### 2.4.1. Root Surface Area Reduction under Osmotic Stress

In this study, the estimated reduction of root surface area of an individual root was related to reduction of diameter and length of root hairs (Table 1, PC2 in Table 2). Reduction of root surface area hampers uptake activity of the plants as available root surface area is directly related to absorption of nutrients, especially phosphorus and water from the soil by the crop plants [28,29]. With confidence, root hairs are valuable tools for nutrient acquisition under all circumstances. Morphology of root hairs differs under different abiotic stresses. For example, Jupp and Newman [30], under drought treatment in *Lolium perenne,* recorded that root hairs disappeared from around 50% and 80% plants after three and eight days, respectively.

### 2.4.2. Increased Total Number of Roots per Tiller

An increase in total number of roots per tiller at 0.6% PEG level compared to control had some similarity to the findings of a previous study of the same group of authors under salinity stress (Table 1, unpublished data). In their study, number of new root appearances at the youngest root bearing phytomer increased compared to control. These results suggested that abiotic stress primarily affects leaf growth and degrades some carbonaceous and nitrogenous compounds like chlorophyll [31,32], and part of C and N released from the bio-molecular de-gradation are possibly transported to the root system for additional root growth.

### 2.4.3. Trait Association: Root Traits with Panicle Traits

Grain yield is a complex and quantitative trait. It is unrealistic to make any straightforward comment that grain yield is directly related to a particular trait only, but in previous studies root hair length was related to grain yield stability [33]. PC2 revealed that root surface area reduction due to treatment effect had a positive association with root hair length and diameter. Root hairs with only 2% of the total carbon required for root construction [34] can contribute total root surface area by many folds, up to 74% of total root surface area of a developed root of a rice plant [35]. In a previous study with barley, longer and denser root hairs yielded a three times higher root surface area [36]. Root hairs not only provide additional surface area to the plants for ion acquisition but also help in rhizosheath formation [28], mediate extra release of organic acids [37,38] and interact with microbial association [39] and, thereby, play an important role in crop yield which deserves further intensive investigation.

### 2.4.4. Varietal Differences

Significant intra and inter-species variation exists for the types and lengths of root hairs [40–43]. Genotypic variations in root hair density along with some other root traits suggested that those traits can be improved by selection (Table 3). PC3 indicated a contrast between panicle traits with the main root axis length for significant varietal difference. Positive and negative PC scores between high yielding wheat varieties, *Prodeep, Shourov, Shotabdi* and *BARI gom27,* and the lowest yielding landrace *Kheri* along with other varieties suggested that high yielding varieties under lower osmotic potential partition comparatively high dry matter to the panicle and less to the roots (see Figures 1 and 2, Table 2). Varietal variation in C, N and dry matter portioning among roots, stem and ear of wheat varieties is evident in the literature [44,45].

## 3. Experimental Section

### 3.1. Plant Culture and Management

The experiment was conducted at the Department of Genetics and Plant Breeding, Bangladesh Agricultural University for the period between August 2013 and February 2014. Seeds of 13 elite wheat varieties were collected from the Wheat Research Centre of the Bangladesh Agricultural Research Institute, Gazipur (Table 4). Seeds were germinated in clean tap water floated in foam net inside the plastic trays in a growth chamber. Around 200 seeds were germinated per tray for each variety in order to select 18 healthy seedlings with synchronized leaf appearance in hydroponic nutrient solution at transplanting. Seed germination process took 5–7 days. Seedlings were transplanted in hydroponic solution following a completely randomized design with three treatments and six replicates per treatment for each of the 13 varieties. An individual plant was considered as a single replicate as root variables were measured in individual plants. There were 18 individual trays for setting up the whole experiment, six trays for each treatment. Each tray contained 13 plants, one from each variety. Plants were cultured at $20 \pm 2$ °C temperature and at $50 \pm 2$ PPFD (photosynthetic photon flux density) light intensity with cool white fluorescent lamps in a 12:12 h day:night cycle. Leaf appearance interval of all varieties was monitored for two weeks and that was calculated at around 10 days (90 °C days). Plants were fed with following nutrient solution: $1mM \cdot NH_4NO_3$, $0.6mM \cdot NaH_2PO_4 \cdot H_2O$, $0.6mM \cdot MgCl_2 \cdot H_2O$, $0.3mM \cdot K_2SO_4$, $0.3mM \cdot CaCl_2 \cdot H_2O$, $50\mu M \cdot H_3BO_3$, $90\mu M \cdot Fe\text{-}EDTA$, $9\mu M \cdot MnSO_4 \cdot 4H_2O$, $0.7\mu M \cdot ZnSO_4 \cdot 7H_2O$, $0.3\mu M \cdot CuSO_4 \cdot 5H_2O$, $0.1\mu M \cdot NaMoO_4 \cdot 2H_2O$ dissolved in water [40]. The nutrient solution was refreshed weekly. All 234 plants were under similar management until they were 60 days old. There were 4–6 live leaves, 3–6 seminal roots and 6–12 adventitious roots per main tiller 60 days after transplantation depending on varieties ($p < 0.001$). Three different polyethylene glycol (PEG 6000, Sigma-Aldrich, Steinheim, Germany) treatments were imposed when the plants were 60 days old. Those were: 0% (control), 0.3% and 0.6% PEG levels. The highest level of PEG 0.6% can potentially create an osmotic potential less than one bar at 20 °C [27]. Even though PEG concentrations are low, however intuitively, these low concentrations of PEG can cause a significant reduction in root hydraulic conductivity, and, therefore, significant water stress, depending on the varieties of wheat.

**Table 4.** List of 13 wheat varieties used for studying root hairs' development and associated traits under PEG-treated hydroponic culture.

| Variety | Seed Source |
|---------|-------------|
| Shotabdi | |
| Shourov | |
| Sufi | |
| Prodeep | |
| Bijoy | |
| BARI gom25 | Wheat Research Centre |
| BARI gom26 | of Bangladesh |
| BARI gom27 | Agricultural Research |
| BARI gom28 | Institute |
| Kheri | |
| Sonalika | |
| Kanchan | |
| Akbar | |

## 3.2. Measurements and Data Collection

Data were recorded on 20 days and 40 days after the PEG treatment by two separate destructive harvests. Three individual plants out of six from each variety × treatment combination were destructively harvested 20 days after treatment, and the remaining three 40 days after treatment. Number of live leaves per main tiller was recorded on the same day. Measurements of root traits were carried out during the destructive harvest. Root measurements include number of root bearing phytomers, number of adventitious roots, mean number of roots per phytomer, number of seminal roots, main axis length at the root bearing phytomer position (Pr) 1, 2, 3, and 4 (youngest root bearing phytomer was considered as reference point), main axis diameter at Pr1, number of root hairs per mm main axis at Pr1, length and diameter of the root hairs at Pr1. Diameter of the main axis, number of root hairs per mm main axis, length and diameter of root hair were measured at 100× magnification under a light microscope. A safranin solution of 0.5% prepared in 50% alcohol was used for staining root hairs. Shoots and roots of the plants at the destructive harvest were dried in an air draft oven for 72 h at 60 °C before recording their dry weights. Panicle length was recorded at the second destructive harvest 40 days after treatment, when plants were 100 days old. Dry weights of the panicles were recorded after 72 h drying at 60 °C in an air draft oven.

## 3.3. Estimation of Root Hair Density and Root Hair Surface Area

To estimate root hair density in no. mm$^{-2}$ main axis, individual main axis was considered as cylinders. Then, the surface area of the main axis per mm length was calculated as: $\pi \times D$ (mm) × 1 mm length. Number of root hairs per mm$^2$ surface area was obtained using the following equation:

$$\text{Root hair density (no. mm}^{-2}) = \frac{Number\ of\ root\ hairs\ counted\ per\ mm\ length\ of\ main\ axis}{Surface\ area\ of\ main\ axis\ for\ per\ mm\ length\ of\ main\ axis} \quad (1)$$

Surface area produced by the root hairs at different PEG levels was calculated from the following equation:

$$\text{Root hair surface area (mm}^2) = \text{Root hair density (no. mm}^{-2}) \times \pi \times \text{RHD} \times \text{RHL} \qquad (2)$$

where RHD and RHL respectively represent root hair diameter and root hair length.

*3.4. Statistical Analyses*

Analysis of variance was carried out following a generalized linear model, using MINITAB 16 statistical software package (Minitab Inc., State College, PA, USA) to find variations among treatments, varieties and treatment × varieties. To test the significant variations among the measured variables, a user defined model was used where the replications within treatment were the error term. Principal component analysis was carried out for some selected traits, the majority of which accounted for either a significant treatment effect or treatment × varieties effect. Analysis of variance of the PC scores was carried out for treatment and variety effects following a generalized linear model.

## 4. Conclusions

One objective of this study was to measure development of root hairs at the reproductive stage under PEG-treated hydroponic culture. This study found that PEG-treated culture decreased the length and diameter of root hairs. Reduction of root hair diameter and root hair length reduced the root surface area of an individual root (Table 2, PC2 in Table 3). In addition, this study revealed a contrast between main axis length and panicle traits with significant differences among varieties, which suggested competition in dry matter partitioning. As low exposure of PEG alters the diameter and length of root hairs, in future experimentation, it will be important to observe how these alterations are associated with growth, development and yield of the wheat varieties under low moisture stress. Moreover, data relevant to root hydraulic conductance or leaf water potential against each of the PEG levels would provide further validation of the experimental results.

## Acknowledgments

This research was supported by the Bangladesh Agriculture University Research System (BAURES) Grant No. 2013/18/BAU.

## Author Contributions

Arif Hasan Khan Robin planned and designed experiments; analyzed data and written the full manuscript. Md. Jasim Uddin and Khandaker Nafiz Bayazid assisted Arif Hasan Khan Robin managing plants and data collection.

## Conflicts of Interest

The authors declare no conflict of interest.

## References

1.   FAOSTAT Data, 2007—Food and Agricultural Commodities Production; FAO: Rome, Italy, 2012. Available online: http://faostat.fao.org (accessed on 4 November 2012).

2.  Anonymous. *Statistical Year Book of Bangladesh-2010, Bangladesh Bureau of Statistics, Planning Division, Ministry of Planning*; Government of the People's Republic of Bangladesh: Dhaka, Bangladesh, 2011.

3.  Hossain, A.; Silva, J.A.T. Wheat production in Bangladesh: Its future in the light of global warming. *AoB Plants* **2013**, *5*, 42, doi:10.1093/aobpla/pls042.

4.  Rahman, A.; Biswas, P.R. Devours resources. *Dhaka Cour.* **1995**, *11*, 7–8.

5.  Shahid, S.; Behrawan, H. Drought risk assessment in the western part of Bangladesh. *Nat. Hazards* **2008**, *46*, 391–413.

6.  Blum, A. Genetic and physiological relationships in plant breeding for drought resistance. *Agric. Water Manag.* **1983**, *7*, 195–205.

7.  Turner, N.C.; Wright, G.C.; Siddique, K.H.M. Adaptation of grain legumes (pulses) to water-limited environments. *Adv. Agron.* **2001**, *71*, 194–233.

8.  Gunasekera, D.; Berkowitz, G.A. Evaluation of contrasting cellular-level acclimation responses to leaf water deficits in three wheat genotypes. *Plant Sci.* **1992**, *86*, 1–12.

9.  Morgan, J.M. Osmoregulation and water stress in higher plants. *Annu. Rev. Plant Physiol.* **1984**, *35*, 299–319.

10. Bray, E.A. Molecular responses to water deficit. *Plant Physiol.* **1993**, *103*, 1035–1040.

11. Manavalan, L.P.; Guttikonda, S.K.; Tran, L.S.P.; Nguyen, H.T. Physiological and molecular approaches to improve drought resistance in soybean. *Plant Cell Physiol.* **2009**, *50*, 1260–1276.

12. Turner, N.C. Adaptation to water deficits: A changing perspective. *Funct. Plant Biol.* **1986**, *13*, 175–190.

13. Turner, N.C.; Jones, M.M. Turgor Maintenance by Osmotic Adjustment: A Review and Evaluation. In *Adaptation of Plants to Water and High Temperature Stress*; Turner, N., Kramer, P., Eds.; John Wiley & Sons, Inc., the University of Michigan: Ann Arbor, MI, USA, 1980; pp. 87–103.

14. Bengough, A.G.; Bransby, M.F.; Hans, J.; McKenna, S.J.; Roberts, T.J.; Valentine, T.A. Root responses to soil physical conditions; growth dynamics from field to cell. *J. Exp. Bot.* **2006**, *57*, 437–447.

15. Bengough, A.G.; McKenzie, B.; Hallett, P.; Valentine, T. Root elongation, water stress, and mechanical impedance: A review of limiting stresses and beneficial root tip traits. *J. Exp. Bot.* **2011**, *62*, 59–68.

16. Brown, L.K.; George, T.S.; Thompson, J.A.; Wright, G.; Lyon, J.; Dupuy, L.; Hubbard, S.; White, P. What are the implications of variation in root hair length on tolerance to phosphorus deficiency in combination with water stress in barley (*Hordeum vulgare* L.)? *Ann. Bot.* **2012**, *110*, 319–328.

17. Crush, J.; Easton, H.; Waller, J.; Hume, D.; Faville, M. Genotypic variation in patterns of root distribution, nitrate interception and response to moisture stress of a perennial ryegrass (*Lolium perenne* L.) mapping population. *Grass Forage Sci.* **2007**, *62*, 265–273.

18. Ehdaie, B.; Layne, A.P.; Waines, J.G. Root system plasticity to drought influences grain yield in bread wheat. *Euphytica* **2012**, *186*, 219–232.

19. Ji, H.; Liu, L.; Li, K.; Xie, Q.; Wang, Z.; Zhao, X.; Li, X. PEG-mediated osmotic stress induces premature differentiation of the root apical meristem and outgrowth of lateral roots in wheat. *J. Exp. Bot.* **2014**, *65*, 4863–4872.

20. Maggio, A.; Hasegawa, P.M.; Bressan, R.A.; Consiglio, M.F.; Joly, R.J. Review: Unravelling the functional relationship between root anatomy and stress tolerance. *Funct. Plant Biol.* **2001**, *28*, 999–1004.

21. Manschadi, A.M.; Christopher, J.; Hammer, G.L.; The role of root architectural traits in adaptation of wheat to water-limited environments. *Funct. Plant Biol.* **2006**, *33*, 823–837.

22. Manschadi, A.M.; Hammer, G.L.; Christopher, J.T. Genotypic variation in seedling root architectural traits and implications for drought adaptation in wheat (*Triticum aestivum* L.). *Plant Soil* **2008**, *303*, 115–129.

23. Palta, J.A.; Chen, X.; Milroy, S.P.; Rebetzke, G.J.; Dreccer, M.F.; Watt, M. Large root systems: Are they useful in adapting wheat to dry environments? *Funct. Plant Biol.* **2011**, *38*, 347–354.

24. Rauf, M.; Munir, M.; ul Hassan, M.; Ahmad, M.; Afzal, M. Performance of wheat genotypes under osmotic stress at germination and early seedling growth stage. *Afr. J. Biotechnol.* **2007**, *6*, 971–975.

25. Dhanda, S.; Sethi, G.; Behl, R. Indices of drought tolerance in wheat genotypes at early stages of plant growth. *J. Agron. Crop Sci.* **2004**, *190*, 6–12.

26. Davidson, D.; Chevalier, P. Influence of polyethylene glycol-induced water deficits on tiller production in spring wheat. *Crop Sci.* **1987**, *27*, 1185–1187.

27. Michel, B.E.; Kaufmann, M.R. The osmotic potential of polyethylene glycol 6000. *Plant Physiol.* **1973**, *51*, 914–916.

28. Haling, R.E.; Brown, L.K.; Bengough, A.G.; Young, I.M.; Hallett, P.D.; White, P.J.; George, T.S. Root hairs improve root penetration, root-soil contact, and phosphorus acquisition in soils of different strength. *J. Exp. Bot.* **2013**, *64*, 3711–3721.

29. Haling, R.E.; Brown, L.K.; Bengough, A.G.; Valentine, T.A.; White, P.J.; Young, I.M.; George, T.S. Root hair length and rhizosheath mass depend on soil porosity, strength and water content in barley genotypes. *Planta* **2014**, *239*, 643–651.

30. Jupp, A.; Newman, E. Morphological and anatomical effects of severe drought on the roots of *Lolium perenne* L. *New Phytol.* **1987**, *105*, 393–402.

31. Khanna-Chopra, R. Leaf senescence and abiotic stresses share reactive oxygen species-mediated chloroplast degradation. *Protoplasma* **2012**, *249*, 469–481.

32. Munné-Bosch, S.; Alegre, L. Die and let live: Leaf senescence contributes to plant survival under drought stress. *Funct. Plant Biol.* **2004**, *31*, 203–216.

33. Gahoonia, T.S.; Nielsen, N.E. Barley genotypes with long root hairs sustain high grain yields in low-P field. *Plant Soil* **2004**, *262*, 55–62.

34. Röhm, M.; Werner, D. Isolation of root hairs from seedlings of Pisum sativum. Identification of root hair specific proteins by *in situ* labeling. *Physiol. Plant.* **1987**, *69*, 129–136.

35. Robin, A.H.K.; Saha, P.S. Morphology of lateral roots of twelve rice cultivars of Bangladesh: Dimension increase and diameter reduction in progressive root branching at the vegetative stage. *Plant Root* **2015**, *9*, 34–42.

36. Gahoonia, T.S.; Nielsen, N.E. Variation in root hairs of barley cultivars doubled soil phosphorus uptake. *Euphytica* **1997**, *98*, 177–182.

37. Gahoonia, T.S.; Nielsen, N.E.; Joshi, P.A.; Jahoor, A. A root hairless barley mutant for elucidating genetic of root hairs and phosphorus uptake. *Plant Soil* **2001**, *235*, 211–219.

38. Narang, R.A.; Bruene, A.; Altmann, T. Analysis of phosphate acquisition efficiency in different Arabidopsis accessions. *Plant Physiol.* **2000**, *124*, 1786–1799.

39. Brown, L.K.; George, T.S.; Barrett, G.E.; Hubbard, S.F.; White, P.J. Interactions between root hair length and arbuscular mycorrhizal colonisation in phosphorus deficient barley (*Hordeum vulgare*). *Plant Soil* **2013**, *372*, 195–205.

40. Robin, A.H.K.; Uddin, M.J.; Afrin, S.; Paul, P.R. Genotypic variations in root traits of wheat varieties at phytomer level. *J. Bangladesh Agric. Univ.* **2014**, *12*, 45–54.

41. Caradus, J. Selection for root hair length in white clover (*Trifolium repens* L.). *Euphytica* **1979**, *28*, 489–494.

42. Haling, R.E.; Richardson, A.E.; Culvenor, R.A.; Lambers, H.; Simpson, R.J. Root morphology, root-hair development and rhizosheath formation on perennial grass seedlings is influenced by soil acidity. *Plant Soil* **2010**, *335*, 457–468.

43. Haling, R.E.; Simpson, R.J.; Delhaize, E.; Hocking, P.J.; Richardson, A.E. Effect of lime on root growth, morphology and the rhizosheath of cereal seedlings growing in an acid soil. *Plant Soil* **2010**, *327*, 199–212.

44. Nicolas, M.E.; Lambers, H.; Simpson, R.J.; Dalling, M.J. Effect of drought on metabolism and partitioning of carbon in two wheat varieties differing in drought-tolerance. *Ann. Bot.* **1985**, *55*, 727–742.

45. Nicolas, M.E.; Simpson, R.J.; Lambers, H.; Dalling, M.J. Effects of drought on partitioning of nitrogen in two wheat varieties differing in drought-tolerance. *Ann. Bot.* **1985**, *55*, 743–754.

# 11

# Interactive Role of Fungicides and Plant Growth Regulator (Trinexapac) on Seed Yield and Oil Quality of Winter Rapeseed

**Muhammad Ijaz [1,2,*], Khalid Mahmood [3,*] and Bernd Honermeier [2]**

[1] College of Agriculture, Bahauddin Zakariya University, Bahadur- Sub Campus Layyah 31200, Pakistan; E-Mail: muhammad.ijaz@bzu.edu.pkM

[2] Institute of Agronomy & Plant Breeding I, Biomedical Research Center Seltersberg (BFS), Justus Liebig University Giessen, Schubertstr. 81, Giessen D-35392, Germany; E-Mail: Bernd.Honermeier@agrar.uni-giessen.de

[3] Department of Agro-ecology, Faculty of Science and Technology, Aarhus University, Aarhus C 8000, Denmark

* Author to whom correspondence should be addressed; E-Mail: khalid.mahmood@agrsci.dk

Academic Editors: Jerry H. Cherney and Peter Langridge

**Abstract:** This study was designed to evaluate the role of growth regulator trinexapac and fungicides on growth, yield, and quality of winter rapeseed (*Brassica napus* L.). The experiment was conducted simultaneously at different locations in Germany using two cultivars of rapeseed. Five different fungicides belonging to the triazole and strobilurin groups, as well as a growth regulator trinexapac, were tested in this study. A total of seven combinations of these fungicides and growth regulator trinexapac were applied at two growth stages of rapeseed. These two stages include green floral bud stage (BBCH 53) and the course of pod development stage (BBCH 65). The results showed that plant height and leaf area index were affected significantly by the application of fungicides. Treatments exhibited induced photosynthetic ability and delayed senescence, which improved the morphological characters and yield components of rape plants at both locations. Triazole, in combination with strobilurin, led to the highest seed yield over other treatments at both experimental locations. Significant effects of fungicides on unsaturated fatty acids of rapeseed oil were observed. Fungicides did not cause any apparent variation in the values of free fatty acids and peroxide of rapeseed oil. Results of our study demonstrate that judicious use of fungicides in rapeseed may help to achieve sustainable farming to obtain higher yield and better quality of rapeseed.

**Keywords:** fungicides; growth regulators; leaf area index; oil quality; rapeseed; seed yield

## 1. Introduction

Different techniques such as osmo-priming, seed treatment, and application of chemicals used to induce growth of plants, are employed to achieve maximum seed yield [1]. Several fungicides also serve this purpose through altering cellular mechanisms of plant growth [2]. After the introduction of various modes of active fungicides, the concept of disease control gained new perspectives due to the positive physiological effects of these chemicals on plants [2]. Triazole and strobilurin treatments, along with plant growth regulators, are associated with various morphological and physiological changes in various plants; including inhibition of plant growth, decrease in inter-nodal elongation, increased chlorophyll content, enlarged chloroplast, thicker leaf tissue, increased root to shoot ratio, delayed senescence, increased antioxidant potentials, and enhancement in alkaloid production [3,4]. For example, triazole fungicides affect the isoprenoid pathway and alter the levels of certain plant hormones by inhibiting gibberellin synthesis [5,6]. Triazole inhibits mono-oxygenases that results in oxidation of ent-kaurene to ent-kaurenoic acid in three steps, which is an early reaction in GA biosynthesis [7]. In winter rapeseed, triazole application reduced the rate of photosynthesis by decreasing the stomatal conductance [8]. Strobilurins, another fungicide group, cause reductions in ethylene concentrations leading to degradation of cytokinins and resulting in delayed senescence [9]. Application of strobilurin fungicides maintain the photosynthetic active green leaf area for a longer period to increase the quantity of assimilates available for grain filling that can result in higher yield [9]. These fungicides were also reported previously to control lodging and to improve seed yield in cereals, but little information exists for the use of these fungicides in combination with plant growth regulators in oil seed crops. The purpose of this study is to evaluate the role of these fungicides, in combination with growth regulator trinexapac, on seed yield and oil quality of winter rapeseed in field conditions.

## 2. Materials and Methods

The experiments were conducted at Giessen (GI) (50°47' N and 8°61' E, 158 m above sea level) and Rauischholzhausen (RH) (50°45' N and 8°39' E, 220 m above sea level) experimental stations. The soil at RH is a loess type, while that at GI is silt clay. The mean air temperature during the growing season was 8.5 °C at GI and 9.7 °C at RH, and total rainfall from August to July were recorded as 660.5 mm and 637.5 mm, respectively. The experiments were set out as randomized complete blocks, with a factorial arrangement and four replications per treatment at each site. Six fungicides belonging to triazole or strobilurin groups, along with growth regulator Moddus (trinexapac), were applied in seven combinations (Table 1) on two winter rapeseed cultivars, "Elektra" and "NK Fair". The treatments were applied at two growth stages of rapeseed. The first application was at the green floral bud stage (BBCH 53) and the second application was at the pod development stage (BBCH 65). The fungicides and trinexapac were applied using a $CO_2$-charged hand boom sprayer equipped with Tee Jet nozzles that delivered 180 L·ha$^{-1}$.

**Table 1.** Various fungicide and growth regulator applications at two developmental stages.

| Stage | Fungicide and Growth Regulator | Dose L·ha$^{-1}$ |
|---|---|---|
| Floral bud development (BBCH 53) | Control | - |
| | Toprex | 0.5 |
| | Toprex | 0.5 |
| | Toprex | 0.35 |
| | Toprex + Moddus | 0.35 + 0.5 |
| | Folicur | 1.0 |
| | Caramba | 1.0 |
| Pod development (BBCH 65) | Control | - |
| | Ortiva | 1.0 |
| | Ortiva | 1.0 |
| | Ortiva | 1.0 |
| | Ortiva | 1.0 |
| | Proline | 0.7 |
| | Cantus | 0.5 |

Triazole:Toprex (Difenoconazole 250 g·L$^{-1}$ + Paclobutrazol 125 g·L$^{-1}$), Folicur (Tebuconazole 251.2 g·L$^{-1}$), roline (Prothioconazole 250 g·L$^{-1}$) and Caramba (Metconazole 60 g·L$^{-1}$); Strobilurin:Ortiva (azoxystrobin 250 g·L$^{-1}$), Cantus (Boscalid 500 g/L) (Cantus is not a strobilurin fungicide but has same mode of action like strobilurin); Growth regulator: Moddus (Trinexapac 222 g·L$^{-1}$ + Ethyl ester 250 g·L$^{-1}$).

The seed-bed was prepared in the autumn prior to the experiment by chisel plowing, followed by two rounds of cultivation during the following spring. Each 5 m × 3 m plot consisted of 12 rows, and the seed was sown at a depth of 2–3cm depth to give a seeding density of 50·m$^{-2}$. Post emergence herbicide Butisan Top (metazachlor 12% + quinmerac 37.5%) was applied to eradicate weeds to avoid weed losses at the rate of 1.8 L·ha$^{-1}$ at both experimental locations. In order to control insect damage Trafo (λ-cyhalothrin) was applied at the rate of 150 g·ha$^{-1}$ at BBCH 57 in Rauischholzhausen, and Biscaya (thiacloprid 240 g·L$^{-1}$) was applied at different growth stages of the crop (BBCH 49 and BBCH 54) at the rate of 300 mL·ha$^{-1}$ in GI. No disease infestation was observed at either experimental station. The previous crop at RH was winter wheat, and at GI was winter barley. A fertilizer dressing of 150kg N·ha$^{-1}$ was applied, half at BBCH 18 and the other half at BBCH 30, whereas 72 kg S·ha$^{-1}$ was applied at BBCH 18. Ammonium nitrate and ammonium sulfate were applied for N and S fertilization.

The leaf area index (LAI) and height of plant stands was measured after one fungicide application per week until maturity. The leaf area index was measured using a Sun Scan canopy analysis system (Delta T Company, CA, USA) [10]. The other recorded traits were the numbers of seeds per main stem, pods per main stem, length of main stem, and pod length. These parameters were assessed from 10 plants per plot across all four replicates at GI. Neither LAI nor any of the morphological traits were assessed in the RH trial. Thousand grain weight (TGW) was obtained by counting two lots of 500 seeds per sample. The extent of lodging was estimated by grading the crop on a 1 to 9 scale (1 for erect and 9 for flat) at BBCH 75. After harvest, the seed moisture content was adjusted to 9% before estimating seed yield.

The Soxhlet method was used to determine seed oil content [11]. For the quantification of fatty acids, the oil was subjected to gas chromatography (Varian CP-3800) equipped with a flame ionization detector (GC-FID) through an OPTIMA-FFAP-Wax column (25 m × 0.32 mm i.d; film thickness 0.25 μm) [12]. The Dumas combustion method [13], utilized by a CHNS analyzer EA1110 type thermo Finnegan device, was used to measure the N content of the sample, and its protein content was obtained by multiplying this value by 6.25. A simple titration was applied to determine the free fatty acid (FFA) content. For this procedure, a 10 g aliquot of oil was dissolved in an equal volume of ethyl ethanol and 50 mL to 10 uL in the presence of phenolphthalein, and titrated against 0.01N NaOH until the solution reached a stable pink endpoint. The FFA content was calculated on either an oleic or a palmitic acid basis. Each sample was titrated in duplicate. The peroxide value (PV) of the oil was also obtained by titration, using 5 g oil dissolved in 30 mL acetic acid: iso-octane (3:2 $v/v$). Following the addition of 0.5 mL saturated potassium iodide (KI), the mixture was titrated with 0.01N sodium thiosulfate, using the starch/iodine reaction as the indicator.

## 3. Statistical Analysis

The experimental data were statistically analyzed using the software package PIAF (Planning Information Analysis Program for Field Trials). A general linear model was assumed, and multiple comparisons were performed using a $t$-test, with a chosen significance level of $p < 0.05$. Mean values were compared using a least significant difference test.

## 4. Results and Discussion

The results of this study indicate that application of fungicides at BBCH 53 and 65 increased leaf area index (LAI) of rapeseed by delaying senescence, in comparison with application of triazole (Toprex) or strobilurin (Ortiva) fungicides, as well as control treatment ,at the later stages of rapeseed in Giessen (Table 2). Maximum LAI was recorded by application of Toprex (paclobutrazole and difenoconazole) at the rate of 0.5 L/ha in combination with Ortiva (azoxystrobin) at all growth stages. This effect can be explained by prolonged photosynthetic duration of green tissues by Ortiva application at BBCH 65 as compared with its combined application with triazole fungicides. Similar results were recorded by Zhang et al. (2010) [4] in wheat, who postulated that Ortiva application delayed senescence by enhancing antioxidative potential and protecting the plants from harmful active oxygen species.

Our triazole treatment (Folicur + Proline) reduced LAI among double-applied treatments at later stages of rapeseed. Zhou and Leoul (1998) [13] reported that application of triazole increased the level of stress hormone abscisic acid (ABA), which favors senescence. Strobilurin performed best as the second application at BBCH 65 to delay senescence compared with triazole.

Triazole and trinexapac are anti-gibberellins that improve stem stability by inhibiting intercalary growth, which reduces the probability of lodging. Combinations of chemicals had higher LAI and reduced lodging more than single applied fungicides (Table 2). Leaf area index is the ratio of green plant material that covers a square meter of land and has a direct influence on crop vigor, root development, carbohydrate storage, and nutrient transport. Healthiness of rape plants was improved with increased LAI and hence lodging was reduced considerably.

Height of plant stands had an inverse relation with lodging, as observed by Armstrong and Nicol (1991) [14]. At both stations sole-applied Ortiva and control resulted in a severely lodged crop with a minimum plant height (PH) in comparison with other treatments (Table 2). Higher plant height (PH) with minimum lodging was recorded for Folicur + Proline over other treatments at both stations. Application of Ortiva in combination with Toprex produced healthy plants which resisted lodging. At Giessen it was observed that by reducing the concentration of Toprex from 0.5 to 0.35 $L \cdot ha^{-1}$ in combination with Ortiva increased lodging. Baylis and Wright (1990) [15] also observed similar results after applying paclobutrazole on winter rapeseed.

**Table 2.** Effect of fungicides and growth regulator on leaf area index (LAI) at GI, lodging (Lodg) and plant height (PH) of two cultivars of rapeseed at GI and RH.

| Treatments | Giessen | | | | | Rauischholzhausen | |
|---|---|---|---|---|---|---|---|
| | Leaf Area Index | | | Lodg (1–9) | PH (cm) | Lodg (1–9) | PH (cm) |
| | BBCH 62 | BBCH 75 | BBCH 80 | BBCH 80 | BBCH 80 | BBCH 80 | BBCH 80 |
| Control | 7.29 | 4.34 | 2.91 d | 5.3 | 147.5 b | 5.8 | 100.3 cd |
| Toprex (Top$_{0.5}$) | 7.08 | 4.59 | 3.64 ab | 4.1 | 160.3 a | 5.1 | 104.2cd |
| Top$_{0.5}$ + Ortiva | 7.78 | 5.07 | 3.99 a | 3.9 | 161.2 a | 5.8 | 105.6 bcd |
| Top$_{0.35}$ + Ortiva | 7.32 | 4.94 | 3.54 bc | 4.5 | 160.3 a | 4.9 | 115.7 abc |
| Top$_{0.35}$ + Mo + Ort. | 6.68 | 4.71 | 3.76 ab | 3.3 | 158.3 a | 6.0 | 108.6 bcd |
| Ortiva | 6.96 | 4.49 | 3.33 c | 4.6 | 154.6 ab | 6.9 | 97.9 d |
| Folicur + Proline | 6.77 | 4.73 | 3.45 bc | 3.6 | 160.4 a | 4.8 | 125.4 a |
| Caramba + Cantus | 6.95 | 4.78 | 3.79 ab | 3.5 | 159.4 a | 4.9 | 121.8 ab |
| Fun. (LSD$_{0.05}$) | ns | ns | 0.36 | - | 7.2 | - | 14.6 |
| NK Fair | 7.01 | 4.70 | 3.72 a | 3.5 | 169.2 a | 4.4 | 125.3 a |
| Elektra | 7.20 | 4.71 | 3.39 b | 4.8 | 146.3 b | 6.6 | 93.4 b |
| Cv. (LSD$_{0.05}$) | ns | ns | 0.18 | - | 3.6 | - | 7.28 |
| Fun. x Cv. (LSD$_{0.05}$) | ns | ns | ns | - | 10.2 | - | ns |

Top$_{0.5}$ = Toprex @ 0.5 $L \cdot ha^{-1}$, Top0.35 = Toprex @ 0.35 $L \cdot ha^{-1}$, Mod = Moddus, Ort = Ortiva, Fun. = Fungicides, ns = non-significant. *a > b > c > c > d.

1000-grain weight (TGW) is an important yield component because large seed is more valuable. Application of fungicides altered TGW significantly at both stations. Heavy-lodged plants of untreated plots attained the lowest TGW at both stations (Table 3). Due to lodging, growing conditions are unfavorable for seed filling because of reduced light, which has a negative impact on photosynthesis [16–18]. Severe lodging interferes with the transport of nutrients and moisture from the soil and, thus, with storage in developing seeds. Incomplete filling results in small seeds with lower oil and protein content and weight. TGW was lower at RH than that of Giessen. This can be explained by severe lodging at RH compared with Giessen.

At both stations, performance of Caramba in combination with Cantus was consistent to enhance TGW, while application of Toprex by itself reduced TGW among fungicidal treatments. Berry and Spink (2009) [19] also reported that Caramba (metconazole) application in winter rapeseed enhanced TGW by increasing leaf area index and developing optimum size of the crop canopy, which helped to reduce lodging. Growth regulator Moddus (trinexapac) treatment attained maximum value of TGW and minimal lodging at Giessen, while at RH the same treatment plots were lodged [17–20]. This was caused by a thunderstorm at the time of maturity. Our results demonstrate a negative relationship between lodging and TGW.

Combined application of triazole fungicides (Folicur + Proline) significantly increased TGW at both stations (Table 2). At the same time this treatment decreased number of pods per main stem and increased number of seeds per stem. The reduction in pod number caused assimilates to accumulate in less pods, leading to higher TGW than that of the control. In both experiments, we found a direct relation of LAI with TGW [21]. Double-applied treatments attained the highest LAI and also maximum TGW than that of the control treatments. Reduction of TGW in the case of cv. NK Fair was associated with higher plant height in comparison with cv. Elektra. Shorter plants received equal and maximum light throughout the canopy and, consequently, large seeds were produced [8].

**Table 3.** Effect of fungicides and growth regulator on TGW, seed yield, and oil content of two cultivars of winter rapeseed at GI and RH.

| Treatments | Giessen | | | Rauischholzhausen | | |
|---|---|---|---|---|---|---|
| | TGW (g) | Seed Yield (dt·ha$^{-1}$) | Oil (%) | TGW (g) | Seed Yield (dt·ha$^{-1}$) | Oil (%) |
| Control | 4.48 *b | 52.0 d | 43.5 | 3.77 d | 53.9 | 42.7 b |
| Toprex (Top$_{0.5}$) | 4.59 ab | 57.5 bc | 43.3 | 3.89 cd | 53.9 | 42.3 b |
| Top$_{0.5}$ + Ortiva | 4.71 a | 62.1 a | 43.9 | 4.11 abc | 55.5 | 43.2 ab |
| Top$_{0.35}$ + Ortiva | 4.63 a | 57.9 b | 45.0 | 4.08 abc | 58.2 | 42.8 b |
| Top$_{0.35}$ + Moddus + Ortiva | 4.71 a | 59.3 ab | 43.9 | 3.92 cd | 54.3 | 43.1 ab |
| Ortiva | 4.61 a | 55.1 cd | 43.7 | 3.95 bcd | 56.5 | 43.2 ab |
| Folicur + Proline | 4.67 a | 57.6 bc | 43.6 | 4.17 ab | 57.2 | 42.2 b |
| Caramba + Cantus | 4.60 ab | 62.1 a | 44.1 | 4.26 a | 58.0 | 44.2 a |
| Fun. (LSD$_{0.05}$) | 0.12 | 3.60 | ns | 0.22 | ns | 1.16 |
| NK Fair | 4.46 b | 58.0 | 45.2 a | 3.73 b | 53.8 b | 43.5 a |
| Elektra | 4.79 a | 57.9 | 42.6 b | 4.31 a | 58.1 a | 42.5 b |
| Cv. (LSD$_{0.05}$) | 0.06 | ns | 0.73 | 0.11 | 0.54 | 0.58 |
| Fun. x Cv. (LSD$_{0.05}$) | ns | ns | ns | ns | ns | 1.63 |

Top$_{0.5}$ = Toprex @ 0.5 L·ha$^{-1}$, Top$_{0.35}$ = Toprex @ 0.35 L·ha$^{-1}$, ns = non-significant.*a > b > c > c > d.

Seed yield and its formation process depend on genetic, environmental, and agronomic factors, including growth regulation and the interaction between them. In this experiment, growth-regulating fungicides altered seed yield significantly at Giessen, while it was unaffected at RH (Table 3). Results showed that TGW was positively correlated with seed yield at both stations. Control treatment exhibited lowest seed yield with minimum TGW at both stations. This yield loss was associated with reduced LAI compared to other treatments. Assimilate production in plants is reduced if the LAI is below the optimum required to capture all light transmitted beyond the flower layer. Maximum seed yield was recorded for the combined application of Caramba and Cantus, corresponding to the highest number of pods and seeds per main stem. Top$_{0.5}$ in combination with Ortiva also attained high seed yield through improving LAI and TGW at Giessen. Tuncturk and Ciftci (2007) [22] also reported that number of seeds per pod, 1000-seed weight, and number of seeds per pod have a direct positive effect on seed yield.

In our study, seed yield was strongly related to the severity of lodging. Ortiva alone-applied plants recorded the lowest seed yield and were susceptible to maximum lodging among fungicidal treatments at Giessen. The increase in individual seed weight was negligible among triazoles. Previous studies found that yield improvements in response to triazole fungicides uniconazole and paclobutrazol were unrelated to changes in individual seed weight [8,14]. Higher LAI in the case of $Top_{0.5}$ + Ortivawas likely a result of more rapid stem elongation and the partitioning of a greater proportion of assimilate to above-ground growth as a result of maximum seed yield at Giessen. It seems plausible that reducing LAI and plant height resulted in a stronger stem. These effects may explain a large reduction in lodging from combined application of Folicur and Proline at both stations.

The value of rapeseed linked to its seed oil content was influenced significantly by the application of fungicides at RH (Table 3). At both stations it was observed that application of only triazole fungicides ($Top_{0.5}$and Folicur + Proline) reduced oil content in comparison with control and other treatments [6,16]. Our results are contradicted by MertTürk et al. (2008) [23] and Butkute et al. (2006) [24], who worked with triazole fungicides Harvesan and Folicur, respectively, and reported that oil content of rapeseed increased significantly after triazole application in comparison with the control. Application of Ortiva and Cantus in combination with triazole fungicides enhanced oil content by extending the seed formation phase which led to increased oil accumulation in the seeds. Including Caramba improved yield associated parameters (number of pods and seeds per main stem) and seed yield, and also resulted in higher oil content.

In the literature it was explained that oil content of rapeseed is influenced by air temperature, especially after flowering. This was confirmed with our results in which oil content of seed samples from RH was 1% lower due to its higher air temperature (9.7 °C) than that of Giessen experimental station (8.5 °C). Hassan et al. (2007) [25] also reported that an increase of 1 °C temperature caused a loss of 1.2% of oil in the rapeseed. Interactions between fungicide and cultivar regarding oil content of rapeseed were significant at Giessen, which showed that cultivars responded differently in oil content after application of fungicides.

Fungicides at RH influenced protein content of seed, whereas no clear variations were observed at the other location (Table 4). Butkute et al. (2006) [24] reported that the application of Folicur on rapeseed enhanced protein content significantly over the control. These results agreed with our findings from RH. Among the fungicidal treatments, Ortiva application improved protein content significantly in comparison to control at RH. Jenkyn et al. (2000) [26] also reported that application of azoxystrobin enhanced protein content in the grains of wheat. An inverse relationship was observed between oil and protein in both experiments. Higher oil and lower protein content was observed at the Giessen experimental station compared to RH experimental station, which might be due to temperature differences. These findings are consistent with those of Pritchard et al. (2000) [27], who recorded increased protein with decreased oil content, and concluded that wetter and cooler spring weather would favor higher oil accumulation, but lower proteins.

Significant interactions were recorded among cultivars and fungicides regarding oil and protein content at RH. Cultivars differed markedly for protein and oil content at both stations and responded differently to fungicide application. NK Fair was a late maturing cultivar and its LAI increased significantly at BBCH 80 compared to cv. Elektra. These results agreed with the findings of Dimov and Möller (2010) [28] that tested modern winter oilseed rape cultivars including cv. NK Fair and cv. Elektra in field experiments under typical German growing conditions.

In the present study, concentration of free fatty acids (FFA) in the oil of rapeseed varied from 0.12 to 0.16%, which was lower than that reported by May et al. (1993) [29] who obtained 0.41 to 0.54% FFA in the oil of rapeseed after application of fungicides. Our experimental data demonstrated that concentration of FFA was significantly influenced by the application of fungicides at RH, while FFA was unaffected at Giessen (Table 4). May *et al.* (1993) [29] reported that FFA was unaffected after application of fungicides, while FFA was significantly influenced by agronomic practices, including low seeding rates, increased nitrogen fertilization, and delayed planting in Ontario grown spring rapeseed. Concentration of FFA correlated with the intensity of lodging. Maximum FFA was recorded from heavily lodged Ortiva alone-treated plants at both stations.

**Table 4.** Effect of fungicides and growth regulator on protein content, free fatty acids (FFA) and peroxides value (PV) of two cultivars of winter rapeseed at GI and RH.

| Treatments | Giessen | | | Rauischholzhausen | | |
|---|---|---|---|---|---|---|
| | Protein (%) | FFA (%) | PV (meq·kg$^{-1}$) | Protein (%) | FFA (%) | PV (meq·kg$^{-1}$) |
| Control | 21.8 | 0.14 | 3.18 *b | 22.2 c | 0.15 ab | 2.46 a |
| Toprex (Top$_{0.5}$) | 22.1 | 0.12 | 3.54 a | 23.1 ab | 0.15ab | 2.34 bc |
| Top$_{0.5}$ + Ortiva | 22.6 | 0.12 | 2.61 c | 22.9 bc | 0.15 ab | 2.44 ab |
| Top$_{0.35}$ + Ortiva | 21.7 | 0.14 | 2.31 d | 22.9 bc | 0.14 b | 2.50 a |
| Top$_{0.35}$ + Moddus + Ortiva | 21.5 | 0.13 | 2.36 d | 23.9 a | 0.15 ab | 2.10 e |
| Ortiva | 21.9 | 0.15 | 2.50 cd | 24.1 a | 0.16 a | 2.43 ab |
| Folicur + Proline | 21.8 | 0.14 | 2.26 d | 23.9 a | 0.15 ab | 2.26 cd |
| Caramba + Cantus | 22.0 | 0.14 | 2.39 cd | 22.8 bc | 0.15 ab | 2.22 d |
| Fun. (LSD$_{0.05}$) | ns | ns | 0.24 | 0.93 | 0.01 | 0.11 |
| NK Fair | 22.7 a | 0.13 | 2.84 | 24.1 a | 0.14 b | 2.19 b |
| Elektra | 21.1 b | 0.13 | 2.45 | 22.1 b | 0.16 a | 2.49 a |
| Cv. (LSD$_{0.05}$) | 0.70 | ns | 0.12 | 0.47 | 0.01 | 0.05 |
| Fun. x Cv. (LSD$_{0.05}$) | ns | 0.03 | 0.34 | 1.32 | 0.01 | ns |

Top$_{0.5}$ = Toprex @ 0.5 L·ha$^{-1}$, Top$_{0.35}$ = Toprex @ 0.35 L·ha$^{-1}$, ns = non-significant. *a > b > c > c > d.

In this study, PV was altered significantly by application of fungicides at both stations. Growth regulator Moddus, and triazole fungicides Folicur and Caramba, treatments reduced PV significantly compared to the control at both stations (Table 4). This may be due to minimal damage of seeds during harvesting from these treatments. These treatments protected plants from severe lodging compared to the control treatment. Lodging caused mechanical damage to seeds during harvesting [15], and damaged seeds led to increased PV. Severely lodged *cv.* Elektra also attained significantly higher PV compared with *cv.* NK Fair at RH.

In this study, it was observed that fungicidal treatments like Caramba + Cantus, which produced maximum oil content also attained higher oleic acid, which agreed with the findings of MertTürk *et al.* (2008) [22], who reported that application of triazole fungicide Harvesan increased oil content, as well as oleic acid compared with control. Linoleic and linolenic acids were significantly affected by application of fungicides. These unsaturated fatty acids slightly increased in concentration by the application of triazole fungicides Top$_{0.5}$ alone at Giessen and Folicur + Proline at RH (Table 5). Results demonstrated that increased oleic acid related to the decreased linoleic acid. This relation of fatty acids can be explained by the activity of the enzyme FAD2 that converts oleic acid to linoleic acid which is in turn, converted to linolenic acid by FAD3.

**Table 5.** Effect of fungicides and growth regulator on unsaturated fatty acids of two cultivars of winter rapeseed at GI and RH.

| Treatments | Giessen | | | Rauischholzhausen | | |
|---|---|---|---|---|---|---|
| | C18:1 | C18:2 | C18:3 | C18:1 | C18:2 | C18:3 |
| Control | 61.3 | 19.5 ab | 9.42 b | 59.0 | 19.8 a | 10.0 cd |
| Toprex (Top$_{0.5}$) | 60.3 | 19.6 a | 9.80 a | 59.8 | 19.3 bc | 9.7 e |
| Top$_{0.5}$ + Ortiva | 60.9 | 19.3 abc | 9.55 ab | 60.0 | 19.2 c | 9.8 cde |
| Top$_{0.35}$ + Ortiva | 61.6 | 18.7 d | 9.52 ab | 60.1 | 19.0 c | 9.8 de |
| Top$_{0.35}$ + Moddus + Ortiva | 61.1 | 19.0 bcd | 9.50 b | 59.5 | 19.1 c | 9.8 de |
| Ortiva | 61.3 | 18.7 d | 9.26 b | 60.3 | 19.7 ab | 10.4 b |
| Folicur + Proline | 60.7 | 18.9 cd | 9.36 b | 59.9 | 19.9 a | 10.7 a |
| Caramba + Cantus | 61.0 | 18.8 cd | 9.42 b | 60.8 | 19.1 c | 10.1 bc |
| Fun. (LSD$_{0.05}$) | ns | 0.49 | 0.29 | ns | 0.31 | 0.26 |
| NK Fair | 62.4* a | 18.5 | 9.26 b | 60.1 a | 19.3 | 10.0 |
| Elektra | 59.7 b | 19.6 | 9.70 a | 59.8 b | 19.5 | 10.1 |
| Cv. (LSD$_{0.05}$) | 0.54 | 0.25 | 0.15 | 0.54 | ns | ns |
| Fun. × Cv. (LSD$_{0.05}$) | ns | ns | ns | ns | 0.45 | 0.37 |

Top$_{0.5}$ = Toprex @ 0.5 L·ha$^{-1}$, Top$_{0.35}$ = Toprex @ 0.35 L·ha$^{-1}$, ns = non-significant; C18:1 = Oleic acid, C18:2 = Linoleic acid, C18:3 = Linolenic acid. *a > b > c > d > e.

We concluded that positive effects on plant growth were observed when triazole and strobilurin fungicides were applied. Moreover, it was shown that application of growth-regulating fungicides can effectively control overlarge canopies in order to reduce lodging and achieve optimally-sized seeds. Positive yield effects were achieved after combined application of triazole at BBCH 53 and strobilurin fungicides at BBCH 65, compared with their sole applications. Quality parameters of rape oil, including oil content, fatty acid profile, free fatty acids, and peroxide value can be influenced by application of fungicides, but are also dependent on weather conditions and cultivar effects.

## Acknowledgments

The authors acknowledge the Higher Education Commission Pakistan and Deutscher Akademischer Austauschdienst (DAAD) Germany for providing funding.

## Authors Contribution

Muhammad Ijaz and Bernd Honermeier designed the study. Muhammad Ijaz prepared the material. Bernd Honermeier helped in conducting field experiments. Muhammad Ijaz, Bernd Honermeier, and Khalid Mahmood, analyzed data through statistically analysis. Muhammad Ijaz,Khalid Mahmood and Bernd Honermeier wrote the manuscript. Bernd Honermeier supervised the project.

## Conflicts of Interest

The authors declare no conflict of interest.

**List of Abbreviations**

| | |
|---|---|
| cm | centimeter |
| Cv. | Cultivar |
| dt | decitones |
| FAO | Food and Agriculture Organization (of the United Nations) |
| FID | Flame Ionization Detector |
| FFA | Free Fatty Acids |
| GC | Gas Chromatography |
| GI | Giessen |
| GSL | Glucosinolates |
| Fun | Fungicides |
| g | gram |
| ha | hectare |
| K | Potassium |
| kg | kilogram |
| LAI | Leaf Area Index |
| LSD | Least Significant Difference |
| mequ | Milliequvalent |
| ns | Non-Significant |
| p | Probability |
| PH | Plant Height |
| PIAF | Planning Information Analysis Program for Field Trials |
| PV | Peroxides value |
| RCBD | Randomized Complete Block Design |
| RH | Rauischholzhausen |
| TGW | 1000-grain weight |

**References**

1.  Kumar, B.; Sing, Y.; Ram, H.; Sarlach, R.S. Enhancing seed yield and quality of Egyptian Clover (*Trifolium alexandrinum* L.) with foliar application of bio-regulators. *Field Crop Res.* **2013**, *146*, 25–30.

2.  Venancio, W.S.; Rodrigues, M.A.T.; Begliomini, E.; Souza, N.L.D. Physiological effects of strobilurin fungicides on plants. *Cienc. Exatas Terra Cienc. Agr. Eng.* **2003**, *9*, 59–68.

3.  Ruske, R.E.; Gooding, M.J.; Jones, S.A. The effects of adding picoxystrobin, azoxystrobin and nitrogen to a triazole programme on disease control, flag leaf senescence, yield and grain quality of winter wheat. *Crop Prot.* **2003**, *22*, 975–987.

4.  Zhang, Y.J.; Zhang, X.; Zhou, M.G.; Chen, C.J.; Wang, J.X.; Wang, H.C.; Zhang, H. Effect of fungicides JS399-19, azoxystrobin, tebuconazole, and carbendazim on the physiological and biochemical indices and grain yield of winter wheat. *Pestic. Biochem. Physiol.* **2010**, *98*, 151–157.

5.  Graebe, J.E. Gibberellin biosynthesis and control. *Ann. Rev. Plant Physiol.* **1987**, *38*, 419–465.

6.  Setia, R.C.; Bhathal, G.; Setia, N. Influence of paclobutrazol on growth and yield of *Brassica carinata* A. Br. *Plant Growth Regul.* **1995**, *16*, 121–127.

7.   Rademacher, W. Growth retardants: Effects on gibberellin biosynthesis and other metabolic pathways. *Ann. Rev. Plant Physiol. Plant Mol. Biol.* **2000**, *51*, 501–531.

8.   Zhou, W.; Ye, Q. Physiological and yield effects of uniconazole on winter rape (*Brassica napus* L.). *Plant Growth Regul.* **1996**, *15*, 69–73.

9.   Ijaz, M.; Honermeier, B. Effect of triazole and strobilurin fungicides on seed yield formation and grain quality of winter rapeseed (*Brassica napus* L.). *Field Crops Res.* **2012**, *130*, 80–86.

10.   Child, R.D.; Evans, D.E.; Allen, J.; Arnold, G.M. Growth responses in oilseed rape (*Brassica napus* L.) to combined applications of the triazole chemicals triapenthenol and tebuconazole and interaction with gibberellin. *Plant Growth Regul.* **1993**, *13*, 203–212.

11.   Jensen, W.B. The origin of the soxhlet extractor. *J. Chem. Educ.* **2007**, *84*, 913–914.

12.   Sepännen, L.; Hiltunen, R. Analysis of fatty acids by gas chromatography and its relevance to research on health and nutrition. *Anal. Chim. Acta* **2002**, *465*, 39–62.

13.   Zhou, W.; Leul, M. Uniconazole-induced alleviation of freezing injury in relation to changes in hormonal balance, enzyme activities and lipid peroxidation in winter rape. *Plant Growth Regul.* **1998**, *26*, 41–47.

14.   Armstrong, E.L.; Nicol, H.I. Reducing height and lodging in rapeseed with growth regulators. *Aust. J. Exp. Agric.* **1991**, *31*, 245–250.

15.   Baylis, A.D.; Wright, T.J. The effects of lodging and a paclobutrazol-chlormequat chloride mixture on the yield and quality of oilseed rape. *Ann. Appl. Biol.* **1990**, *116*, 287–295.

16.   Baylis, A.D.; Hutley-Bull, P.D. The effects of a paclobutrazol based growth regulator on the yield, quality and ease of management of oilseed rape. *Ann. Appl. Biol.* **1991**, *118*, 445–452.

17.   Berding, N.; Hurney, A.P. Flowering and lodging, physiological-based traits affecting cane and sugar yield. *Field Crops Res.* **2005**, *92*, 261–275.

18.   Rolston, R.; Trethewey, J.; Chynoweth, R.; Mccloy, B. Trinexapac-ethyl delays lodging and increases seed yield in perennial ryegrass seed crops. *N. Zeal. J. Agric. Res.* **2010**, *53*, 403–406.

19.   Berry, P.M.; Spink, J.H. Understanding the effect of a triazole with anti-gibberellin activity on the growth and yield of oilseed rape (*Brassica napus*). *J. Agric. Sci.* **2009**, *147*, 273–285.

20.   Rajala, A.; Peltonen-Sainio, P.; Onnela, M.; Jackson, M. Effects of applying stem shortening plant growth regulators to leaves on root elongation by seedlings of wheat, oat and barley: Mediation by ethylene. *J. Plant Growth Regul.* **2002**, *38*, 51–59.

21.   Faraji, A. Quantifying factors determining seed weight in open pollinated and hybrid oilseed rape (*Brassica napus* L.) cultivars. *Crop Breed. J.* **2011**, *1*, 41–54.

22.   Tuncturk, M.; Ciftci, V. Relationships between yield and some yield components in rapeseed (*Brassica napus* ssp. *Oleifera* L.) cultivars by using correlation and path analysis. *Pak. J. Bot.* **2007**, *39*, 81–84.

23.   Mert-türk, F.; Gül, M.K.; Egesel, C.Ö. Nitrogen and fungicide applications against *Erysiphe cruciferarum* affect quality components of oilseed rape. *Mycopathol.* **2008**, *165*, 27–35.

24.   Butkute, B.; Sidlauskas, G.; Brazauskiene, I. Seed yield and quality of winter oilseed rape as affected by nitrogen rates, sowing time and fungicide application. *Commun. Soil Sci. Plant Anal.* **2006**, *37*, 272–274.

25.   Hassan, F.U.; Manaf, A.; Qadir, G.; Basra, S.M.A. Effects of sulphur on seed yield, oil, protein and glucosinolates of canola cultivars. *Int. J. Agric. Biol.* **2007**, *3*, 504–508.

26. Jenkyn, J.F.; Bateman, G.L.; Gutteridge, R.J.; Edwards, S.G. Effect of foliar sprays of azoxystrobin on take-all in wheat. *Ann. Appl. Biol.* **2000**, *137*, 99–106.

27. Prithchard, F.M.; Eagles, A.; Norton, R.M.; Salisbury, P.A.; Nicolas, M. Environmental effects on seed composition of Victorian canola. *Aust. J. Exp. Agric.* **2000**, *40*, 679–685.

28. Dimov, Z.; Möllers, C. Genetic variation for saturated fatty acid content in a collection of European winter oilseed rape material (*Brassica napus*). *Plant Breed.* **2010**, *129*, 82–86.

29. May, W.E.; Hume, D.J.; Hale, B.A. Effects of agronomic practices on free fatty acid levels in the oil of Ontario-grown spring canola. *Can. J. Plant Sci.* **1993**, *74*, 267–274.

12

# Role of Arbuscular Mycorrhizal Fungi in the Nitrogen Uptake of Plants: Current Knowledge and Research Gaps


**Heike Bücking * and Arjun Kafle**

Biology and Microbiology Department, South Dakota State University, Brookings, SD 57007, USA;
E-Mail: arjun.kafle@sdstate.edu

* Author to whom correspondence should be addressed; E-Mail: heike.bucking@sdstate.edu

Academic Editors: Anne Krapp and Bertrand Hirel

**Abstract:** Arbuscular mycorrhizal (AM) fungi play an essential role for the nutrient uptake of the majority of land plants, including many important crop species. The extraradical mycelium of the fungus takes up nutrients from the soil, transfers these nutrients to the intraradical mycelium within the host root, and exchanges the nutrients against carbon from the host across a specialized plant-fungal interface. The contribution of the AM symbiosis to the phosphate nutrition has long been known, but whether AM fungi contribute similarly to the nitrogen nutrition of their host is still controversially discussed. However, there is a growing body of evidence that demonstrates that AM fungi can actively transfer nitrogen to their host, and that the host plant with its carbon supply stimulates this transport, and that the periarbuscular membrane of the host is able to facilitate the active uptake of nitrogen from the mycorrhizal interface. In this review, our current knowledge about nitrogen transport through the fungal hyphae and across the mycorrhizal interface is summarized, and we discuss the regulation of these pathways and major research gaps.

**Keywords:** ammonium; arginine; interface; mycorrhiza; nitrate; nutrient transport; organic nitrogen; periarbuscular membrane; phosphate; plant microbe interaction

# 1. Introduction

The arbuscular mycorrhizal (AM) symbiosis plays a key role for the nutrient uptake of more than 60% of land plants, including many important crop species such as wheat, corn, and soybean [1]. AM fungi are ubiquitous and can account for up to 50% of the microbial biomass in soils [2]. The extraradical mycelium (ERM) of the fungus acts as an extension of the root system and increases the uptake of phosphate (P), nitrogen (N), sulfur, and magnesium but also of trace elements, such as copper and zinc. In addition, AM fungi also provide non-nutritional benefits to their host plant, as they improve the resistance of plants against several abiotic (drought, salinity, heavy metals) and biotic (pathogens, herbivores) stresses [3]. AM fungi therefore play a key role in the survival and fitness of plants and act as "ecosystem engineers" of plant communities [4]. However, the benefits for the plant are not free of charge, and the plant transfers between 4% to 22% of its assimilated carbon (C) to the AM fungus [5]. It has been suggested that the substantial C costs are responsible for the variability of mycorrhizal growth responses that have been described. These growth responses can range from highly beneficial to detrimental and follow a mutualism to parasitism continuum [6–8].

AM fungi belong to the fungal phylum Glomeromycota, and are obligate biotrophs that are unable to complete their life cycle without the carbon supply from their host. This dependency on the host and the observation that many host plants suppress their mycorrhizal colonization particularly under high nutrient supply conditions has led to the overall assumption that the host plant is in control of the symbiosis [9]. However, the long co-evolution of about 400 to 450 million years for both partners in the AM symbiosis also allowed the fungus to improve its strategies to control the nutrient transport to the host despite its obligate biotrophic life cycle [10,11]. It has been suggested that carbon to nutrient exchange in the AM symbiosis is controlled by biological market dynamics and that reciprocal reward mechanisms ensure a "fair trade" between both partners in the AM symbiosis [10].

AM fungi and their plant partners form a complex network of many-to-many interactions; each host plant is colonized by communities of AM fungi and fungal individuals colonize multiple host plants simultaneously and interconnect plants by common mycorrhizal networks (CMNs). These many-to-many interactions allow both partners in the symbiosis to choose among multiple trading partners but also force both partners to compete with other partners for nutrient or carbon resources [12,13]. CMNs play a key role for the long distance transport of nutrients, water, stress chemicals and allelochemicals and allow the interconnected host plants to "communicate" with other plants within their CMN [14–18]. CMNs have also been discussed as a pathway for the transport of N from donor to recipient plants [19]. It is clear that CMNs affect the survival and fitness, behavior and competitiveness of the plants and fungi that are linked via these networks, but our current understanding about how the nutrient or infochemical allocation among plants within a CMN is controlled, is very limited [11,20].

The contribution of AM fungi to the N nutrition of their host plant is still under debate [21], and it has been suggested that higher N contents in mycorrhizal plants are just a consequence of an improved supply with P [22]. However, it is clear that AM fungi transfer N to their host, and plants are able to take up N from the mycorrhizal interface. This review summarizes our current knowledge about N uptake, metabolism, and transport in the AM symbiosis and discusses major research gaps in our understanding.

## 2. Mycorrhizal Roots Have Two Uptake Pathways for Nutrients

Mycorrhizal roots have two uptake pathways for nutrients: the plant uptake pathway (PP) and the mycorrhizal uptake pathway (MP; Figure 1). The PP involves the uptake of nutrients via high- or low affinity uptake transporters in the epidermis or root hairs. Particularly for nutrients with a low mobility in the soil (e.g., P), the uptake via the PP is often limited by the development of depletion zones around the roots. By contrast, the MP involves the uptake by high affinity nutrient transporters in the ERM, followed by the translocation along the hyphae to the intraradical mycelium (IRM) in the root cortex, and the uptake from the mycorrhizal interface by mycorrhiza-inducible plant uptake transporters [23]. However, a plant is simultaneously colonized by communities of AM fungi that can differ in their efficiency with which their MP contributes to the total uptake of nutrients by the plant. The uptake and transport of nutrients via both pathways and their contribution to the nutrient supply of the plant has so far primarily been studied for P [21,23–25], but both pathways also play a role in the N uptake by plants.

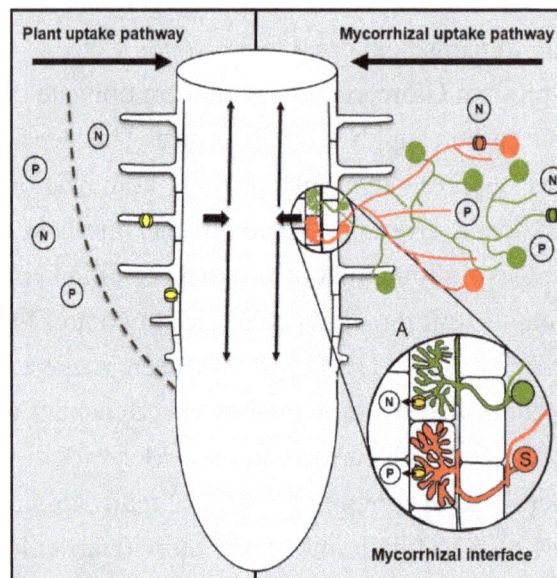

**Figure 1.** Plant uptake and mycorrhizal uptake pathway. Plants can take up nutrients by transporters that are located in epidermis or root hairs (yellow symbols) or via the mycorrhizal uptake pathway that comprises the uptake of nutrients by fungal transporters in the extraradical mycelium (red or green symbols), the transport through the hyphae from the ERM to the IRM (see mycorrhizal interface), and the uptake from the mycorrhizal interface by mycorrhiza-inducible plant transporters in the periarbuscular membrane (orange symbols). Indicated by the red and green fungal structures is the colonization of one host root by multiple fungal species that can differ in their efficiency with which they are able to take up nutrients from the soil and transfer these nutrients to their host.

Ectomycorrhizal roots are enclosed by a more or less densely arranged fungal sheath that can represent a significant apoplastic barrier and restricts the nutrient uptake of ectomycorrhizal roots via the PP [26,27]. However, in contrast to ectomycorrhizal roots, AM roots are structurally unaltered and can theoretically use both pathways for nutrient uptake. Previously, it was generally believed that the PP is not affected by the symbiosis and that both uptake pathways act additively. This led to the overall assumption that the uptake via the MP was negligible in cases in which the mycorrhizal growth responses were neutral and the mycorrhizal plants did not differ in their P contents from non-mycorrhizal plants. However, there is increasing evidence that this view is oversimplified, and that mycorrhizal plants differ in their nutrient acquisition strategy from non-mycorrhizal plants [21,23,28].

Multiple studies have demonstrated that even in non-responsive plants the MP can contribute significantly to the P uptake of the plant [24,29]. Plant P transporters that are involved in the uptake via the PP are down-regulated in response to the AM symbiosis [30,31], while mycorrhiza-specific transporters that are involved in the P uptake from the mycorrhizal interface are induced [32–34]. The suppression of the PP by the AM symbiosis can also lead to negative mycorrhizal growth responses when the reduction in the P uptake via the PP is not fully compensated for by an increase in the P uptake via the MP. This can cause an overall reduction in total P uptake and lead to P deficiency of the plant [21]. AM fungi differ in their capability to suppress the PP [31]. A strong suppression of the PP will shift the ratio between the two uptake pathways towards the MP and will result in a higher mycorrhizal dependency of the host.

Whether the N uptake via the PP is affected by the AM symbiosis is currently unknown. However, similar to what has previously been described for P, there are indications that plant ammonium ($NH_4^+$) or nitrate ($NO_3^-$) transporters are down-regulated [35,36], while mycorrhiza-inducible N transporters are up-regulated by the AM symbiosis [36–39]. Arbuscules are prematurely degraded when the mycorrhiza-inducible P transporter *PT4* is not expressed [40]. Recently, it was shown that the arbuscular life span in *Medicago* plants is not only influenced by *PT4* but also by the mycorrhiza-inducible $NH_4^+$ transporter *AMT2;3* [41]. There is increasing evidence that suggests that the N and P uptake and transport are tightly linked in the AM symbiosis and that the transport of both nutrients controls mycorrhizal functioning (see also below). More studies are necessary to understand the role that both uptake pathways play for the N uptake, and how the activity of both pathways is regulated.

*Contribution of the Arbuscular Mycorrhizal Symbiosis to Plant Nutrition*

While the positive effect of the AM symbiosis on P nutrition is long known [3,21], the contribution of AM fungi to the N nutrition of their host plant is still under debate [23]. The mobility of the inorganic N sources nitrate ($NO_3^-$) and ammonium ($NH_4^+$) in the soil is relatively high, and the rhizosphere is less likely to become N depleted. Unlike for P, where the access of the ERM to P sources beyond the root depletion zone is clearly advantageous, it is expected that the ERM does not increase the access to the N resources in the soil. Recent studies, however, suggest that the $NH_4^+$ uptake system of AM fungi has a five times higher affinity for $NH_4^+$ than typical uptake systems of plants, what would enable the fungus to take up $NH_4^+$ from the soil even under low N supply conditions [42].

In the literature, negative [43], neutral [44] or positive [45,46] effects of the AM symbiosis on N nutrition have been reported (for review [47]), and several authors postulated that an improved N status of AM plants is only the consequence of an improved P nutrition [22]. Hawkins and George [44], for example, reported that the hyphal N supply was not sufficient to sustain an adequate N nutrition of a host plant under N limitation. By contrast, other studies have shown that AM fungi can increase the N acquisition of plants compared to non-mycorrhizal controls that were supplied with additional P [48]. Recent results demonstrated that the ability of AM fungi to improve the N nutrition of their host is relatively widespread within the Glomeromycota, but that there is a high intraspecific diversity between different isolates of one AM fungal morphospecies [46]. The authors reported that out of the 31 fungal isolates that were tested, the colonization with six fungal isolates led on average to a mycorrhizal growth response of 191% and a 2.4-fold increase in the N tissue concentration of *Medicago sativa*. These positive effects were independent from the contribution of these fungal isolates to P supply. There is increasing evidence that a pathway for N through the fungal hyphae to the host plant exists, even if the percentage contribution to total N nutrition of the host plant can vary considerably and is context dependent [23,46]. In AM root organ cultures, 21% of the total N in the roots were taken up by the ERM [49]; and in similar experiments even higher proportions were observed [50]. In maize, 75% of the N in the leaves were taken up by the ERM of an AM fungus [51].

In several plant species mycorrhiza-inducible $NO_3^-$ [52] or $NH_4^+$ [36,38,39,41,53], transporters have been identified that are able to facilitate the uptake of N sources from the mycorrhizal interface. Recent results demonstrated that both P and N are important determinants for the AM symbiosis and that the colonization of the plant host is controlled by feedback mechanisms between both nutrients [12]. For example, both, P and N starvation of the plant induce a nutrient stress transcriptome that is favorable for AM colonization. Under P and N stress, plant defense genes are down-regulated, while genes that are involved in the strigolactone biosynthesis are up-regulated [54]. Strigolactone serves as an important signal for AM fungi in the soil and stimulates hyphal branching during the presymbiotic growth stage [55]. High P availabilities often reduce the AM colonization of the plant, but N starvation triggers a signal that promotes AM colonization and reverses the inhibitory effects of high P availabilities on AM colonization [41,56]. However, the recovery of AM colonization did not lead to increased N levels in these plants [56]. When the fungal P transport to *Medicago truncatula* plants is inhibited by the suppression of the mycorrhiza-inducible plant P transporter *MtPt4*, arbuscules are prematurely degraded, and the symbiosis is not maintained by the host [40]. However, AM fungi can escape premature degeneration of their arbuscules in *MtPt4* mutants under N starvation [41,57], and it has been demonstrated that the expression of mycorrhiza-induced plant P transporters are not critical for the AM symbiosis as long as the plant is grown under low N conditions [41]. All these results clearly demonstrate that N is transferred through the fungal hyphae to the host, and that N transport to the host may play a critical role for the maintenance and the nutrient efficiency of the AM symbiosis.

## 3. Nitrogen Uptake by Arbuscular Mycorrhizal Fungi

### 3.1. Uptake of Inorganic N Sources

Nitrate ($NO_3^-$) is the dominant form of N that is available to plants and fungi in most agricultural soils, whereas ammonium ($NH_4^+$) predominates in many undisturbed or very acidic soils, where $NO_3^-$ can be almost entirely absent. The ERM of AM fungi can take up $NH_4^+$ [58–62] and $NO_3^-$ [62,63], but $NH_4^+$ is generally preferred, because it is energetically more efficient than $NO_3^-$ [49,50,62,64] (see also Table 1). Based on pH changes induced by the ERM, when hyphae were supplied with $NO_3^-$ or $NH_4^+$, it has been hypothesized that the $NO_3^-$ uptake by the hyphae is active and coupled to an $H^+$-symport mechanism, while $NH_4^+$ is taken up by an antiport mechanism with a net $H^+$ efflux [65,66]. Similar to plants and other fungi, AM fungi have high-affinity and low-affinity uptake systems for $NH_4^+$. The low

$K_m$ value of $2.53 \pm 0.25$ µM for the high affinity uptake system of *Rhizophagus irregularis* demonstrates that the fungus is able to take up $NH_4^+$ from soils with very low concentrations [42]. $K_m$ values of high affinity $NH_4^+$ uptake systems of plants are typically higher than the estimated $K_m$ value for the fungal $NH_4^+$ uptake system [42,67,68].

**Table 1.** Demonstrated metabolic pathways of nitrogen (N) in arbuscular mycorrhizal (AM) fungi. Shown are the fungal species or the tissues that were tested. Fungal species were renamed according to the reclassification of the Glomeromycota [69]. Species abbreviations and authorities: *Acla—Acaulospora laevis* (Gerd. & Trappe), *Clet—Claroideoglomus etunicatum* (W. N. Becker & Gerd.; previously *Glomus etunicatum)*, *Fumo—Funneliformis mosseae* (T. H. Nicolson & Gerd., previously *Glomus mosseae*), *Gima—Gigaspora margarita* (W. N. Becker & I. R. Hall), *Giro—Gigaspora rosea* (T. H. Nicolson & N. C. Schenck), *Glfa—Glomus fasciculatum* (Gerd. & Trappe), *Glma—Glomus macrocarpus* (Tul. & C. Tul.), *Glve—Glomus versiforme* (P. Karst., S. M. Berch), *Rhcl—Rhizophagus clarus* (T. H. Nicolson & N. C. Schenck, previously *Glomus clarum*), *Rhir—Rhizophagus irregularis* (N.C. Schenck & G. S. Sm.), *Rhma—Rhizophagus manihotis* (R. H. Howeler, Sieverd. & N. C. Schenck).

| Pathway/Enzyme | Fungal Species | Tissue | | | | References |
|---|---|---|---|---|---|---|
| | | Spore | ERM | AM | IRM | |
| **Inorganic N uptake** | | | | | | |
| $NH_4^+$ uptake | *Fumo, Rhir, Acla, Gima* | + | + | nd | nd | [50,58,60–62,64,70,71] |
| $NH_4^+$ transporter | *Rhir* | + | + | + | + | [53,72,73] |
| $NO_3^-$ uptake | *Glfa, Rhir, Fumo* | + | + | nd | nd | [62–65,70,71] |
| $NO_3^-$ transporter | *Rhir* | + | + | + | + | [71,72,74–76] |
| Nitrate permease | *Rhir* | + | + | + | + | [76] |
| **Organic N uptake** | | | | | | |
| | *Acla, Rhir, Gima* | nd | +/− | nd | nd | [60] |
| | *Fumo* | nd | + | + | nd | [45,58] |
| | | nd | − | nd | nd | [77] |
| Amino acid transporter | | + | + | + | + | [72] |
| Amino acid permease | *Fumo* | − | + | + | nd | [78] |
| Peptide transporter | *Rhir* | nd | + | + | + | [79] |
| Alanine | *Rhir* | − | nd | nd | nd | [71] |
| Arginine | *Rhir* | − | + | nd | nd | [50,71] |
| Cysteine | *Rhir* | − | + | nd | nd | [80] |
| Glycine | *Rhir, Fumo, un* | + | + | nd | nd | [62,71] |
| Glutamate | *Fumo, Rhir* | + | + | nd | nd | [62,71] |
| Glutamine | *Fumo, Rhir* | + | + | nd | nd | [71,81] |
| Methionine | *Rhir* | − | + | nd | nd | [80] |
| Ornithine | *Rhir* | + | nd | nd | nd | [71] |
| Urea [4] | *Gima* | + | + | nd | nd | [71,82] |

**Table 1.** *Cont.*

| Pathway/Enzyme | Fungal Species | Tissue | | | | References |
|---|---|---|---|---|---|---|
| | | Spore | ERM | AM | IRM | |
| **Nitrate reduction** | | | | | | |
| Nitrate reductase[1] | *Glfa, Rhir, Fumo, Glma, Glsp* | + | nd | + | nd | [72,83–86] |
| Nitrate reductase (NADH) | *Un, Fumo* | + | + | + | nd | [52,87–89] |
| Nitrate reductase (NADPH) | *Rhir* | nd | + | + | + | [52,90] |
| Nitrite reductase | *Rhir* | nd | + | + | + | [72,84] |
| **GDH pathway** | | | | | | |
| GDH[1] | *Clet* | nd | nd | + | + | [91] |
| GDH (NADH) | *Fumo, Rhir* | + | + | + | + | [71,72,92] |
| GDH (NADPH) | Glfa | nd | nd | +/− | nd | [83,93] |
| **GS/GOGAT pathway** | | | | | | |
| Glutamine synthetase | *Glfa, Fumo, Rhir* | + | + | + | + | [71,72,74,75,81,83,84,93,94] |
| Glutamate synthase (NADH) | Rhir | + | + | + | + | [64,72,74,75,84,94] |
| Glutamate synthase (NADPH) | Glfa, Rhir | nd | nd | + | + | [72,83] |
| **Amino acid biosynthesis** | | | | | | |
| Transaminases | *Glfa* | nd | nd | + | nd | [83,95] |
| Asparagine synthase | *Rhir* | + | + | + | + | [72] |
| **Arginine biosynthesis** | | | | | | |
| Carbamoyl-P synthase | *Rhir* | + | + | + | + | [72,74,75] |
| Argininosuccinate synthase | *Rhir* | + | + | + | + | [72,74,75] |
| Argininosuccinate lyase | *Rhir* | + | + | + | + | [72,74,75] |
| **Arginine breakdown** | | | | | | |
| Arginase | *Rhir* | + | + | + | + | [72,74,75] |
| Urease | *Rhcl, Rhir* | + | + | + | + | [72,74,75] |
| Urease accessory protein | *Rhir* | + | + | + | + | [71,72,75,94] |
| Ornithine aminotransferase | *Rhir* | + | + | + | + | [71,72,74,75,94] |
| **Polyamine biosynthesis** | | | | | | |
| Ornithine decarboxylase | *Rhir* | + | + | + | + | [72,74,75] |
| **N uptake from interface** | **Plant species** | | | | | |
| Plant NH4+ transporter | *Medicago truncatula* | na | na | + | na | [39,41] |
| | *Lotus japonicus* | na | na | + | na | [38] |
| | *Glycine max* | na | na | + | na | [36] |
| | *Sorghum bicolor* | na | na | + | na | [37] |

Abbreviations: AM—arbuscular mycorrhizal roots; ERM—extraradical mycelium; IRM—intraradical mycelium; na—not applicable; nd—not determined.

Despite the fungal preference for $NH_4^+$, Hawkins and George [70] reported that when $NH_4^+$ was the sole N source for mycorrhizal plants, root and shoot biomass, hyphal length densities and N transport via the hyphae to the plant were lower than after $NO_3^-$ supply. Valentine *et al.* [96] also reported, that $NH_4^+$ fed AM roots had lower numbers of vesicles and arbuscules than $NO_3^-$ supplied roots. An excess of $NH_4^+$ as sole N source is often considered to be toxic for plants, and inhibits root growth [97]. $NH_4^+$ assimilation occurs in the root, while the assimilation of $NO_3^-$ is predominantly foliar [98,99]. When $NH_4^+$ is the sole N source, the assimilation of $NH_4^+$ could increase the consumption of C skeletons in the root, and reduce the carbon availability for the fungus. Hawkins and George [70] found that $NH_4^+$ reduced the hyphal length in the soil, but not the number of arbuscules, and assumed that high concentrations of $NH_4^+$ could also have a direct deleterious effect on the ERM.

Transcriptome studies revealed the expression of several fungal $NH_4^+$ and $NO_3^-$ transporters in spores, ERM, and IRM [72]. The expression of *GintAMT1*, an $NH_4^+$ transporter of the AM fungus *Rhizophagus irregularis* (previously *Glomus intraradices*) is induced by low additions of $NH_4^+$ to the medium (in the presence of relatively high concentrations of $NO_3^-$ in the medium) but suppressed under high $NH_4^+$ supply. This suggests that the expression of this transporter is substrate inducible, and is regulated by the $NH_4^+$ supply and fungal $NH_4^+$ status [73]. By contrast, the $NH_4^+$ transporter *GintAMT2* is constitutively expressed in the ERM under N limiting conditions, and transitory induced when different N sources are added to the N deficient ERM [53]. The differential localization of high transcript levels of these transporters in colonized roots suggests that both transporters may differ in their role for N uptake and transport. The high expression levels of *GintAMT1* in the ERM indicate that this transporter could be primarily involved in the $NH_4^+$ acquisition of fungal hyphae from the soil. By contrast, *GintAMT2* is particularly expressed in the IRM, suggesting that this transporter could play a role in the re-uptake of $NH_4^+$ by the fungus from the symbiotic interface [53].

An exogenous supply of $NO_3^-$ stimulates the expression of a fungal $NO_3^-$ transporter in the ERM of *Rhizophagus irregularis* (also known as *Rhizophagus intraradices* or previously *Glomus intraradices*) [74]. However, the expression of this $NO_3^-$ transporter is repressed by an increase in the internal levels of $NH_4^+$ or of a downstream metabolite, such as glutamine [75]. Nitrate transporters in many organisms are suppressed when more preferred N sources, such as $NH_4^+$ become available. This process is known as N catabolite repression, and this regulatory system also controls the expression of $NH_4^+$ transporters in other fungi [100]. In ectomycorrhizal fungi, the expression of high affinity $NH_4^+$ transporters is suppressed by high levels of intracellular glutamine [101], and $NO_3^-$ transporter proteins in the yeast *Hansenula polymorpha* are degraded in response to glutamine [102]. GATA transcription factors have been shown to be involved in N catabolite repression and in the promoter sequence of the $NH_4^+$ transporter *GintAMT2* two GATA core sequences have been identified [53]. The expression profiles of AM fungal $NO_3^-$ and $NH_4^+$ transporters under different N availabilities suggest that N catabolite repression could also operate in AM fungi and could control the N uptake by AM fungi from the soil.

## 3.2. Uptake of Organic N by Hyphae

Organic N can represent in many soils a significant proportion of total soil N [103], but the obligate biotrophic life cycle of AM fungi has led to the overall assumption that AM fungi are unable to utilize organic N sources. Several studies, however, have demonstrated that the hyphae of AM fungi grow into organic patches of varying complexity and transfer $^{15}N$ from these organic patches to their host plant [104,105], and that this can lead to higher plant N contents [106]. When the fungus had access to organic patches that were labelled with $^{15}N$ and $^{13}C$, the fungal ERM got only enriched with $^{15}N$, but not with $^{13}C$. This confirms that AM fungi do not have saprophytic capabilities and that the fungus acquires $^{15}N$ from these organic patches likely as a decomposition product [105]. However, even if AM fungi themselves do not act as decomposers, AM fungi accelerate the N mineralization from organic matter [107] and affect the carbon flow through soil microbial communities during decomposition [108].

Plant roots release fixed carbon into their rhizosphere, and this carbon flux into the root rhizosphere acts as an important trigger for decomposition processes and nutrient cycling in soils [109]. AM fungi form an extensive network of hyphae in soils (ERM), and the mycorrhizosphere, the large interface between the fungal hyphae and the soil, could play a similar role than the rhizosphere for decomposition and nutrient cycling in soils. The ERM of the AM fungus acts as a conduit between decomposing microbial communities and the host plant, and provides decomposers with plant-derived carbon inputs and transfers decomposition products to the plant. The mycorrhizosphere represents in soils an important ecological niche and provides nutritionally favorable conditions for diverse microbial communities, and it has been suggested that the presence of AM fungal hyphae plays an important role in the bacterial community assembly during decomposition [110].

Compared to ectomycorrhizal fungi, the capability of AM fungi to utilize organic N sources is considered to be relatively small but also AM fungi are able to take up organic N sources from the soil, such as amino acids (Table 1). Free amino acids can represent an important N source in soils, and AM fungi can take up several amino acids, e.g., aspartic acid, serine [111], glycine, glutamic acid [62,112], glutamine [81], cysteine or methionine [80]. Some amino acids are also taken up by germinating spores during the presymbiotic growth stage of the fungus [71]. Transcriptome studies have shown that fungal amino acid transporters are expressed in the ERM [72]. Cappellazzo and co-authors characterized an amino acid permease of the AM fungus *Funnelliformis mosseae* (previously *Glomus mosseae*) and found that this amino acid transporter binds preferably neutral, nonpolar and hydrophobic amino acids [78]. While some amino acids such as glycine, serine and glutamine competed strongly with proline for this transporter, positively charged amino acids such as arginine, histidine and lysine were only poor competitors. The mycorrhizal colonization of *Sorghum bicolor* led to an increase in the uptake of phenylalanine, methionine, asparagine, tryptophane, and cysteine, but also increased the uptake of the charged amino acids arginine, lysine, and histidine [113]. Other studies demonstrated that arginine can be taken up by the ERM and that arginine is not metabolized during its transport from the ERM to the IRM [50]. This suggests that multiple transporters are involved in the amino acid uptake by the ERM.

Recently, the putative dipeptide transporter *RiPTR2* has been identified in the AM fungus *Rhizophagus irregularis* [79]. Short peptides can represent a greater proportion of N in soils than free amino acids, and this transporter facilitated in complementation assays the uptake of several dipeptides. This suggests that this transporter could play a role in the fungal uptake of small peptides from the soil. However, the transcript levels of this transporter were higher in the IRM than in the ERM, suggesting that *RiPTR2* could also play a role in the reabsorption of peptides from the interfacial apoplast [79].

## 4. Nitrogen Assimilation and Transport by Arbuscular Mycorrhizal Fungi

### 4.1. Nitrate Reduction in Arbuscular Mycorrhizal Fungi

After $NO_3^-$ is taken up by plants or fungi, $NO_3^-$ is first reduced to nitrite by a nitrate reductase. Both NADH and NADPH can act as reductant for nitrite formation [52], but, in AM fungi, nitrate reductase activity is mainly driven by NADPH [90]. Nitrate reductase activity in both roots [44,83,84,87] and shoots [83,87,95] of AM plants is generally higher than in non-mycorrhizal control plants (for more information, see also Table 1). In nonmycorrhizal plants, $NO_3^-$ reduction primarily takes place in the leaves, while in mycorrhizal plants $NO_3^-$ is predominantly reduced in the roots [88,90]. A gene that encodes a fungal nitrate reductase is expressed in spores, ERM and IRM, but the transcript levels are particularly high in the IRM. This could indicate that $NO_3^-$ that is not directly assimilated in the ERM can also be reduced in the fungal tissue within the host root [72,85,90].

The second step in N assimilation is the conversion of nitrite into $NH_4^+$ by nitrite reductase. A gene encoding a putative fungal nitrite reductase of the AM fungus *Rhizophagus irregularis* shows particularly high expression levels in spores and in the IRM [72]. This is consistent with the high transcript levels of a fungal nitrate reductase in the IRM of AM roots [72]. The expression of *tbnir1*, the nitrite reductase of the ectomycorrhizal (ECM) fungus *Tuber borchii*, is induced by $NO_3^-$ but suppressed when more preferred N sources, such as $NH_4^+$ or glutamine become available [114]. It has been demonstrated in ECM roots that the activity of the fungal nitrite reductase controls the expression of the plant nitrite reductase gene. When the plant is colonized with a fungal wild type strain, the expression of the plant nitrite reductase is repressed, but the expression increased in roots that were colonized with nitrate reductase deficient fungal strains [115]. This is consistent with the finding that, in mycorrhizal roots, the transcript levels of plant nitrate reductase are lower than in non-mycorrhizal control plants [52,90] and suggests that the expression of the plant nitrite reductase is suppressed by the transfer of reduced N compounds from the fungus to the host [115].

## 4.2. Nitrogen Assimilation into Amino Acids in Arbuscular Mycorrhizal Fungi

Since excess N in its reduced form $NH_4^+$ is toxic for cells due to its ability to uncouple respiration, $NH_4^+$ is rapidly assimilated into amino acids [116]. Two pathways can be involved in the assimilation of $NH_4^+$ by fungi: the NAD(P)-glutamate dehydrogenase (GDH) or the glutamine synthetase—glutamate synthase (GS-GOGAT) pathway. GDH uses $NH_4^+$ and 2-oxoglutarate to produce glutamate (Glu), while in the GS-GOGAT pathway, first glutamine (Gln) is produced from Glu and $NH_4^+$ by glutamine synthetase, and then Gln and 2-oxoglutarate are converted by glutamate synthase (also known as glutamine oxoglutarate aminotransferase) into two molecules of Glu. In the cytoplasm of ascomycetes and basidiomycetes activities of NAD and NADP-dependent GDH can be detected [117]. However, Smith *et al.* [93] found no increased GDH activity in AM roots, and also in the ERM of *Rhizophagus irregularis* no GDH activity could be detected [49]. A putative NAD-dependent GDH gene of *R. irregularis* is down-regulated in the ERM in response to either $NO_3^-$ or $NH_4^+$ supply, suggesting that GDH plays in AM fungi a more catabolic role rather than a role in N assimilation [49,71,83,94].

$NH_4^+$ is predominately assimilated via the GS-GOGAT pathway in AM fungi [64,81,83,94]. GS and GOGAT activities have been shown to be significantly higher in roots and shoots of AM plants [83,84,93,95] (see also Table 1), and the decrease in the levels of free amino acids in the ERM of the AM fungus *R. irregularis* after incubation with the GOGAT inhibitor albizzine confirms the activity of a functional GS-GOGAT pathway in AM fungi [64]. Tian and co-authors [74] identified two different functional GS isoforms of *R. irregularis*, *GiGS1* and *GiGS2*. *GiGS1* has a lower $K_m$ than *GiGS2* and is constitutively expressed at high levels in the ERM, while *GiGS2* is strongly induced by an addition of $NO_3^-$ to the ERM. This suggests that *GiGS1* is the main functional enzyme for N assimilation at low N availabilities, and that *GiGS2* may play a more significant role for N assimilation under high N supply conditions [74]. Breuninger and co-authors [81] also found that the fungal GS genes of *Funneliformis mosseae* (*GmGln1*) and *R. irregularis* (*GiGln1*) are constitutively expressed but that the GS activities in the ERM are modulated in response to different N availabilities. Based on this observation, the authors concluded that the fungal GS activity is not controlled on a transcriptional level, but is subjected to post-transcriptional regulation [81]. However, the GS transcript that was examined by these authors has a closer sequence similarity to *GiGS1* (see above), and only *GiGS2* has been shown to be strongly up-regulated after $NO_3^-$ supply [74].

Consistent with the GS-GOGAT pathway, Gln becomes highly labelled when $^{15}NH_4^+$ is supplied to AM fungi, and represents one of the major N sinks [50,71,83,95,118]. Gln plays a central role in N metabolism (1) as key N donor; (2) as precursor of many essential metabolites such as nucleic acids, amino sugars, and other amino acids, such as histidine, tyrosine and asparagine; and (3) as key effector for N metabolite repression and regulator of genes involved in N metabolism [67,101,102,119,120]. Due to these important functions, the free levels of Gln in AM fungi are tightly controlled [71]. In addition to Gln, Glu, asparagine (Asn), aspartate (Asp) and alanine (Ala) are abundant free amino acids in germinating spores [71], in the ERM [64] or in AM roots [50,95,118]. Ornithine, serine and glycine are also detectable but in much lower concentrations [64].

Arginine levels (Arg) were not determined in several of the earlier investigations [64,84,95], but more recent studies demonstrate that Arg is the most abundant free amino acid and can represent more than 90% of the total free amino acids in the ERM [118]. Arg levels of up to 200 nM·mg$^{-1}$ dry weight have been reported in the ERM [50]. Due to its low carbon-to-nitrogen ratio of 6:4, Arg plays an important role for N storage and N transfer from the ERM to the IRM [50,121]. For example, Arg serves as a N storage molecule in quiescent spores and its catabolic breakdown during spore germination provides the N and C skeletons for the biosynthesis of other amino acids or proteins for the presymbiotic growth of the AM fungus [71].

### 4.3. Nitrogen Transport from the Extraradical Mycelium to the Intraradical Mycelium

The ERM takes up N from the soil in a significant distance from the root, and transfers N from the ERM to the IRM. The transport of N through the hyphae of the AM symbiosis can be very fast and flux rates similar to those of P have been observed [121,122]. Fungal vacuoles often contain polyphosphates (polyP) and basic amino acids in equimolar concentrations [123–125], and it has been suggested that N could move in the form of Arg with fungal polyP from the ERM to the IRM [50,94,121]. PolyP are negatively charged polyanions and the basic amino acid Arg could serve together with other cations such as $K^+$ and $Mg^{2+}$ as counter charge and contribute to the required charge balance [126]. Studies on the N transport in the AM symbiosis of *Agropyron repens*, however, also suggest that in addition to Arg, other amino acids such as Gln or Glu could be involved in the translocation of N from the ERM to the IRM [127].

Gene expression studies are consistent with the biosynthesis of Arg in the ERM. Shortly after $NO_3^-$ supply, the transcript levels of a fungal carbamoyl-phosphate synthetase (CPS), argininosuccinate synthase (ASS), and argininosuccinate lyase (AL) are induced in the ERM (Figure 2). All of these enzymes are involved in Arg biosynthesis; CPS catalyzes the formation of carbamoylphosphate from $CO_2$, ATP, and $NH_3$, which is converted together with ornithine to citrulline and $P_i$ by an ornithine transcarbamoylase (OTC). Citrulline and aspartate are converted to argininosuccinate (AS) by ASS, and AL converts argininosuccinate to fumarate and Arg. In contrast, in the IRM a fungal arginase (CAR1) and urease (URE) are up-regulated that are involved in the catabolic breakdown of Arg [74]. The biosynthesis of Arg in the ERM followed by the subsequent breakdown of Arg in the IRM are spatially separated but synchronized processes, and confirm the function of the anabolic (ERM) and catabolic part (IRM) of the urea cycle in the AM symbiosis (Figure 2). The synchronization of these processes suggests that Arg plays an important role in the N translocation from the ERM to the IRM [50,74,121].

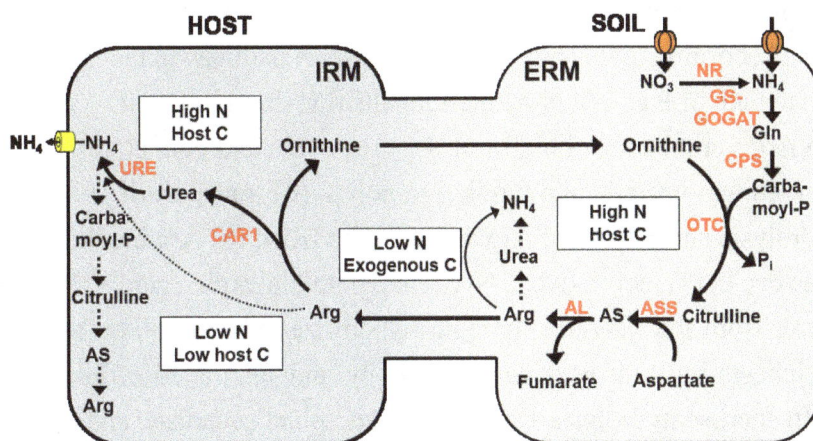

**Figure 2.** Function and regulation of the anabolic and catabolic arm of the urea cycle in the ERM or the IRM, respectively. The activities of the following enzymes are indicated in red: nitrate reductase/nitrite reductase (NR), glutamine synthetase—glutamate synthase (GS-GOGAT), carbamoyl-phosphate synthetase (CPS), ornithine transcarbamoylase (OTC), argininosuccinate synthase (ASS), argininosuccinate lyase (AL), arginase (CAR1), and urease (URE). Favored pathways are indicated by bold lines, suppressed pathways by dotted lines. White boxes provide additional information about the putative regulation of these pathways.

The C supply of the host plant plays a significant role in the regulation of these processes [12,75]. An increase in the C supply from the host plant led to an up-regulation of genes that are involved in Arg biosynthesis (AL, ASS, CPS), but to a down-regulation of a fungal urease (URE) in the ERM [75]. This will lead to increasing levels of Arg in the ERM (stimulation in biosynthesis and reduced catabolic breakdown) and stimulate the transfer of Arg from the ERM to the IRM. In contrast, in the IRM, an increase in the fungal arginase and urease activity and a down-regulation of genes involved in Arg biosynthesis were observed. This will favor the catabolic breakdown of Arg but prevent the re-assimilation of the released $NH_4^+$ in the IRM [75]. This demonstrates that the host plant with its C supply is able to regulate fungal gene expression and to trigger fungal N transport. However, when the fungus had access to an exogenous C source (acetate supply to the ERM), the transcript levels of genes involved in N assimilation and Arg biosynthesis were reduced, but the expression of a fungal arginase was induced in the ERM. This could indicate that the Arg transport from the ERM to the IRM is suppressed when the fungus has access to an exogenous carbon source and is not solely dependent on its host for its carbon supply [75].

The active hydrolysis of polyP will release $P_i$ and Arg (for subsequent breakdown into $NH_3$ and $NH_4^+$ by the catabolic arm of the urea cycle) into the fungal cytoplasm of the IRM and will facilitate the efflux of P and of N into the mycorrhizal interface. AM and ECM fungi regulate the nutrient transport to the host by the accumulation or remobilization of polyP, and it has been shown that the carbon supply of the host plant triggers polyP hydrolysis [10,75,128,129]. The regulation of polyP and Arg biosynthesis in the ERM and of polyP and Arg breakdown in the IRM would allow the fungus to control its P and N transport across the mycorrhizal interface.

### 4.4. Nitrogen Transport across the Mycorrhizal Interface

In Figure 3, a model of the current N transport pathway and its associations to the P and C flux in the AM symbiosis is shown. According to this model, the N transport pathway in the AM symbiosis includes the following steps: (1) uptake of P and of N by the fungal ERM through $P_i$, $NO_3^-$ or $NH_4^+$ transporters; (2) N assimilation into Arg via the anabolic arm of the urea cycle and conversion of $P_i$ into polyp; (3) transport of Arg into the fungal vacuole and binding to polyp; (4) transport of polyP from the ERM to the IRM; (5) polyP hydrolysis and release of Arg and $P_i$ in the IRM; (6) Arg breakdown to $NH_4^+$ via the catabolic arm of the urea cycle; (7) facilitated $P_i$, $NH_4^+$, and potential amino acid efflux into the interfacial apoplast; (8) plant uptake from the interface through mycorrhiza-inducible $P_i$ or $NH_4^+$ transporters; (9) stimulation in photosynthesis by improved nutrient supply and facilitated efflux of sucrose into the interfacial apoplast; (10) sucrose hydrolysis via an apoplastic plant invertase, and uptake of hexoses by the AM fungus through fungal monosaccharide transporters.

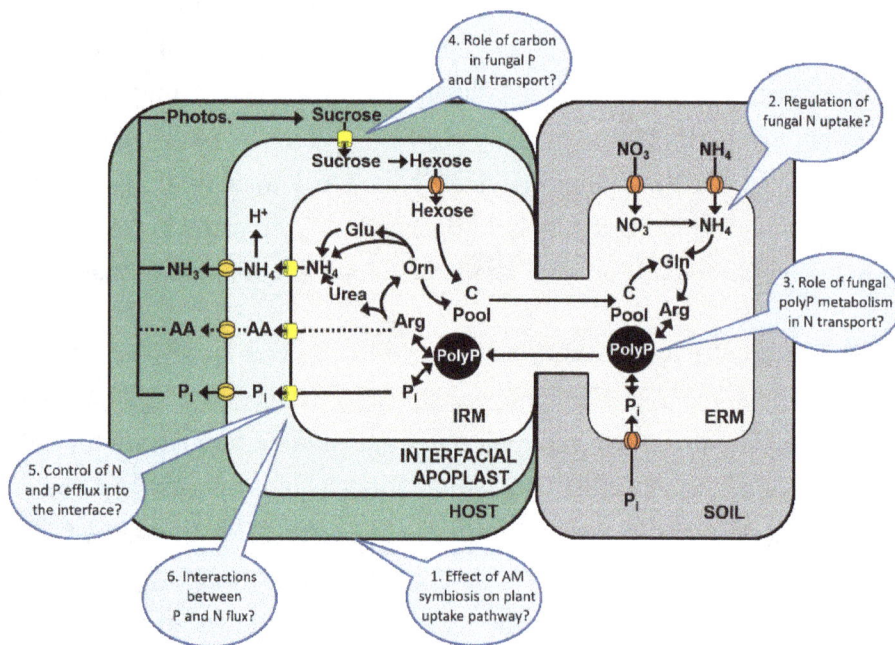

**Figure 3.** Model of the N and P transport in the AM symbiosis. The model is based on previous models presented in the literature [50,74,75,94]. The model shows the nutrient uptake by the fungal ERM through $P_i$, $NO_3^-$ or $NH_4^+$ transporters (**red**), N assimilation into Arg and the conversion of $P_i$ into polyP, the transport of polyP from the ERM to the IRM, polyP hydrolysis and the release of Arg and $P_i$ in the IRM, Arg breakdown to $NH_4^+$ via the catabolic arm of the urea cycle, and the $P_i$, $NH_4^+$, and potential amino acid efflux (**yellow**) into the interfacial apoplast, and the plant uptake from the interface through mycorrhiza-inducible $P_i$ or $NH_4^+$ transporters. The blue ovals indicate gaps in our current understanding of these processes.

Previous models suggested that N is transferred from the fungus to the host in form of amino acids, but this would lead to a substantial reflux of C from the obligately biotrophic fungus to the host. The uptake of organic N from the mycorrhizal interface would also require the presence of e.g., amino acid transporters in the periarbuscular membrane that could facilitate an uptake of organic N from the mycorrhizal interface. In *Lotus japonicus* a mycorrhiza-inducible plant amino acid transporter (*LjLHT1.2*) has been identified that shows a high expression in arbusculated cells but also in non-colonized cells of the root cortex. The expression of this amino acid transporter in the periarbuscular membrane could indicate that this transporter is involved in the uptake of amino acids from the mycorrhizal interface [130]. However, the expression of this transporter in non-colonized cells suggests that the transporter could also play a role for the re-absorption of amino acids by the cortical cells.

Based on the elucidation of the N transport pathway [74,75,94] (see above and Figure 3), and the identification of several mycorrhiza-inducible $NH_4^+$ transporters in the periarbuscular membrane [36–38], newer models of N transport in the AM symbiosis now conclude that $NH_4^+$ is the main form in which N is transferred into the mycorrhizal interface and is taken up by the host. Studies after labeling with $^{13}C$ also indicate that the resulting $CO_2$ or $HCO_3^-$ from the breakdown of Arg is not transferred to the plant and remains within the fungal compartment [50,94].

The release of nutrients from free living fungi is normally regarded as slow, and membrane transport processes are generally favoring fungal re-absorption [131]. Therefore, it is likely that conditions at the mycorrhizal interface exist that facilitate the efflux of nutrients from the fungus or reduce the level of competing fungal uptake systems. It is currently unknown how the N transport from the fungal hyphae into the mycorrhizal interface is facilitated, but several different mechanisms have been discussed. For example, in arbusculated cells and in the ERM of maize roots, two functional fungal aquaporins have been identified that are particularly expressed under drought stress [132]. Aquaporins are integral membrane channels that facilitate the concentration gradient-driven water transport through the plasma membrane. However, in addition to their function as water channels, aquaporins have also been shown to facilitate the diffusion across membranes of low molecular weight neutral solutes such as glycerol, ammonia, and carbon dioxide [133]. This has led to the hypothesis that aquaporins could play a role in the leakage of $NH_4^+$ into the interface [134]. The continuous breakdown of Arg in the IRM and the re-assimilation of $NH_4^+$ in the root cells will lead to a concentration gradient that could facilitate the movement of $NH_4^+$ into the mycorrhizal interface, and reduce the re-absorption of $NH_4^+$ by the AM fungus.

Plant $NH_4^+$ transporters in the periarbuscular membrane compete with fungal transporters for the $NH_4^+$ that has been released into the interfacial apoplast. The expression of fungal $NH_4^+$ transporters in arbusculated cells indicates that the fungus is also able to re-absorb released $NH_4^+$, and the fungus could modulate the amount of N that is delivered to the host by the expression of these transporters in the IRM [53]. Several plant $NH_4^+$ transporters have been identified that are induced in arbusculated cells during the AM symbiosis and that have been implicated in the $NH_4^+$ uptake from the mycorrhizal interface [37–39,41]. One of these mycorrhiza-inducible $NH_4^+$ transporters has been identified in *Lotus japonicus* (*LjAMT2;2*) [38]. Interestingly, when *LjAMT2;2* is expressed in oocytes of *Xenopus laevis* the uptake of $NH_4^+$ does not result in a flow of current. Based on this observation, the authors concluded that the transporter binds $NH_4^+$ externally but transfers uncharged $NH_3$ across the periarbuscular membrane. It is unlikely that this high-affinity transporter could recruit $NH_3$ from the interfacial apoplast, because $NH_4^+$ should be the dominant form of $NH_3/NH_4^+$ in the interfacial apoplast due to the assumed low pH conditions in this compartment (*pKa* of protonation 9.25) [38].

Recently, two plant $NH_4^+$ transporters of *Medicago truncatula* (*AMT2;3* and *AMT2;4*) have been identified that play a role for the arbuscular life span [41]. Arbuscules are prematurely degraded in mutants in which *pt4*, the mycorrhiza-inducible plant P transporter that is critical for the uptake of P from the mycorrhizal interface, is not expressed [40,57]. This premature degeneration of arbuscules is suppressed by an expression of *AMT2;3* when the plant is grown under N stress, but not by *AMT2;4*. However, only *AMT2;4* has been shown to be a functional $NH_4^+$ transporter and was able to support in complementation assays the growth of a yeast $NH_4^+$ transporter mutant [41]. This suggests that *AMT2;3* and *AMT2;4* differ in their function, and that *AMT2;3* could play more a sensing or signaling function that is not present in other AMT transporters [41]. It has been hypothesized that some mycorrhiza-inducible nutrient transporters that are localized in the periarbuscular membrane could act as transceptors, transporter-like proteins with a receptor function [135].

## 5. Conclusions

In recent years, significant progress has been made in the elucidation of the N transport pathway and the identification of fungal or plant genes that are involved in the N transport from the ERM via the IRM to the host. In this review, we summarized our current understanding of N transport and its regulation in the AM symbiosis that is primarily based on $^{13}C$ and $^{15}N$ labeling studies and more recent data on fungal and plant gene expression. However, despite the progress that has been made, many critical questions are still unanswered and should be addressed in future studies. Some of these major research gaps are also indicated in Figure 3, and are listed below.

### 5.1. Effect of the Arbuscular Mycorrhizal Symbiosis on the Plant Uptake Pathway

There is increasing evidence that the AM symbiosis changes the nutrient uptake strategy of the plant, and that P transporters of the plant uptake pathway are downregulated in response to the AM symbiosis [31]. It has been shown that even in non-responsive plants, the uptake via the mycorrhizal pathway can dominate the overall P uptake by the plant [29]. Whether and how the plant uptake pathway for N is affected by the colonization with AM fungi is currently unknown, but there is evidence that some plant N transporters are downregulated in colonized roots [35,36].

### 5.2. Regulation of Fungal N Uptake

Fungal N uptake systems have a high affinity for the uptake of N from the soil [42], but how the expression of these transporters is regulated is largely unknown. The transcript levels of some of these transporters are substrate inducible, and regulated by the $NH_4^+$ supply and/or fungal $NH_4^+$ status [73]. However, our current understanding of the N transport pathways in the AM symbiosis, is still limited by the fact that these transport pathways were only studied in a small number of fungal species (see also Table 1), and by the lack of molecular tools in AM research. There is evidence that there is a high intraspecific diversity in the efficiency with which fungi are able to improve the N nutrition of their host [46]. So far, the transcriptome and genome of only one AM fungal species is known [72,76], and attempts to develop fungal mutants for research studies were unsuccessful.

## 5.3. Role of Fungal PolyP Metabolism in N Transport

PolyP play a significant role for the P storage in fungal hyphae, for the long distance transport of P and other nutrients from the ERM to the IRM, and in the regulation of the nutrient transport to the host [10,126,129]. There are indications that the polyP metabolism may play a role in the efficiency with which AM fungi are able to provide nutritional benefits to their host [46]. The processes that are involved in fungal $P_i$ homeostasis and polyP biosynthesis or remobilization have so far mainly been studied in yeasts (e.g., [136]), but the regulatory mechanisms that control the rates of polyP turnover in mycorrhizal fungal hyphae are largely unknown.

## 5.4. Role of Carbon in Fungal P and N Transport

Some studies have demonstrated that the carbon supply of the host acts as an important trigger for P and N transport in the AM symbiosis [10,12,75]. However, there are also other reports in which the carbon to nutrient resource exchange was not directly correlated [13]. Currently, we have only limited information how the carbon to nutrient exchange across the mycorrhizal interface is regulated [47]. It has been shown that the compatibility between host plants and AM fungi plays a significant role for the nutritional benefits in the AM symbiosis [25], but it is currently unknown why some AM fungi are only beneficial to certain host plant species, despite their ability to colonize a variety of host plants.

## 5.5. Control of Fungal N and P Efflux into the Interface

The processes by which nutrients are released into the mycorrhizal interface, or the mechanisms that control this efflux of nutrients through the fungal plasma membrane, are currently unknown. However, a better knowledge of these processes is key to understand the variability or context-dependency of nutritional benefits or mycorrhizal growth responses that have been described for the AM symbiosis.

## 5.6. Interactions between P and N Flux

There is a growing body of evidence that suggests that there are interactions between the P and N transport in the AM symbiosis [121] and that the maintenance of the AM symbiosis is more determined by the sum of nutritional benefits that are provided via the mycorrhizal interface [54,56,57,137]. It has been speculated that the transport of P or $NH_4^+$ through the respective mycorrhiza-inducible transporters in the periarbuscular membrane is not only important for the uptake of these nutrients from the mycorrhizal interface but also initiates signaling that controls the arbuscular life span [41]. Currently, the processes by which plants control the symbiosis in order to maximize their nutritional gains or control their C costs are largely unknown. These considerations are also complicated by the fact that, in addition to P and N, other nutrients are also transferred from the fungus to the host and that the symbiosis also provides a variety of non-nutritional benefits to host plants [3]. However, the transport of other nutrients in the AM symbiosis is under-researched and it is currently unknown how other nutrient fluxes or non-nutritional benefits contribute to the C costs of the symbiosis for the host plant.

## Acknowledgments

We thank the National Science Foundation (IOS award 1051397) and the South Dakota Soybean Research and Promotion Council for financial support.

## Author Contributions

The manuscript was primarily written by Heike Bücking with contributions by Arjun Kafle.

## Conflicts of Interest

The authors declare no conflict of interest.

## References

1.    Wang, B.; Qiu, Y.L. Phylogenetic distribution and evolution of mycorrhizae in land plants. *Mycorrhiza* **2006**, *16*, 299–363.

2.    Ryan, M.H.; Graham, J.H. Is there a role for arbuscular mycorrhizal fungi in production agriculture? *Plant Soil* **2002**, *244*, 263–271.

3.    Smith, S.E.; Read, D.J. *Mycorrhizal Symbiosis*, 3rd ed.; Academic Press: New York, NY, USA, 2008.

4.    Cameron, D.D. Arbuscular mycorrhizal fungi as (agro)ecosystem engineers. *Plant Soil* **2010**, *333*, 1–5.

5.    Wright, D.P.; Read, D.J.; Scholes, J.D. Mycorrhizal sink strength influences whole plant carbon balance of *Trifolium repens* L. *Plant Cell Environ.* **1998**, *21*, 881–891.

6.    Smith, F.A.; Smith, S.E. How useful is the mutualism-parasitism continuum of arbuscular mycorrhizal functioning? *Plant Soil* **2013**, *363*, 7–18.

7.    Johnson, N.C.; Graham, J.H. The continuum concept remains a useful framework for studying mycorrhizal functioning. *Plant Soil* **2013**, *363*, 411–419.

8.    Johnson, N.C.; Graham, J.H.; Smith, F.A. Functioning of mycorrhizal associations along the mutualism-parasitism continuum. *New Phytol.* **1997**, *135*, 575–585.

9.    Smith, S.E.; Smith, F.A. Fresh perspectives on the roles of arbuscular mycorrhizal fungi in plant nutrition and growth. *Mycologia* **2012**, *104*, 1–13.

10.   Kiers, E.T.; Duhamel, M.; Beesetty, Y.; Mensah, J.A.; Franken, O.; Verbruggen, E.; Fellbaum, C.R.; Kowalchuk, G.A.; Hart, M.M.; Bago, A.; *et al.* Reciprocal rewards stabilize cooperation in the mycorrhizal symbiosis. *Science* **2011**, *333*, 880–882.

11.   Bücking, H.; Mensah, J.A.; Fellbaum, C.R. Common mycorrhizal networks and their effect on the bargaining power of the fungal partner in the arbuscular mycorrhizal symbiosis. *Comp. Integr. Biol.* **2015**, in press.

12.   Fellbaum, C.R.; Mensah, J.A.; Cloos, A.J.; Strahan, G.D.; Pfeffer, P.E.; Kiers, E.T.; Bücking, H. Fungal nutrient allocation in common mycelia networks is regulated by the carbon source strength of individual host plants *New Phytol.* **2014**, *203*, 645–656.

13. Walder, F.; Niemann, H.; Natarajan, M.; Lehmann, M.F.; Boller, T.; Wiemken, A. Mycorrhizal networks: Common goods of plants shared under unequal terms of trade. *Plant Physiol.* **2012**, *159*, 789–797.

14. Babikova, Z.; Gilbert, L.; Bruce, T.J.A.; Birkett, M.; Caulfield, J.C.; Woodcock, C.; Pickett, J.A.; Johnson, D. Underground signals carried through common mycelial networks warn neighbouring plants of aphid attack. *Ecol. Lett.* **2013**, *16*, 835–843.

15. Teste, F.P.; Veneklaas, E.J.; Dixon, K.W.; Lambers, H. Is nitrogen transfer among plants enhanced by contrasting nutrient-acquisition strategies? *Plant Cell Environ.* **2015**, *38*, 50–60.

16. Gorzelak, M.A.; Asay, A.K.; Pickles, B.J.; Simard, S.W. Inter-plant communication through mycorrhizal networks mediates complex adaptive behaviour in plant communities. *AOB Plants* **2015**, doi:10.1093/aobpla/plv050.

17. Barto, E.K.; Hilker, M.; Muller, F.; Mohney, B.K.; Weidenhamer, J.D.; Rillig, M.C. The fungal fast lane: Common mycorrhizal networks extend bioactive zones of allelochemicals in soils. *PLoS ONE* **2011**, *6*, 7.

18. Weremijewicz, J.; Janos, D.P. Common mycorrhizal networks amplify size inequality in *Andropogon gerardii* monocultures. *New Phytol.* **2013**, *198*, 203–213.

19. He, X.; Xu, M.; Qiu, G.Y.; Zhou, J. Use of $^{15}$N stable isotope to quantify nitrogen transfer between mycorrhizal plants. *J. Plant Ecol.* **2009**, *2*, 107–118.

20. Babikova, Z.; Johnson, D.; Bruce, T.J.A.; Pickett, J.; Gilbert, L. Underground allies: How and why do mycelial networks help plants defend themselves? *Bioassays* **2013**, *36*, 21–26.

21. Smith, S.E.; Jakobsen, I.; Grønlund, M.; Smith, F.A. Roles of arbuscular mycorrhizas in plant phosphorus nutrition: Interactions between pathways of phosphorus uptake in arbuscular mycorrhizal roots have important implications for understanding and manipulating plant phosphorus acquisition. *Plant Physiol.* **2011**, *156*, 1050–1057.

22. Reynolds, H.L.; Hartley, A.E.; Vogelsang, K.M.; Bever, J.D.; Schultz, P.A. Arbuscular mycorrhizal fungi do not enhance nitrogen acquisition and growth of old-field perennials under low nitrogen supply in glasshouse culture. *New Phytol.* **2005**, *167*, 869–880.

23. Smith, S.E.; Smith, F.A. Roles of arbuscular mycorrhizas in plant nutrition and growth: New paradigms from cellular to ecosystem scales. *Annu. Rev. Plant Biol.* **2011**, *62*, 227–250.

24. Smith, S.E.; Smith, F.A.; Jakobsen, I. Mycorrhizal fungi can dominate phosphate supply to plants irrespective of growth responses. *Plant Physiol.* **2003**, *133*, 16–20.

25. Smith, S.E.; Smith, F.A.; Jakobsen, I. Functional diversity in arbuscular mycorrhizal (AM) symbioses: The contribution of the mycorrhizal P uptake pathway is not correlated with mycorrhizal responses in growth or total P uptake. *New Phytol.* **2004**, *162*, 511–524.

26. Bücking, H.; Kuhn, A.J.; Schröder, W.H.; Heyser, W. The fungal sheath of ectomycorrhizal pine roots: An apoplastic barrier for the entry of calcium, magnesium, and potassium into the root cortex? *J. Exp. Bot.* **2002**, *53*, 1659–1669.

27. Behrmann, P.; Heyser, W. Apoplastic transport through the fungal sheath of *Pinus sylvestris/Suillus bovinus* ectomycorrhizae. *Bot. Acta* **1992**, *105*, 427–434.

28. Smith, S.E.; Facelli, E.; Pope, S.; Smith, F.A. Plant performance in stressful environments: Interpreting new and established knowledge of the roles of arbuscular mycorrhizas. *Plant Soil* **2010**, *326*, 3–20.

29. Li, H.Y.; Smith, S.E.; Holloway, R.E.; Zhu, Y.G.; Smith, F.A. Arbuscular mycorrhizal fungi contribute to phosphorus uptake by wheat grown in a phosphorus-fixing soil even in the absence of positive growth responses. *New Phytol.* **2006**, *172*, 536–543.

30. Chiou, T.J.; Liu, H.; Harrison, M.J. The spatial expression patterns of a phosphate transporter (*MtPt1*) from *Medicago truncatula* indicate a role in phosphate transport at the root/soil interface. *Plant J.* **2001**, *25*, 281–293.

31. Grunwald, U.; Guo, W.B.; Fischer, K.; Isayenkov, S.; Ludwig-Müller, J.; Hause, B.; Yan, X.L.; Küster, H.; Franken, P. Overlapping expression patterns and differential transcript levels of phosphate transporter genes in arbuscular mycorrhizal, $P_i$-fertilised and phytohormone-treated *Medicago truncatula* roots. *Planta* **2009**, *229*, 1023–1034.

32. Harrison, M.J.; Dewbre, G.R.; Liu, J. A phosphate transporter from *Medicago truncatula* involved in the acquisition of phosphate released by arbuscular mycorrhizal fungi. *Plant Cell* **2002**, *14*, 2413–2429.

33. Paszkowski, U.; Kroken, U.; Roux, C.; Briggs, S.P. Rice phosphate transporters include an evolutionarily divergent gene specifically activated in arbuscular mycorrhizal symbiosis. *Proc. Natl. Acad. Sci. USA* **2002**, *99*, 13324–13329.

34. Xu, G.H.; Chague, V.; Melamed-Bessudo, C.; Kapulnik, Y.; Jain, A.; Raghothama, K.G.; Levy, A.A.; Silber, A. Functional characterization of *LePt4*: A phosphate transporter in tomato with mycorrhiza-enhanced expression. *J. Exp. Bot.* **2007**, *58*, 2491–2501.

35. Burleigh, S.H. Relative quantitative rt-pcr to study the expression of plant nutrient transporters in arbuscular mycorrhizas. *Plant Sci.* **2001**, *160*, 899–904.

36. Kobae, Y.; Tamura, Y.; Takai, S.; Banba, M.; Hata, S. Localized expression of arbuscular mycorrhiza-inducible ammonium transporters in soybean. *Plant Cell Physiol.* **2010**, *51*, 1411–1415.

37. Koegel, S.; Lahmidi, N.A.; Arnould, C.; Chatagnier, O.; Walder, F.; Ineichen, K.; Boller, T.; Wipf, D.; Wiemken, A.; Courty, P.E. The family of ammonium transporters (AMT) in *Sorghum bicolor*: Two AMT members are induced locally, but not systemically in roots colonized by arbuscular mycorrhizal fungi. *New Phytol.* **2013**, *198*, 853–865.

38. Guether, M.; Neuhauser, B.; Balestrini, R.; Dynowski, M.; Ludewig, U.; Bonfante, P. A mycorrhizal-specific ammonium transporter from *Lotus japonicus* acquires nitrogen released by arbuscular mycorrhizal fungi. *Plant Physiol.* **2009**, *150*, 73–83.

39. Gomez, S.K.; Javot, H.; Deewatthanawong, P.; Torres-Jerez, I.; Tang, Y.; Blancaflor, E.B.; Udvardi, M.K.; Harrison, M.J. *Medicago truncatula* and *Glomus intraradices* gene expression in cortical cells harboring arbuscules in the arbuscular mycorrhizal symbiosis. *BMC Plant Biol.* **2009**, *9*, 10.

40. Javot, H.; Penmetsa, R.V.; Terzaghi, N.; Cook, D.R.; Harrison, M.J. A *Medicago truncatula* phosphate transporter indispensable for the arbuscular mycorrhizal symbiosis. *Proc. Natl. Acad. Sci. USA* **2007**, *104*, 1720–1725.

41. Breuillin-Sessoms, F.; Floss, D.S.; Gomez, S.K.; Pumplin, N.; Ding, Y.; Levesque-Tremblay, V.; Noar, R.D.; Daniels, D.A.; Bravo, A.; Eaglesham, J.B.; *et al.* Suppression of arbuscule degeneration in *Medicago truncatula* phosphate transporter 4 mutants is dependent on the ammonium transporter 2 family protein AMT2;3. *Plant Cell* **2015**, doi:10.1105/tpc.114.131144.

42. Pérez-Tienda, J.; Valderas, A.; Camañes, G.; García-Agustín, P.; Ferrol, N. Kinetics of $NH_4^+$ uptake by the arbuscular mycorrhizal fungus *Rhizophagus irregularis*. *Mycorrhiza* **2012**, *22*, 485–491.

43. George, E.; Marschner, H.; Jakobsen, I. Role of arbuscular mycorrhizal fungi in uptake of phosphorus and nitrogen from soil. *Crit. Rev. Biotechnol.* **1995**, *15*, 257–270.

44. Hawkins, H.J.; George, E. Effect of plant nitrogen status on the contribution of arbuscular mycorrhizal hyphae to plant nitrogen uptake. *Physiol. Plant.* **1999**, *105*, 694–700.

45. Saia, S.; Benitéz, E.; Garcia-Garrido, J.M.; Settanni, L.; Amato, G.; Giambalvo, D. The effect of arbuscular mycorrhizal fungi on total plant nitrogen uptake and nitrogen recovery from soil organic material. *J. Agric. Sci.* **2014**, *152*, 370–378.

46. Mensah, J.A.; Koch, A.M.; Antunes, P.M.; Hart, M.M.; Kiers, E.T.; Bücking, H. High functional diversity within arbuscular mycorrhizal fungal species is associated with differences in phosphate and nitrogen uptake and fungal phosphate metabolism. *Mycorrhiza* **2015**, *25*, 533–546.

47. Corrêa, A.; Cruz, C.; Ferrol, N. Nitrogen and carbon/nitrogen dynamics in arbuscular mycorrhiza: The great unknown. *Mycorrhiza* **2015**, *25*, 499–515.

48. Azcon-Aguilar, C.; Alba, C.; Montilla, M.; Barea, J.M. Isotopic ($^{15}N$) evidence of the use of less available N forms by VA mycorrhizas. *Symbiosis* **1993**, *15*, 39–48.

49. Toussaint, J.P.; St-Arnaud, M.; Charest, C. Nitrogen transfer and assimilation between the arbuscular mycorrhizal fungus *Glomus intraradices* Schenck & Smith and RI t-DNA roots of *Daucus carota* l. In an vitro compartmented system. *Can. J. Microbiol.* **2004**, *50*, 251–260.

50. Jin, H.; Pfeffer, P.E.; Douds, D.D.; Piotrowski, E.; Lammers, P.J.; Shachar-Hill, Y. The uptake, metabolism, transport and transfer of nitrogen in an arbuscular mycorrhizal symbiosis. *New Phytol.* **2005**, *168*, 687–696.

51. Tanaka, Y.; Yano, K. Nitrogen delivery to maize via mycorrhizal hyphae depends on the form of N supplied. *Plant Cell Environ.* **2005**, *28*, 1247–1254.

52. Hildebrandt, U.; Schmelzer, E.; Bothe, H. Expression of nitrate transporter genes in tomato colonized by an arbuscular mycorrhizal fungus. *Physiol. Plant.* **2002**, *115*, 125–136.

53. Pérez-Tienda, J.; Testillano, P.S.; Balestrini, R.; Fiorilli, V.; Azcón-Aguilar, C.; Ferrol, N. GintAmt2, a new member of the ammonium transporter family in the arbuscular mycorrhizal fungus *Glomus intraradices*. *Fungal Genet. Biol.* **2011**, *48*, 1044–1055.

54. Bonneau, L.; Huguet, S.; Wipf, D.; Pauly, N.; Truong, H.N. Combined phosphate and nitrogen limitation generates a nutrient stress transcriptome favorable for arbuscular mycorrhizal symbiosis in *Medicago truncatula*. *New Phytol.* **2013**, *199*, 188–202.

55. Besserer, A.; Puech-Pagès, V.; Kiefer, P.; Gomez-Roldan, V.; Jauneau, A.; Roy, S.; Portais, J.C.; Roux, C.; Bécard, G.; Séjalon-Delmas, N. Strigolactones stimulate arbuscular mycorrhizal fungi by activating mitochondria. *PLoS Biol.* **2006**, *4*, 1239–1247.

56. Nouri, E.; Breuillin-Sessoms, F.; Feller, U.; Reinhardt, D. Phosphorus and nitrogen regulate arbuscular mycorrhizal symbiosis in *Petunia hybrida*. *PLoS ONE* **2014**, *9*, e90841.

57. Javot, H.; Penmetsa, R.V.; Breuillin, F.; Bhattarai, K.K.; Noar, R.D.; Gomez, S.K.; Zhang, Q.; Cook, D.R.; Harrison, M.J. *Medicago truncatula MtPt4* mutants reveal a role for nitrogen in the regulation of arbuscule degeneration in arbuscular mycorrhizal symbiosis. *Plant J.* **2011**, *68*, 954–965.

58. Ames, R.N.; Reid, C.P.P.; Porter, L.K.; Cambardella, C. Hyphal uptake and transport of nitrogen from two [15]N-labelled sources by *Glomus mosseae*, a vesicular-arbuscular mycorrhizal fungus. *New Phytol.* **1983**, *95*, 381–396.

59. Barea, J.M.; Azcón-Aguilar, C.; Azcón, R. Vesicular-arbuscular mycorrhiza improve both symbiotic $N_2$ fixation and N uptake from soil as assessed with a [15]N technique under field conditions. *New Phytol.* **1987**, *106*, 717–725.

60. Frey, B.; Schüepp, H. Acquisition of nitrogen by external hyphae of arbuscular mycorrhizal fungi associated with *Zea mays* L. *New Phytol.* **1993**, *124*, 221–230.

61. Johansen, A.; Jakobsen, I.; Jensen, E.S. Hyphal transport by a vesicular-arbuscular mycorrhizal fungus of N applied to the soil as ammonium or nitrate. *Biol. Fertil. Soils* **1993**, *16*, 66–70.

62. Hawkins, H.J.; Johansen, A.; George, E. Uptake and transport of organic and inorganic nitrogen by arbuscular mycorrhizal fungi. *Plant Soil* **2000**, *226*, 275–285.

63. Tobar, R.; Azcón, R.; Barea, J.M. Improved nitrogen uptake and transport from [15]N-labelled nitrate by external hyphae of arbuscular mycorrhiza under water-stressed conditions. *New Phytol.* **1994**, *126*, 119–122.

64. Johansen, A.; Finlay, R.D.; Olsson, P.A. Nitrogen metabolism of external hyphae of the arbuscular mycorrhizal fungus *Glomus intraradices*. *New Phytol.* **1996**, *133*, 705–712.

65. Bago, B.; Vierheilig, H.; Piché, Y.; Azcón-Aguilar, C. Nitrate depletion and pH changes induced by the extraradical mycelium of the arbuscular mycorrhizal fungus *Glomus intraradices* grown in monoxenic cultures. *New Phytol.* **1996**, *133*, 273–280.

66. Bago, B.; Azcón-Aguilar, C. Changes in the rhizospheric pH induced by arbuscular mycorrhiza formation in onion (*Allium cepa* L.). *Z. Pflanzenernähr. Bodenkd.* **1997**, *160*, 333–339.

67. Howitt, S.M.; Udvardi, M.K. Structure, function and regulation of ammonium transporters in plants. *Biochim. Biophys. Acta* **2000**, *1465*, 152–170.

68. D'Apuzzo, E.; Rogato, A.; Simon-Rosin, U.; El Alaoui, H.; Barbulova, A.; Betti, M.; Dimou, M.; Katinakis, P.; Marquez, A.; Marini, A.M.; *et al.* Characterization of three functional high-affinity ammonium transporters in *Lotus japonicus* with differential transcriptional regulation and spatial expression. *Plant Physiol.* **2004**, *134*, 1763–1774.

69. Schüßler, A.; Walker, C. *The Glomeromycota. A species list with new families and new genera*; Libraries at the Royal Botanic Garden Edinburgh, The Royal Botanic Garden Kew, Botanische Staatssammlung Munich, and Oregon State University: Gloucester, UK, 2010.

70. Hawkins, H.J.; George, E. Reduced [15]N-nitrogen transport through arbuscular mycorrhizal hyphae to *Triticum aestivum* L. supplied with ammonium *vs.* nitrate nutrition. *Ann. Bot.* **2001**, *87*, 303–311.

71. Gachomo, E.; Allen, J.W.; Pfeffer, P.E.; Govindarajulu, M.; Douds, D.D.; Jin, H.R.; Nagahashi, G.; Lammers, P.J.; Shachar-Hill, Y.; Bücking, H. Germinating spores of *Glomus intraradices* can use internal and exogenous nitrogen sources for de novo biosynthesis of amino acids. *New Phytol.* **2009**, *184*, 399–411.

72. Tisserant, E.; Kohler, A.; Dozolme-Seddas, P.; Balestrini, R.; Benabdellah, K.; Colard, A.; Croll, D.; da Silva, C.; Gomez, S.K.; Koul, R.; *et al.* The transcriptome of the arbuscular mycorrhizal fungus *Glomus intraradices* (DAOM 197198) reveals functional tradeoffs in an obligate symbiont. *New Phytol.* **2012**, *193*, 755–769.

73. Lopez-Pedrosa, A.; González-Guerrero, M.; Valderas, A.; Azcón-Aguilar, C.; Ferrol, N. *GintAmt1* encodes a functional high-affinity ammonium transporter that is expressed in the extraradical mycelium of *Glomus intraradices*. *Fungal Genet. Biol.* **2006**, *43*, 102–110.

74. Tian, C.; Kasiborski, B.; Koul, R.; Lammers, P.J.; Bücking, H.; Shachar-Hill, Y. Regulation of the nitrogen transfer pathway in the arbuscular mycorrhizal symbiosis: Gene characterization and the coordination of expression with nitrogen flux. *Plant Physiol.* **2010**, *153*, 1175–1187.

75. Fellbaum, C.R.; Gachomo, E.W.; Beesetty, Y.; Choudhari, S.; Strahan, G.D.; Pfeffer, P.E.; Kiers, E.T.; Bücking, H. Carbon availability triggers fungal nitrogen uptake and transport in arbuscular mycorrhizal symbiosis. *Proc. Natl. Acad. Sci. USA* **2012**, *109*, 2666–2671.

76. Tisserant, E.; Malbreil, M.; Kuo, A.; Kohler, A.; Symeonidi, A.; Balestrini, R.; Charron, P.; Duensing, N.; Frey, N.F.D.; Gianinazzi-Pearson, V.; *et al.* Genome of an arbuscular mycorrhizal fungus provides insight into the oldest plant symbiosis. *Proc. Natl. Acad. Sci. USA* **2014**, *111*, 563–563.

77. Hodge, A.; Robinson, D.; Fitter, A.H. An arbuscular mycorrhizal inoculum enhances root proliferation in, but not nitrogen capture from, nutrient-rich patches in soil. *New Phytol.* **2000**, *145*, 575–584.

78. Cappellazzo, G.; Lanfranco, L.; Fitz, M.; Wipf, D.; Bonfante, P. Characterization of an amino acid permease from the endomycorrhizal fungus *Glomus mosseae*. *Plant Physiol.* **2008**, *147*, 429–437.

79. Belmondo, S.; Fiorilli, V.; Pérez-Tienda, J.; Ferrol, N.; Marmeisse, R.; Lanfranco, L. A dipeptide transporter from the arbuscular mycorrhizal fungus *Rhizophagus irregularis* is upregulated in the intraradical phase. *Front. Plant Sci.* **2014**, *5*, 436.

80. Allen, J.W.; Shachar-Hill, Y. Sulfur transfer through an arbuscular mycorrhiza. *Plant Physiol.* **2009**, *149*, 549–560.

81. Breuninger, M.; Trujillo, C.G.; Serrano, E.; Fischer, R.; Requena, N. Different nitrogen sources modulate activity but not expression of glutamine synthetase in arbuscular mycorrhizal fungi. *Fungal Genet. Biol.* **2004**, *41*, 542–552.

82. Nakano, A.; Takahashi, K.; Koide, R.T.; Kimura, M. Determination of the nitrogen source for arbuscular mycorrhizal fungi by $^{15}N$ application to soil and plants. *Mycorrhiza* **2001**, *10*, 267–273.

83. Cliquet, J.B.; Stewart, G.R. Ammonia assimilation in *Zea mays* L. infected with a vesicular-arbuscular mycorrhizal fungus *Glomus fasciculatum*. *Plant Physiol.* **1993**, *101*, 865–871.

84. Subramanian, K.S.; Charest, C. Arbuscular mycorrhizae and nitrogen assimilation in maize after drought and recovery. *Physiol. Plant.* **1998**, *102*, 285–296.

85. Kaldorf, M.; Zimmer, W.; Bothe, H. Genetic evidence for the occurence of assimilatory nitrate reductase in arbuscular mycorrhizal and other fungi. *Mycorrhiza* **1994**, *5*, 23–28.

86. Ho, I.; Trappe, J.M. Nitrate reducing capacity of two vesicular-arbuscular mycorrhizal fungi. *Mycologia* **1975**, *67*, 886–888.

87. Oliver, A.J.; Smith, S.E.; Nicholas, D.J.D.; Wallace, W. Activity of nitrate reductase in *Trifolium subterraneum*: Effects of mycorrhizal infection and phosphate nutrition. *New Phytol.* **1983**, *94*, 63–79.

88. Vázquez, M.; Barea, J.; Azcón, R. Impact of soil nitrogen concentration on *Glomus* spp.-*Sinorhizobium* interactions as affecting growth, nitrate reductase activity and protein content of *Medicago sativa*. *Biol. Fertil. Soils* **2001**, *34*, 57–63.

89. Hawkins, H.J.; Cramer, M.D.; George, E. Root respiratory quotient and nitrate uptake in hydroponically grown non-mycorrhizal and mycorrhizal wheat. *Mycorrhiza* **1999**, *9*, 57–60.

90. Kaldorf, M.; Schmelzer, E.; Bothe, H. Expression of maize and fungal nitrate reductase genes in arbuscular mycorrhiza. *MPMI* **1998**, *11*, 439–448.

91. Nemec, S. Histochemical characterization of *Glomus etunicatus* infection of *Citrus limon* fibrous roots. *Can. J. Bot.* **1981**, *59*, 609–617.

92. MacDonald, R.M.; Lewis, M. The occurence of some acid phosphatases and dehydrogenases in the vesicular-arbuscular mycorrhizal fungus *Glomus mosseae*. *New Phytol.* **1978**, *80*, 135–141.

93. Smith, S.E.; St John, B.J.; Smith, F.A.; Nicholas, D.J.D. Activity of glutamine synthetase and glutamate dehydrogenase in *Trifolium subterraneum* L. and *Allium cepa* L.: Effects of mycorrhizal infection and phosphate nutrition. *New Phytol.* **1985**, *99*, 211–227.

94. Govindarajulu, M.; Pfeffer, P.E.; Jin, H.R.; Abubaker, J.; Douds, D.D.; Allen, J.W.; Bücking, H.; Lammers, P.J.; Shachar-Hill, Y. Nitrogen transfer in the arbuscular mycorrhizal symbiosis. *Nature* **2005**, *435*, 819–823.

95. Faure, S.; Cliquet, J.B.; Thephany, G.; Boucaud, J. Nitrogen assimilation in *Lolium perenne* colonized by the arbuscular mycorrhizal fungus *Glomus fasciculatum*. *New Phytol.* **1998**, *138*, 411–417.

96. Valentine, A.J.; Osborne, B.A.; Mitchell, D.T. Form of inorganic nitrogen influences mycorrhizal colonisation and photosynthesis of cucumber. *Sci. Hortic.* **2002**, *92*, 229–239.

97. Liu, Y.; Lai, N.; Gao, K.; Chen, F.; Yuan, L.; Mi, G. Ammonium inhibits primary root growth by reducing the length of meristem and elongation zone and decreasing elemental expansion rate in the root apex in *Arabidopsis thaliana*. *PLoS ONE* **2013**, *8*, e61031.

98. Cramer, M.D.; Lewis, O.A.M. The influence of $NO_3^-$ and $NH_4^+$ nutrition on the carbon and nitrogen partitioning characteristics of wheat (*Triticum aestivum*) and maize (*Zea mays*) plants. *Plant Soil* **1993**, *154*, 289–300.

99. Cramer, M.D.; Lewis, O.A.M.; Lips, S.H. Inorganic carbon fixation and metabolism in maize roots as affected by nitrate and ammonium nutrition. *Physiol. Plant.* **1993**, *89*, 632–639.

100. Cooper, T.G.; Sumrada, R.A. What is the function of nitrogen catabolite repression in *Saccharomyces cerevisiae*? *J. Bacteriol.* **1983**, *155*, 623–627.

101. Javelle, A.; Morel, M.; Rodriguez-Pastrana, B.R.; Botton, B.; Andr,, B.; Marini, A.M.; Brun, A.; Chalot, M. Molecular characterization, function and regulation of ammonium transporters (Amt) and ammonium-metabolizing enzymes (GS, NADP-GDH) in the ectomycorrhizal fungus *Hebeloma cylindrosporum*. *Mol. Microbiol.* **2003**, *47*, 411–430.

102. Navarro, F.J.; Machín, F.; Martín, Y.; Siverio, J.M. Down-regulation of eukaryotic nitrate transporter by nitrogen-dependent ubiquitinylation. *J. Biol. Chem.* **2006**, *281*, 13268–13274.

103. Schulten, H.-R.; Schnitzer, M. The chemistry of soil organic nitrogen: A review. *Biol. Fertil. Soils* **1998**, *26*, 1–15.

104. Leigh, J.; Hodge, A.; Fitter, A.H. Arbuscular mycorrhizal fungi can transfer substantial amounts of nitrogen to their host plant from organic material. *New Phytol.* **2009**, *181*, 199–207.

105. Hodge, A.; Fitter, A.H. Substantial nitrogen acquisition by arbuscular mycorrhizal fungi from organic material has implications for N cycling. *Proc. Natl. Acad. Sci. USA* **2010**, *107*, 13754–13759.

106. Thirkell, J.D.; Cameron, D.D.; Hodge, A. Resolving the "nitrogen paradox" of arbuscular mycorrhizas: Fertilization with organic matter brings considerable benefits for plant nutrition and growth. *Plant Cell Environ.***2015**, doi:10.1111/pce.12667.

107. Atul-Nayyar, A.; Hamel, C.; Hanson, K.; Germida, J. The arbuscular mycorrhizal symbiosis links N mineralization to plant demand. *Mycorrhiza* **2009**, *19*, 239–246.

108. Herman, D.J.; Firestone, M.K.; Nuccio, E.; Hodge, A. Interactions between an arbuscular mycorrhizal fungus and a soil microbial community mediating litter decomposition. *FEMS Microbiol. Ecol.* **2012**, *80*, 236–247.

109. Finzi, A.C.; Abramoff, R.Z.; Spiller, K.S.; Brzostek, E.R.; Darby, B.A.; Kramer, M.A.; Phillips, R.P. Rhizosphere processes are quantitatively important components of terrestrial carbon and nutrient cycles. *Glob. Chang. Biol.* **2015**, *21*, 2082–2094.

110. Nuccio, E.E.; Hodge, A.; Pett-Ridge, J.; Herman, D.J.; Weber, P.K.; Firestone, M.K. An arbuscular mycorrhizal fungus significantly modifies the soil bacterial community and nitrogen cycling during litter decomposition. *Environ. Microbiol.* **2013**, *15*, 1870–1881.

111. Cliquet, J.B.; Murray, P.J.; Boucaud, J. Effect of the arbuscular mycorrhizal fungus *Glomus fasciculatum* on the uptake of amino nitrogen by *Lolium perenne*. *New Phytol.* **1997**, *137*, 345–349.

112. Whiteside, M.D.; Digman, M.A.; Gratton, E.; Treseder, K.K. Organic nitrogen uptake by arbuscular mycorrhizal fungi in a boreal forest. *Soil Biol. Biochem.* **2012**, *55*, doi:10.1016/j.soilbio.2012.1006.1001.

113. Whiteside, M.D.; Garcia, M.O.; Treseder, K.K. Amino acid uptake in arbuscular mycorrhizal plants. *PLoS ONE* **2012**, *7*, e47643.

114. Guescini, M.; Zeppa, S.; Pierleoni, R.; Sisti, D.; Stocchi, L.; Stocchi, V. The expression profile of the *Tuber borchii* nitrite reductase suggests its positive contribution to host plant nitrogen nutrition. *Curr. Genet.* **2007**, *51*, 31–41.

115. Bailly, J.; Debaud, J.C.; Verner, M.C.; Plassard, C.; Chalot, M.; Marmeisse, R.; Fraissinet-Tachet, L. How does a symbiotic fungus modulate expression of its host-plant nitrite reductase? *New Phytol.* **2007**, *175*, 155–165.

116. Temple, S.J.; Vance, C.P.; Gantt, S.J. Glutamate synthase and nitrogen assimilation. *Trends Plant Sci.* **1998**, *3*, 51–56.

117. Botton, B.; Chalot, M. Nitrogen assimilation: Enzymology in ectomycorrhizas. In *Mycorrhiza*; Varma, A., Hock, B., Eds.; Springer-Verlag: Berlin, Germany, 1995; pp. 325–363.

118. Rolin, D.; Pfeffer, P.E.; Douds, D.D.; Farrell, H.M.; Shachar-Hill, Y. Arbuscular mycorrhizal symbiosis and phosphorus nutrition: Effects on amino acid production and turnover in leek. *Symbiosis* **2001**, *30*, 1–14.

119. Forde, B.G. Nitrate transporters in plants: Structure, function and regulation. *Biochim. Biophys. Acta* **2000**, *1465*, 219–235.

120. Marzluf, G.A. Genetic regulation of nitrogen metabolism in the fungi. *Microbiol. Mol. Biol. Rev.* **1997**, *61*, 17–32.

121. Cruz, C.; Egsgaard, H.; Trujillo, C.; Ambus, P.; Requena, N.; Martins-Loucao, M.A.; Jakobsen, I. Enzymatic evidence for the key role of arginine in nitrogen translocation by arbuscular mycorrhizal fungi. *Plant Physiol.* **2007**, *144*, 782–792.

122. Cox, G.; Moran, K.J.; Sanders, F.; Nockolds, C.; Tinker, P.B. Translocation and transfer of nutrients in vesicular-arbuscular mycorrhizas. III. Polyphosphate granules and phosphorus translocation. *New Phytol.* **1980**, *84*, 649–659.

123. Cramer, C.L.; Davis, H.H. Polyphosphate-cation interaction in the amino acid-containing vacuole of *Neurospora crassa. J. Biol. Chem.* **1984**, *259*, 5152–5157.

124. Cramer, C.L.; Vaughn, L.E.; Davis, R.H. Basic amino acids and inorganic polyphosphates in *Neurospora crassa*: Independent regulation of vacuolar pools. *J. Bacteriol.* **1980**, *142*, 945–952.

125. Westenberg, B.; Boller, T.; Wiemken, A. Lack of arginine- and polyphosphate-storage pools in a vacuole-deficient mutant (end1) of *Saccharomyces cerevisiae. FEBS Lett.* **1989**, *254*, 133–136.

126. Bücking, H.; Heyser, W. Elemental composition and function of polyphosphates in ectomycorrhizal fungi - an X-ray microanalytical study. *Mycol. Res.* **1999**, *103*, 31–39.

127. George, E.; Häussler, K.U.; Vetterlein, D.; Gorgus, E.; Marschner, H. Water and nutrient translocation by hyphae of *Glomus mosseae. Can. J. Bot.* **1992**, *70*, 2130–2137.

128. Bücking, H.; Heyser, W. Uptake and transfer of nutrients in ectomycorrhizal associations: Interactions between photosynthesis and phosphate nutrition. *Mycorrhiza* **2003**, *13*, 59–68.

129. Bücking, H.; Shachar-Hill, Y. Phosphate uptake, transport and transfer by the arbuscular mycorrhizal fungus *Glomus intraradices* is stimulated by increased carbohydrate availability. *New Phytol.* **2005**, *165*, 899–912.

130. Guether, M.; Volpe, V.; Balestrini, R.; Requena, N.; Wipf, D.; Bonfante, P. *LjLht1.2*—A mycorrhiza-inducible plant amino acid transporter from *Lotus japonicus. Biol. Fertil. Soils* **2011**, *47*, 925–936.

131. Beever, R.E.; Burns, D.J.W. Phosphorus uptake, storage and utilization by fungi. *Adv. Bot. Res.* **1980**, *8*, 127–219.

132. Li, T.; Hu, Y.J.; Hao, Z.P.; Li, H.; Wang, Y.S.; Chen, B.D. First cloning and characterization of two functional aquaporin genes from an arbuscular mycorrhizal fungus *Glomus intraradices. New Phytol.* **2013**, *197*, 617–630.

133. Maurel, C.; Plassard, C. Aquaporins: For more than water at the plant–fungus interface? *New Phytol.* **2011**, *190*, 815–817.

134. Aroca, R.; Bago, A.; Sutka, M.; Paz, J.A.; Cano, C.; Amodeo, G.; Ruiz-Lozano, J.M. Expression analysis of the first arbuscular mycorrhizal fungi aquaporin described reveals concerted gene expression between salt-stressed and nonstressed mycelium. *Mol. Plant Microb. Interact.* **2009**, *22*, 1169–1178.

135. Xie, X.A.; Huang, W.; Liu, F.C.; Tang, N.W.; Liu, Y.; Lin, H.; Zhao, B. Functional analysis of the novel mycorrhiza-specific phosphate transporter *AsPt1* and *Pht1* family from *Astragalus sinicus* during the arbuscular mycorrhizal symbiosis. *New Phytol.* **2013**, *198*, 836–852.

136. Secco, D.; Wang, C.; Shou, H.; Whelan, J. Phosphate homeostasis in the yeast *Saccharomyces cerevisiae*, the key role of the SPX domain-containing proteins. *FEBS Lett.* **2012**, *586*, 289–295.

137. Carbonnel, S.; Gutjahr, C. Control of arbuscular mycorrhiza development by nutrient signals. *Front. Plant Sci.* **2014**, *5*, 462.

# 13

# The Role of Canadian Agriculture in Meeting Increased Global Protein Demand with Low Carbon Emitting Production

**James A. Dyer** [1,†,*] **and Xavier P.C. Vergé** [2,†]

[1]  Contract Researcher, 122 Hexam Street, Cambridge, Ontario, N3H 3Z9, Canada

[2]  Contract Researcher; 2055, Carling avenue, #1016, Ottawa, Ontario, K2A 1G6, Canada;
    E-Mail: xavier_vrg@yahoo.fr

[†]  These authors contributed equally to this work.

[*]  Author to whom correspondence should be addressed; E-Mail: jamesdyer@sympatico.ca

Academic Editors: Paul C. Struik and Yantai Gan

**Abstract:** Although the demand on agriculture to produce food could double by 2050, changing diets will expand the global demand for protein even faster. Canadian livestock producers will likely expand in response to this market opportunity. Because of the high greenhouse gas (GHG) emissions from animal protein production, the portion of this protein demand that can be met by pulse crops must be considered. The protein basis for GHG emission intensity was assessed for 2006 using a multi-commodity GHG emissions inventory model. Because arable land is required for other agricultural products, protein production and GHG emissions were also assessed on the basis of the land use. GHG emissions per unit of protein are one or two orders of magnitude higher for protein from livestock, particularly ruminants, than for protein from pulses. The protein production from pulses was moderately higher per unit of land than the protein from livestock. This difference was greater when soybeans were the only pulse in the comparison. Protein from livestock, especially ruminants, resulted in much higher GHG emissions per unit of land than the protein from pulses. A shift towards more protein from pulses could assure a better global protein supply and reduce GHG emissions associated with that supply.

**Keywords:** livestock and pulse protein; greenhouse gas emissions; daily protein intake; performance indicators

## 1. Introduction

The global demand for protein is rising rapidly. This is largely due to more disposable income in emerging economies and changing dietary preferences in much of the developed world [1]. Dire predictions have been made about a global food crisis for 2050, with global population exceeding nine billion and the effect of more extreme weather on farm productivity around the globe [2–5]. Global food security is exacerbated by inequitable distribution between rich and poor countries [6]. Nowhere is this inequality more apparent than in access to protein, where grains are increasingly being fed to meat animals, while the world's poorest people struggle with chronic hunger and poor nutrition. The need to address climate change is the added imperative that pushes food production into a perfect storm situation [7]. As the global community awakens to the reality that all countries and all sectors must do whatever they can to reduce greenhouse gas (GHG) emissions [8], all agricultural products can expect to be evaluated on the basis of their carbon footprints (CF). Since livestock account for 14.5% of global GHG emissions [9], consuming fewer livestock products could significantly lower these emissions [10,11].

This CF scrutiny will be particularly intense for livestock products, which are estimated to be responsible for 18% of greenhouse gases [12]. Dyer *et al.* argued and demonstrated that the fairest way to assess the CF of livestock production was by protein-based GHG emission intensity [13]. This paper focusses specifically on protein production from both plant and animal sources in Canada as a distinct challenge from food security. It determines how much plant sources can help Canada produce more protein with lower GHG emissions. This paper will quantify the CF of these products and describe three indicators suited to this task. The diversity of protein sources and differences in livestock types and production systems among the Canadian provinces result in significant differences among the CF of protein from those provinces. Therefore, a second goal was to determine how these spatial differences affect the protein CF distribution across the Canadian provinces.

## 2. Background

Protein is a macronutrient necessary for the proper growth and function of the human body [14]. Although there is some debate over the amount of protein a person needs, a deficiency in protein leads to muscle atrophy and impaired functioning of the human body. Whitbread [14] recommended that the current daily intake for protein should be 46 grams for women aged 19–70 and 56 grams for men aged 19–70. Using food balance sheet data from the Statistics Division from the Food and Agriculture Organization of the United Nations, the ChartsBin Statistics Collector Team [15] defined world daily intake per capita of protein as 77 g, ranging from 100 g in the developed world to as low as 55 g/day in the developing world.

While small grain cereals (particularly hard red wheat) produce proteins, this type of protein is considered to be incomplete, because it lacks some of the amino acids that are found in animal proteins and which are essential to the human diet [16,17]. Legumes (or pulses), however, produce complete proteins. Therefore, plant and animal agriculture can be compared on the basis of their respective supply of "complete" protein. Beans and other legumes are a critical source of protein in many parts of the world [14,18]. They are an inexpensive food, high in fiber, calcium and iron.

Trends and patterns in GHG emissions from livestock and field crops in Canada are often discussed simply on an east-west basis, because agriculture west of the Great Lakes is dominated by the Prairie Provinces, while east of Lake Superior, it is the north shore of Lakes Erie and Ontario and the Saint Lawrence River basin that are the dominant farming regions. To discuss regional differences, the provinces were grouped so that the Atlantic Provinces (AP) (treated as one province), Quebec (QC) and Ontario (ON) were defined as Eastern Canada, and Manitoba (MN), Saskatchewan (SA), Alberta (AB) and British Columbia (BC) were defined as Western Canada. Given the size of Canada and the additional small, but distinct, farming regions in the coastal provinces, these trends and patterns also need to be assessed on a sub-provincial basis, as well as a provincial level. However, reporting results on a sub-provincial scale was beyond the scope of this paper.

## 3. Materials and Methods

### 3.1. Selecting the Performance Indicators

The assessment described in this paper is an extension of the previous assessment by Dyer *et al.* [10], which compared the five major livestock industries on the basis of GHG emissions per unit of protein, but excluded any plant source protein. Furthermore, the animal protein-based GHG emission intensity indicator has been applied to the Canadian sheep industry [19]. This paper will describe the first application of this indicator to compare regions on the basis of total protein supply. The second goal of this paper integrated the plant and animal sources of protein production in each province to show the distribution of protein types across the Canadian provinces. Thus, the assessment described below was applied to all means of producing protein in each province.

Ensuring global food security, especially under anticipated climate change impacts on farmland, will also require that a minimum of land capable of growing food-quality carbohydrates is diverted to protein production. Therefore, the indicators used in this assessment should track GHG emission intensity on the basis of both protein production and land use, as well as protein production on the basis of land use. Thus, these indicators include the land use basis for GHG emission intensity (Indicator 1) and all protein production (Indicator 2), as well as the protein-based intensity for GHG emissions (Indicator 3). The dimensions of these three indicators are the GHG emissions and protein production per ha and the GHG emissions per unit of protein production. To quantify these indicators, an automated computer modelling system was assembled capable of integrating the diversity of factors that define the CF of livestock and pulse crop production.

### 3.2. Performance Modelling Methodology

The Unified Livestock Industry and Crop Emissions Estimation System (ULICEES), which is a spreadsheet-based inventory model, was used in the assessment of livestock GHG emissions. Since ULICEES has been described in detail elsewhere [20], only the main concepts used in inter-commodity assessment are described in this paper. ULICEES was created by assembling the five groups of livestock-specific GHG computations from the Canadian beef, dairy, pork, poultry and sheep (lamb) industries [19,21–24] in one spreadsheet file. As well as the direct emissions from livestock, these calculations account for GHG emissions from producing the crops that feed each livestock population and the storage practices and characteristics of manure from each livestock type. ULICEES integrated $N_2O$ and fossil $CO_2$ emissions with $CH_4$ emissions from livestock. ULICEES was initially applied to 2001 [20], the most recent year with livestock diet survey data [25]. ULICEES was updated to 2006 [19], the latest year with a complete set of input census data. In order to quantify the indirect, as well as the direct GHG emissions from livestock, the concept of the livestock crop complex (LCC), the land that supports feed production in the livestock industry, was a critical component of ULICEES [20]. ULICEES used the Intergovernmental Panel on Climate Change (IPCC) Tier 2 methodology for enteric methane and the Tier 1 methodology modified for Canadian conditions for $N_2O$ emissions [26,27].

For the assessment of GHG emissions from just plant protein, the farm fieldwork fossil fuel energy and emissions (F4E2) model [28] was applied to pulse crops. Since pulses are nitrogen-fixing crops and are assumed to not require N-fertilizer, there should be negligible $N_2O$ emissions [27]. Although a range of legumes are cultivated in Canada, only annual legumes, or pulses, are considered in this assessment. The F4E2 output files were also integrated with ULICEES to determine the fossil $CO_2$ emissions associated with livestock production [20]. The F4E2 model was based on farm machinery management coefficients from the American Society of Agricultural Engineers (ASAE) [23] and farm machinery management equations [29,30]. The database generated by the 1996 Farm Energy Use Survey (FEUS) [31] was used to verify the F4E2 model and to calibrate several non-fieldwork terms in the farm energy budget.

The pulse crops that were destined for animal feed had to be excluded from the comparison of proteins from pulses to avoid double counting those pulses that are already counted indirectly for their contribution to animal protein production. Since those pulse crops were used for animal feed, they would, therefore, contribute to the protein production by livestock. The livestock diet data used in ULICEES identified these crops as soybeans and dry peas [20]. To correct for this potential double counting of protein production, the areas in those pulse crops that were designated by ULICEES as being in the LCC were subtracted from the crop area statistics. Because ULICEES identifies LCC areas by diet requirements and yields, rather than empirically, more land was identified in these two crops in some provinces than was in the crop statistics. In these cases, the areas that would produce food-quality pulses were assumed to be zero. Whereas this does not guarantee that the soybeans and dry peas outside the LCC were not eventually consumed by livestock, they were at least sold on the market and into another province or the USA.

There were several limitations of this assessment. First, it did not credit ruminant livestock production with any sequestration of soil carbon. That was because this assessment was limited to one year and did not consider any land use transfers among the protein sources over time. Although ULICEES can accommodate this carbon sink term, it only does so for land in the LCC that has been shifted into or out of perennial ground cover (forage) [20]. The second limitation of this study was that it did not take into account the processing energy of any of the agricultural products considered in this paper. A full life cycle assessment of each of the commodities considered in this assessment, such as done for the Canadian dairy industry [32], would have been too involved to include in this paper. This assessment also excluded the potential non-CF risks associated with livestock in Canada [33]. The energy budget for pulses also did not consider the potential of vegetable oils. Soy oil, for example, is one of the feedstocks for biodiesel production [34] and is a popular cooking oil [35].

The processing of soy into tofu can raise the CF of the soybean crop by a factor of four, due to the need to remove most of the oil and ferment the soy meal [36]. However, soybeans are consumed in their whole food form (cooked whole beans *versus* soy protein alone) throughout Asia [37]. Therefore, besides giving soybeans a higher market value (which was beyond the scope of this paper), converting them into soymilk or tofu is not essential to the consumption of soybeans throughout the world. Therefore, including tofu production in the CF of soybean protein from a theoretical perspective is unnecessary in the context of this paper.

### 3.3. Defining Protein Conversion Factors

Table 1 provides the protein conversions that were used in this paper for both plant and animal protein sources. Except for lamb, all of the protein conversion factors (PCF) from live weight (LW) were taken from Dyer *et al.* [13]. The PCF for lamb was taken from Dyer *et al.* [19]. The PCF for sheep was lower than those for the other three meat animals. Two of these factors had to be expressed as weight of protein per head because the primary products from those animals, milk and eggs, are not their carcasses. The USDA [38] provided the protein contents of one dozen medium-sized eggs as 0.53 kg and of fluid milk as 3.2% by weight [13,21]. For the annual milk yield, the national average for 2006 of 7405 kg/head [21] was used, while the annual egg yield was 186 per layer [21]. The protein from culled dairy cows and unwanted calves was attributed to the beef industry [19], because the slaughtered dairy cows cannot be tracked in the census statistics.

**Table 1.** Protein conversion factors for five plant (legume) sources and five livestock type sources. LW, live weight. head, one individual producing animal. DM, dry matter.

| Plant Protein | | Animal Protein | | |
|---|---|---|---|---|
| Pulse Crops | % of DM Yield | Livestock | % of LW | Kg(protein)/head |
| Dry peas | 7.7 | Beef | 8.3 | |
| Soybeans | 35.0 | Dairy | | 237 |
| Lentils | 9.0 | Sheep | 6.4 | |
| Chick peas | 8.9 | Pork | 9.8 | |
| White beans | 9.7 | Layers | | 1.03 |
| Coloured beans | 8.9 | Broilers | 10.1 | |

On the pulse side, PCFs were expressed as the percent of dry matter yield. Using the methodology described by the USDA [38] and nutrition tables published on the USDA website [12,14,34], PCF values were derived for each of the pulses. The highest PCF was for soybeans, which is three- to four-times the PCF of the other pulse PCF (Table 1). This gives soybeans a particularly important role in potential alternate protein sources. The other five pulses have PCFs that are reasonably close to each other. Based on this difference in protein content, the following protein production assessment was applied separately to soybeans and the other pulses.

## 3.4. Defining the Area Inputs for Protein Production

The design of the assessment process described in this paper depended largely on how land use was distributed among the agricultural products being considered. Figures 1–3 summarize the land use inputs to ULICEES and F4E2 that drove the calculations in the three indicators.

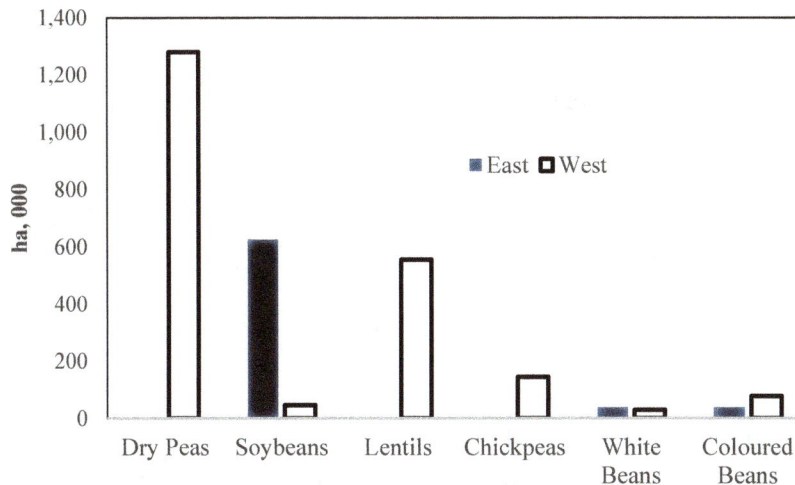

**Figure 1.** Areas (ha × $10^3$) in the pulses (annual legume crops) that were not used to support livestock production in Eastern and Western Canada in 2006.

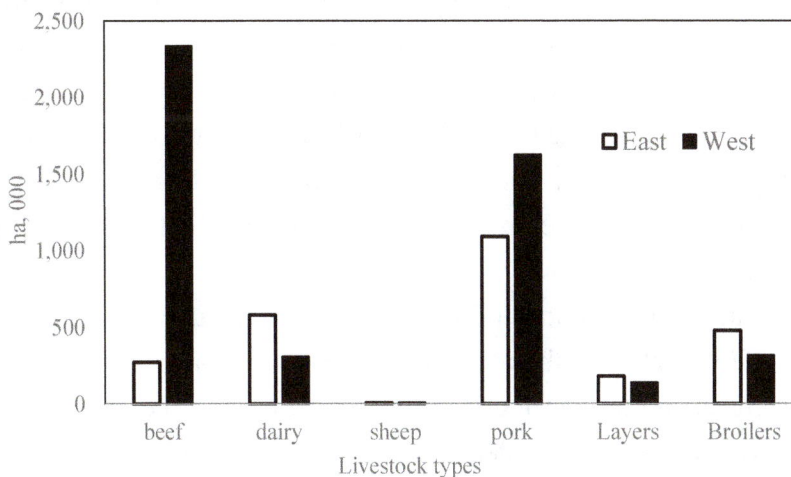

**Figure 2.** Areas (ha×$10^3$) in annual crops (including grains, oilseeds and pulses) used to support different types of livestock in Eastern and Western Canada in 2006.

In Figure 1, the general east-west differences among pulse production in Canada indicated which of those crops warranted the closest attention in this assessment. Six pulses were grown on a reportable scale in Canada in 2006, including dry peas, soybeans, lentils, chickpeas, white beans and coloured beans. Soybeans are the predominant pulse crop in the east. No lentils or chickpeas are grown on a reportable scale in the east. In the west, pulse production is more diversified among the six pulses, although areas in dry peas were twice as much as were in lentils, the next highest area.

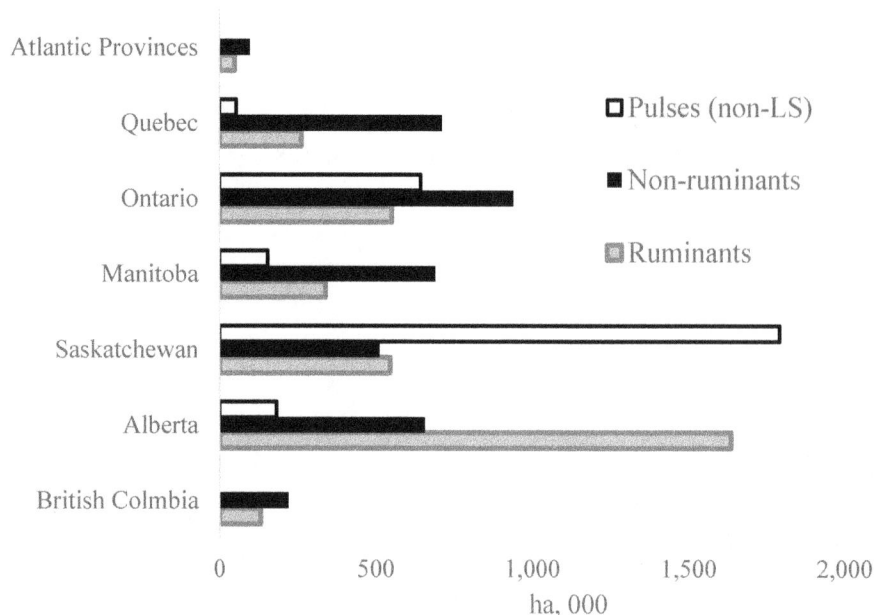

**Figure 3.** Areas in annual crops (ha $\times$ $10^3$) in each province of Canada used to grow or support three generalized groups of protein sources in 2006.

Figure 2 gives an east-west breakdown of the annual crop areas used to produce livestock protein, similar to Figure 1 for pulse protein. Figure 2 implies that the ruminant livestock industries are smaller than they really are because the land in perennial forage was not included in the comparison of land use. This was because this assessment only considered the land base that could be used to produce any of the plant or animal proteins. Consequently, it was assumed that none of the land that was growing perennial forage in 2006 could be used to grow pulse crops. This masks the full magnitude of dairy in relation to pork and the two poultry products as a land user, particularly in Eastern Canada. Nevertheless, the dominant role of western beef and, more importantly, the amount of land capable of growing annual field crops are still apparent.

Figure 3 shows the provincial distributions of all of the land in annual crops that supported protein production in Canada in 2006, including the areas that support ruminants, non-ruminants and pulses. The pulse areas were designated as non-livestock feed use (non-LS) crops because (as explained above) those pulses that support livestock were excluded. Because the areas in pulses are small relative to the crop areas that support livestock in Canada, the soybean areas and areas in non-soybean pulses were recombined into one land use category in Figure 4. The areas attributed to ruminant protein production suggest that the resource use by this livestock system is lower than is really the case because of this exclusion of land in perennial forage.

## 4. Results

### 4.1. Area Inputs for Livestock and Pulse Protein Production

In Figure 3, the areas in non-LS pulses were only comparable in size to the areas that support the Canadian livestock industry in Ontario and Saskatchewan. This was mainly due to the dominance of soybeans in Ontario and dry peas in Saskatchewan. Alberta was notable for having much more annual crop land devoted to ruminant livestock due to the large beef population and number of beef feedlots operating in that province [19]. The annual crop land devoted to non-ruminant livestock was relatively evenly distributed across all provinces, although it was the five provinces other than Saskatchewan and Alberta where non-ruminants account for the largest shares of the land that supports protein production.

## 4.2. Livestock GHG Emissions and Protein Production

Given the importance of enteric methane to the livestock CF [12], livestock commodities were grouped in Table 2 according to whether they were ruminants (which emit copious amounts of enteric methane) and non-ruminants (from which enteric methane is minimal). Table 2 gives a generalized summary of GHG emissions from the Canadian livestock industry for Eastern and Western Canada during 2006. Thus, Table 2 shows the emission estimates from ULICEES [20] for each livestock group and region for $CH_4$, $N_2O$ and fossil $CO_2$. Table 2 also shows the estimates from ULICEES for the protein produced by all livestock grouped as ruminants and non-ruminants and on an east-west basis.

**Table 2.** Total greenhouse emissions (GHG), grouped as being from either ruminant or non-ruminant livestock, and disaggregated by GHG type ($CH_4$, $N_2O$ and fossil $CO_2$), as well as protein production ($t \times 10^3$) from livestock farms in Eastern and Western Canada during 2006.

| | $CH_4$ | $N_2O$ | Fossil $CO_2$ | All GHG | Protein |
|---|---|---|---|---|---|
| | \multicolumn{4}{c}{Tg $CO_2$eq} | t,000 |
| | | | Eastern Canada | | |
| Ruminants [1] | 7.57 | 4.04 | 1.46 | 13.07 | 226 |
| Non-ruminants [2] | 2.10 | 2.21 | 1.00 | 5.31 | 292 |
| All livestock | 9.67 | 6.25 | 2.46 | 18.38 | 518 |
| | | | Western Canada | | |
| Ruminants [1] | 19.41 | 8.50 | 2.14 | 30.06 | 274 |
| Non-ruminants [2] | 2.05 | 1.14 | 0.84 | 4.03 | 204 |
| All livestock | 21.47 | 9.64 | 2.98 | 34.09 | 478 |
| | | | Canada | | |
| Ruminants [1] | 26.98 | 12.54 | 3.60 | 43.12 | 500 |
| Non-ruminants [2] | 4.16 | 3.35 | 1.84 | 9.34 | 496 |
| All livestock | 31.14 | 15.89 | 5.44 | 52.46 | 996 |

[1] Includes beef, dairy and sheep (lamb); [2] includes pork, layers (eggs) and broilers.

The two largest sources of GHG in Table 2 were the two methane terms from ruminants, which accounted for 58% of all GHG emissions in the Eastern Canadian ruminants and 65% of the western ruminants. Methane accounted for 40% and 51% of GHG from the non-ruminants in Eastern and Western Canada, respectively. Nitrous oxide emissions accounted for 31% and 28% of all GHG emissions from ruminants in Eastern and Western Canada, respectively, and 42% and 28% of all GHG emissions in Eastern and Western Canada, respectively, from non-ruminants. Fossil $CO_2$ emissions accounted for 11% and 7% from ruminants in the east and west and 19% and 21% from non-ruminants in the east and west, respectively.

## 4.3. Pulse GHG Emissions and Protein Production

Figure 4 shows the distribution of pulses across the provinces and illustrates the significance of treating soybeans separately. The GHG emissions and protein production from soybeans was quite distinct from the other pulses among the provinces in 2006. Ontario and Saskatchewan were the only two provinces with significant pulse protein production. Almost all of the soybeans were grown in Ontario, while no reportable soybeans were grown in Saskatchewan. The 2006 Ontario soybean crop emitted a third as much $CO_2eq$ as did the Saskatchewan pulse crops, but provided almost three-times as much protein.

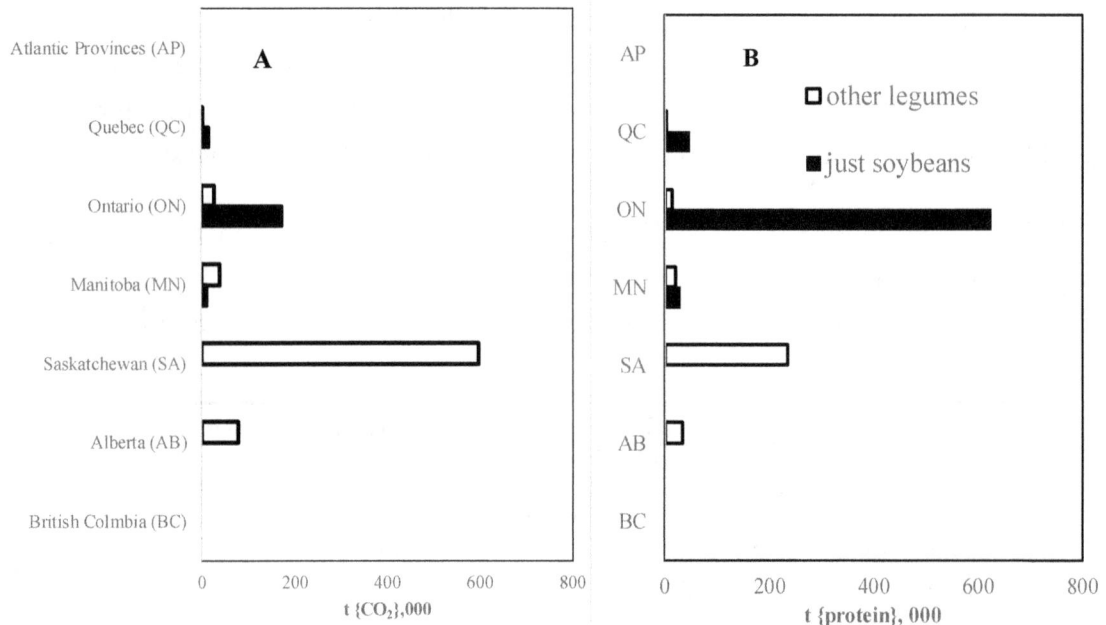

**Figure 4.** Fossil $CO_2$ emissions (**A**) and protein production (**B**) from pulses (annual legume crops) grown in each province of Canada in 2006.

## 4.4. Results of Three Indicators for Pulse and Protein Production

Table 3 summarizes the three performance indicators used in this paper for both plant and animal protein production in Eastern and Western Canada in 2006. The combined plant-animal protein production intensities (both pulse and livestock sources) are shown in Table 4 on a provincial basis for 2006. It should be noted that the direction of values considered beneficial for both Tables 3 and 4 is from low to high in Indicator 2 and from high to low in Indicators 1 and 3. Therefore, in Indicator 2, high numbers mean good, rather than poor, performance.

Table 3 shows that soybeans produced four- to eight-times as much protein per ha as either livestock group in 2006, whereas the area-based protein production of the other pulses was roughly the same order of magnitude as both livestock groups. Excluding soybeans, pulses and the two eastern livestock production systems have similar land use intensities for proteins (Column 3). The land use intensities of non-soy pulses exceeded both of the western livestock production systems (Column 4). For the other two performance indicators, plant protein production showed a dramatic advantage over animal protein (Columns 1, 2, 5 and 6). This was particularly true for soybeans, where ruminant protein emitted

40–50-times as much $CO_2$eq per ha as did soy protein. For the other pulses compared to ruminant animal protein, this emission rates ratio was roughly 35 to one. Even for the other pulses compared to non-ruminant animal protein, the animal protein production had five- to eight-times as high a $CO_2$eq emission rate per ha as did plant protein production. These animal to plant ratios were even more dramatic for $CO_2$eq emission rates per unit of protein, with the ratio of ruminant to soy protein being roughly two orders of magnitude.

**Table 3.** Greenhouse gas and land use indicators for protein production in Eastern and Western Canada in 2006.

| Protein Sources: | $tCO_2$/ha | | Kg (protein)/ha | | $tCO_2$/t (protein) | |
|---|---|---|---|---|---|---|
| **Animal:** | **East** | **West** | **East** | **West** | **East** | **West** |
| Ruminants | 15.17 | 11.33 | 263 | 103 | 57.77 | 109.83 |
| Non-ruminants | 3.13 | 1.82 | 167 | 83 | 18.79 | 21.97 |
| **Plant:** | **East** | **West** | **East** | **West** | **East** | **West** |
| Soybeans | 0.30 | 0.26 | 1077 | 630 | 0.28 | 0.42 |
| Other legumes | 0.41 | 0.34 | 207 | 139 | 1.98 | 2.46 |

**Table 4.** Assessment of the integrated protein production on a provincial basis for 2006 in Canada using the three indicators described in this paper, including, the land use basis [1] for GHG emission intensity and all protein [2] production and the protein-based intensity for GHG emissions.

| Provinces [3] | AP | QC | ON | MN | SA | AB | BC |
|---|---|---|---|---|---|---|---|
| Reporting units | | | | | | | |
| | Land [3] use intensity of GHG emissions | | | | Indicator 1 | | |
| $tCO_2$eq/ha | 9.3 | 7.8 | 4.4 | 5.2 | 3.4 | 6.7 | 13.0 |
| | Land use intensity of protein production | | | | Indicator 2 | | |
| T (protein)/ha | 0.23 | 0.27 | 0.42 | 0.15 | 0.12 | 0.09 | 0.32 |
| | Protein-based intensity for GHG emissions | | | | Indicator 3 | | |
| $tCO_2$eq/t (protein) | 40.4 | 28.6 | 10.5 | 35.2 | 27.9 | 77.6 | 40.3 |

[1] For just land growing annual crops in the livestock crop complex (LCC) and growing non-LS pulses; [2] includes only complete, or animal equivalent, protein; [3] provinces defined using the abbreviations given in Figure 4. Atlantic Provinces: AP; Quebec: QC; Ontario: ON; Manitoba: MN; Saskatchewan: SA; Alberta: AB; British Columbia: BC.

To assess the results from Table 4, provincial rankings over the three indicators were determined. For Indicator 1, the low-to-high rankings were SA, ON, MN, AB, QC, Atlantic Provinces (AP) and, finally, BC. For Indicator 2, the high-to-low rankings were ON, BC, QC, AP, MN, SA and then AB. For Indicator 3, the low-to-high rankings were ON, SA, QC, MN, BC, AP and then AB. Ontario ranked second in Indicator 1 and first in Indicators 2 and 3, giving this province the best overall ranking among provinces. Alberta ranked fourth, seventh and seventh over the three indicators, respectively, making it the overall worst performing province. With rankings of sixth, fourth and sixth over the three indicators, the Atlantic Provinces were the second worst performers overall. Quebec and Manitoba were quite consistently mid-range performers over the three indicators, with rankings of third, fifth and fourth for Manitoba and fifth, third and third for Quebec. The rankings were somewhat inconsistent for Saskatchewan and British Columbia, with rankings of first, sixth and second for Saskatchewan, and seventh, second and fifth for British Columbia over the three indicators.

## 5. Discussion

### 5.1. Evaluating Indicator Performance

Although Table 2 showed that ruminants emit far more $CH_4$ than non-ruminants, especially considering that the greatest LW mass of livestock is in beef cattle, the differences between ruminants and non-ruminants were not as much as would be expected from just considering enteric methane. This was because the stored manure from the non-ruminants typically emitted larger quantities of $CH_4$ than did the stored manure from the ruminants (largely due to more liquid manure storage by non-ruminant producers). The east-west differences in $N_2O$ emissions between the ruminants and non-ruminants also indicate more liquid manure storage in non-ruminant operations, since dry manure storage typically emits more $N_2O$ than liquid manure storage. For livestock, and particularly for ruminants in Western Canada, fossil $CO_2$ emissions were the lowest of the three GHG emissions.

It can be seen in the first two columns of Table 3 (or Indicator 1) that the CF of proteins from pulses was very small compared to the CF of livestock protein. Table 2 partly demonstrates why this would be so. For ruminants, fossil $CO_2$ (the only GHG emitted by pulse production) is around 10% or less of the whole GHG emissions budget of ruminant livestock and 20% of the non-ruminant livestock GHG emissions budget. In addition, the fossil $CO_2$ component of the livestock GHG emissions budget includes the manufacturing of nitrogen fertilizer [20], which accounts for roughly 40% of the fossil $CO_2$ emissions of the fossil energy and $CO_2$ budget of field crops in the LCC. In contrast, fertilizer supply was not included in (and does not contribute to) the $CO_2$ emissions budget for producing the pulse protein. Without including the fertilizer supply term in the livestock CF, the fossil $CO_2$ per ha cost of growing feed grains would be reasonably close to the fossil $CO_2$ per ha cost of growing pulses. In Table 3, pulses also showed a higher rate of protein production per ha (Indicator 2). However, much of this advantage was due to the soybean PCF compared to the other pulses.

Some of the discrepancy among the rankings of provinces for the three indicators in Table 4 can be explained by the differences in the two sub-figures in Figure 4. In Figure 4A, soybean production in Ontario is less than half that of the other pulses grown in Saskatchewan, but the protein production from Ontario soybeans was almost three-times as much as the protein from the other pulses in Saskatchewan. This was due to the much higher protein component (PCF) for soybeans compared to the PCF of the dry peas and lentils grown in Saskatchewan (Table 1). Similarly, the ratios of different annual crop areas of ruminants to non-ruminants among provinces in Figure 3 (taking into account the difference in CF for the two livestock groups) and of western beef in Figure 2 also help to explain why there were some discrepancies in the rankings among the three indicators in Table 4.

The ordinal assessment (rankings) of the three indicators in Table 4 points to Ontario being the most effective protein-producing province in 2006. Saskatchewan should be considered the second best protein-producing province, if allowance is made for the slower growing conditions in the semi-arid cool agro-climate of Saskatchewan compared to BC and the eastern provinces. Alberta was the poorest performing province with the lowest ranks in two of the three indicators. Figure 3 showed that two of the three provinces that ranked lower than Alberta in Indicator 1, BC and AP, had no reportable areas in pulses, while Quebec had quite small areas in pulses. Although Alberta had a large beef herd in 2006, a large portion of these cattle were in feedlots with high energy, low roughage diets, which lowered the enteric methane emissions [22].

Some caution is needed in that this assessment did not consider the processing energy to bring the respective protein to an edible condition for humans. None of the protein sources considered here are edible in their raw state. The top market value of soy products, for example, may be partly offset by the extensive preparation needed to convert it to tofu [36,39]. On the other hand, all of the animal products require refrigeration on their way to market and the dinner table. However, the life cycle assessment for the dairy industry by Vergé *et al.* [30] found that 90% of the GHG emissions happened before the milk left the farm. Caution is also needed in the equivalency between plant and animal proteins, because this interpretation is largely qualitative. While it is possible to quantify the respective percentages of complete proteins from carcasses and pulses, it seems to be beyond the current state of knowledge what fractions of those two respective protein sources consist of the nine essential, or critical, amino acids that make proteins complete.

Another caution is that the results reported in this paper are specific to Canada. As soybean production expands around the world in response to growing markets, such as China, the displacement of beef cattle by soybeans can have a range of impacts depending on the type of land being shifted from the beef industry to pulse production [4,40,41]. For example, tofu production in European studies has a higher CF than tofu produced from North American soybeans, because the soybeans imported to Europe are largely from Brazil and cause deforestation of the Amazon [5,38].

### 5.2. Implications for Protein Demand

The last step in this assessment was to interpret protein production estimates in human terms. With a million t of protein from livestock (Table 2) and another million t from pulses (Figure 4B) in Canada and using an approximate requirement for daily intake of protein per capita, a rough estimate of the number of protein person years (PPY), or the number of people per year that can be provided with adequate protein, can be calculated. A rough mid-range value of 70 g of complete protein per capita per day was taken from the daily intake values given above. This calculation suggests that the actual 2006 protein production in Canada could have provided in the order of 78 million PPY, with about half coming from pulses and half from livestock.

From this perspective, it is useful to consider how the conversion of all of the annual crop land in the LCC to pulses might have changed protein production and the CF of that protein. If the 2006 annual crops area in the LCC for all of Canada were taken out of livestock production and reseeded to a combined mix of all of the 2006 pulses, then the protein from this conversion, combined with the protein actually produced from non-LS pulses in 2006, would have provided 141 million PPY from Canada. The mix of pulses across Canada in 2006 resulted in 70% of the non-LS pulse protein being from soybeans and 30% from the other four pulses. If that LCC area were reseeded into just the other pulses (no soybeans), then the combined 2006 protein supply from Canada would have been 80 million PPY.

If that LCC area were reseeded into just soybeans, the combined total 2006 protein supply from Canada would have been 340 million PPY.

The CF of this hypothetical all-pulse protein supply scenario would be roughly the same intensity as the actual 2006 protein production shown in Table 3, or 0.33 t $CO_2$/ha. There would be an appreciable difference in the two pulse protein sources, since soybeans yield more protein per weight of crop than the other pulses (Table 1). The protein-based GHG emission intensities (Indicator 3) in Table 3 for the other pulses were four- to six-times higher than for soybeans. However, when compared to livestock, Indicator 3 shows that both sources of pulse protein have orders of magnitude lower protein-based CF values than either the ruminant or non-ruminant protein sources. Caution is needed in interpreting the results of the assessment in this generalized manner, given the potential margins of error in the ULICEES-F4E2 modelling system and the wide range of recommendations in the literature regarding quantity and quality of daily protein intake. In spite of these limitations, the degree of differences between the pulse and livestock protein-based CF cannot be discounted.

Should land growing feed grains for beef be converted to non-LS pulses (as in the PPY scenarios), there remains the potential for the cattle being displaced to be converted back to a more perennial forage diet. While that would mean increased enteric methane emissions from these displaced cattle [22], it could also mean increased soil carbon sequestration, if the land growing that perennial forage were taken from land currently growing annual crops [20]. The PPY scenario did not take these indirect impacts into account, because it was assumed that the cattle being displaced by expanded pulse production would be eliminated. Otherwise, new land has to be brought into the PPY scenario, thus making it too speculative to be of any policy value.

Although hay and pasture are a major resource in ruminant production, potential land use changes were not a factor in this assessment until the displacement of feed grain by non-LS pulse was considered in the PPY scenarios. This transfer of land out of feed grains was at the expense of feedlot cattle, a major part of the Canadian beef industry [20,22]. Thus, as the feedlot cattle go to market, they would not be replaced, and the beef industry would shift back towards cattle being sent directly to market after being raised mainly on hay and pasture. Hence, the amount of hay and pasture would remain mostly unchanged while only the areas in feed grains would decline. A coincidental increase in food demand could also make feed grains more costly and the less intensive pasture-grown beef more profitable than feedlot-finished beef. In contrast to Alexander *et al.* [5], who defined land displacement as the migration of activities to another place, a cattle population shrinkage was assumed, since the new pulse protein would push beef out of part of its market. There remains the possibility that the displaced feedlot cattle could be transferred to other land. However, this secondary land use change would also result in the pulse protein being an additional protein supply, rather than livestock protein being replaced by pulse protein (as required in the final PPY scenario).

## 6. Conclusions

The conclusions drawn in this paper relied on all three of the indicators described above. Hence, Indicator 3, the protein-based GHG emission intensity, is most effective when used in conjunction with the two land use indicators. The provinces that ranked best under Indicator 3 were those with the largest areas in non-LS pulses, which is consistent with the lower CF of pulse protein than animal protein.

The province ranking the lowest, Alberta, depended heavily on beef and had a relatively small area in pulses. The two coastal provinces (AP and BC) were almost as low as Alberta, which was to be expected, since neither had any reportable non-LS pulse areas.

The results of this protein comparison identified soybeans as having the lowest CF and as being the most effective land use for protein production. In spite of far greater areas in pulses in Saskatchewan, the lower PCF of dry peas and lentils allowed Ontario, the leading soybean producer, to be the highest protein-producing province in 2006. However, caution is needed when an analysis points to just one crop as a wonder crop or super food. There are both nutritional and environmental risks from creating a global-scale dependence on just one crop. On the other hand, the benefits of soybeans derived from this assessment did not include the additional value of that crop as a potential source of biodiesel feedstock [34] or a cooking oil [35], a consideration that would have lowered its net CF even further.

Much of the methodology described in this paper could also be applied in other countries with developed agriculture. In Canada, the two provinces with the lowest CF for protein production, Ontario and Saskatchewan, both on an area and protein basis, also had the largest pulse production. This result suggests that developed countries can reduce the CF of their capacities to produce protein by increasing their consumption and export of pulse protein rather than by producing more livestock. Such extrapolation to other countries, however, should be limited to those regions that do not depend heavily on extensive grazing, because cattle raised primarily on grass have a different CF than cattle raised in the more intensive beef operations typical of North America. In many parts of the world, the soil under the grazing land may not be suitable for annual cultivation, and converting such land to annual crops could lead to serious land degradation.

The two issues identified at the start of this paper, global protein supply and a lower CF for protein production, can both be achieved through a shift in land use towards pulses and away from livestock. Although this analysis also suggests that ruminants are less efficient than non-ruminants with respect to Indicator 3, it should be cautioned that only ruminants can convert perennial forage to protein. However, keeping cattle on low roughage diets (as in feedlots) negates much of the ruminant advantage. The three indicators, particularly 2 and 3, address the global challenges of demand for protein in the human diet and minimizing the GHG emissions from protein production. While expressing the protein production in terms of satisfying global nutritional requirements went beyond the assessment of the three indicators discussed in Tables 3 and 4, this interpretation has potential food policy value. Inside the farm gate, findings, such as described in this paper, should link with the LCA of whole life cycles when more specific protein products are being assessed.

Although this interpretation is not likely to influence the market strategies of the livestock industries, it should help to position Canada's role in the global protein supply chain. The PPY concept may be a useful index for evaluating this supply chain on a global scale. The low CF of Canadian soybeans relative to soybeans imported from places where deforestation is a big factor in their CF [38] (such as Brazil) should also help to establish this position for Canada. Since this paper was not aimed at marketing of Canadian pulses, tofu and other meat analogues manufactured from pulses could be removed from the CF calculation of pulses and the scope of this assessment. As a protein supplement to food aid, pulse protein enjoys another advantage over livestock protein, because it can be shipped dry, stored without refrigeration and is usually ready to eat right after boiling.

It must be acknowledged that it is relatively easy to find literature and Internet sources that extol the potential environmental and social benefits of replacing livestock protein with pulse protein. It is somewhat surprising, however, just how much the pulse proteins out-performed the livestock protein as rated by all three indicators within the farm gate. However, the true value of this assessment was not in repeating what some might see as the obvious, but in quantifying the process into a repeatable and scalable package. The level of integration in this modelling system brought all of the variables into one quantitative process (inventory model) that could be applied directly, or rebuilt, in many other countries. Although only 2006 was assessed in this paper, having one computational package will facilitate temporal flexibility by allowing the findings for 2006 to be compared to past or future years (when those data become available) or to hypothetical years that reflect policy scenarios.

In undertaking the second goal of this paper, the disaggregation to provinces, the spatial scalability of this methodology was able to highlight the impact of the inter-provisional diversity of protein sources in Canada. The analysis described in this paper could be applied on a sub-provincial scale, such as ecodistricts and census agricultural regions (CAR). Eventually, international treaties encourage all countries to adopt measures that will mitigate GHG emissions in all sectors. Implementation of such measures for Canadian agriculture will require communication with farming communities on sub-provincial scales (either ecodistricts or CARs). Assuming that the provincial scale inputs used in this assessment are available at these finer scales, this assessment could be repeated at those scales.

## Author Contributions

James Dyer was responsible for the design of the study, the generation of the pulse performance data, integration of the livestock and pulse protein estimates, the interpretation of results and for the manuscript preparation. Xavier Vergé was responsible for the ULICEES model, generating the estimates for livestock GHG emissions and contributions to writing the paper.

## Conflicts of Interest

The authors declare no conflict of interest.

## References

1.  Jacob, B. Global Hunger for Protein Fuels Food-Industry Deals. *The Wall Street Journal*, 11 June 2014. Available online: http://www.wsj.com/articles/global-hunger-for-protein-fuels-food-industry-deals-1402444464 (accessed on 31 August 2015).
2.  Kelly, T. Impending Crisis: Earth to Run Out of Food by 2050. *Time*, 7 December 2010. Available online: http://newsfeed.time.com/2010/12/07/impending-crisis-earth-to-run-out-of-food-by-2050/ (accessed on 31 August 2015).
3.  Ritter, K. Climate Change Could Push 100 Million into Extreme Poverty By 2030. *The Associated Press*. Available online: http://www.huffingtonpost.ca/2015/11/09/climate-change-100-million-poverty-world-bank_n_8509444.html (accessed 10 November 2015).

4.  Smith, P.; Haberl, H.; Popp, A.; Erb, K.-H.; Lauk, C.; Harper, R.; Tubiello, F.N.; de Siqueira P.A.; Jafari, M.; Sohi, S.; *et al.* How much land-based greenhouse gas mitigation can be achieved without compromising food security and environmental goals? *Glob. Chang. Biol.* **2013**, *19*, 2285–2302, doi:10.1111/gcb.12160.

5.  Alexander, P.; Rounsevell, M.D.A.; Dislich, C.; Dodson, J.R.; Engström, K.; Moran, D. Drivers for global agricultural land use change: The nexus of diet, population, yield and bioenergy. *Glob. Environ. Chang.* **2015**, *35*, 138–147, doi:10.1016/j.gloenvcha.2015.08.011.

6.  D'Odorico, P.; Carr, J.A.; Laio, F.; Ridolfi, L.; Vandoni, S. Feeding humanity through global food trade. *Earth Future* **2014**, *2*, 458–469, doi:10.1002/2014EF000250.

7.  Cheadle B. Obama's Climate-Change Talk Stands in Stark Contrast to Canadian Party Leaders. *The Canadian Press*, 1 September 2015. Available online: http://www.cbc.ca/news/world/obama-s-climate-change-talk-stands-in-stark-contrast-to-canadian-party-leaders-1.3211523 (accessed on 1 September 2015).

8.  Green, Low-Emission and Climate-Resilient Development Strategies (United Nations Development Programme (UNDP)), 2015. Available online: http://www.undp.org/content/undp/en/home/ourwork/environmentandenergy/focus_areas/climate_strategies.html (accessed on 31 August 2015).

9.  Gerber, P.J.; Steinfeld, H.; Henderson, B.; Mottet, A.; Opio, C.; Dijkman, J.; Falcucci, A.; Tempio, G. *Tackling Climate Change Through Livestock—A Global Assessment of Emissions and Mitigation Opportunities*; Food and Agriculture Organization of the United Nations (FAO): Rome, Italy, 2013; p. 139.

10. Goodland, R. The Overlooked Climate Solution: Joint Action by Governments, Industry, and Consumers. *J. Hum. Secur.* **2010**, *6*. Available online: http://search.informit.com.au/document Summary;dn=332506353716599;res=IELHSS (accessed on 6 November 2015).

11. Ripple, W.J.; Smith, P.; Haberl, H.; Montzka, S.A.; McAlpine, C.; Boucher, D.H. Ruminants, climate change and climate policy. *Nat. Clim. Chang. Opin. Comment* **2014**, *4*, 2–4.

12. The Role of Livestock in Climate Change. Available online: http://www.fao.org/agriculture/lead/themes0/climate/en/ (accessed on 31 August 2015).

13. Dyer, J.A.; Vergé, X.P.C.; Desjardins, R.L.; Worth, D.E. The protein-based GHG emission intensity for livestock products in Canada. *J. Sustain. Agric.* **2010**, *34*, 618–629, doi:10.1080/1044004 6.2010.493376.

14. Whitbread, D. Top 10 Foods Highest in Protein You Can't Miss. *Health Alicious Ness*, 2011. Available online: http://www.healthaliciousness.com/articles/foods-highest-in-protein.php (accessed on 21 August 2015).

15. ChartsBin Statistics Collector Team (CBSCT). Daily Protein Intake Per Capita. Available online: http://chartsbin.com/view/1155 (accessed on 30 August 2015).

16. Mack, S. How Many Amino Acids Does the Body Require? *Living Strong*, Last Updated February 2014. Available online: http://www.livestrong.com/article/463106-how-many-amino-acids-does-the-body-require/ (accessed on 30 August 2015).

17. Understanding Our Bodies: Amino Acids Are Important. Available online: http://nutrition wonderland.com/2009/07/understanding-our-bodies-amino-acids/ (accessed on 30 August 2015).

18. Whitbread, D. Beans and Legumes with the Most Protein. *HealthAliciousNess*, 2011. Available online: http://www.healthaliciousness.com/articles/beans-legumes-highest-protein.php#percent-protein (accessed on 21 August 2015).

19. Dyer, J.A.; Vergé, X.P.C.; Desjardins, R.L.; Worth, D.E. A comparison of the greenhouse gas emissions from the sheep industry with beef production in Canada. *Sustain. Agric. Res.* **2014**, *3*, 65–75, doi:10.5539/sar.v3n3p65.

20. Vergé, X.P.C.; Dyer, J.A.; Worth, D.; Smith, W.N.; Desjardins, R.L.; McConkey, B.G. A greenhouse gas and soil carbon model for estimating the carbon footprint of livestock production in Canada. *Animals* **2012**, *2*, 437–454, doi:10.3390/ani20x000x.

21. Vergé, X.P.C.; Dyer, J.A.; Desjardins, R.L.; Worth, D. Greenhouse gas emissions from the Canadian dairy industry during 2001. *Agric. Syst.* **2007**, *94*, 683–693, doi:10.1016/j.agsy.2007.02.008.

22. Vergé, X.P.C.; Dyer, J.A.; Desjardins, R.L.; Worth, D. Greenhouse gas emissions from the Canadian beef industry. *Agric. Syst.* **2008**, *98*, 126–134, doi:10.1016/j.agsy.2008.05.003.

23. Vergé, X.P.C.; Dyer, J.A.; Desjardins, R.L.; Worth, D. Greenhouse gas emissions from the Canadian pork industry. *Livest. Sci.* **2009**, *121*, 92–101, doi:10.1016/j.livsci.2008.05.022.

24. Vergé, X.P.C.; Dyer, J.A.; Desjardins, R.L.; Worth, D. Long Term trends in greenhouse gas emissions from the Canadian poultry industry. *J. Appl. Poult. Res.* **2009**, *18*, 210–222, doi:10.3382/japr.2008–00091.

25. McLaughlin, E.M.; Alain, B. *Livestock Feed Requirements Study 1999–2001*; Catalogue No. 23–501-XIE; Statistics Canada: Ottawa, ON, Canada, 2003.

26. 2006 IPCC Guidelines for National Greenhouse Gas Inventories, Volume 4 Agriculture, Forestry and Other Land Use. Available online: http://www.ipcc-nggip.iges.or.jp/public/2006gl/vol4.html (accessed on 16 September 2015).

27. Rochette, P.; Worth, D.E.; Lemke, R.L.; McConkey, B.G.; Pennock, D.J.; Wagner-Riddle, C.; Desjardins, R.L. Estimation of N2O emissions from agricultural soils in Canada. I. Development of a country-specific methodology. *Can. J. Anim. Sci.* **2008**, *88*, 1–14.

28. Dyer, J.A.; Desjardins, R.L. Simulated farm fieldwork, energy consumption and related greenhouse gas emissions in Canada. *J. Sustain. Agric.* **2003**, *34*, 618–629, doi:10.1080/10440046.2010.493376.

29. Finner, M.F. *Farm Field Machinery*, 2nd ed.; American Printing and Publishing Inc. and Agricultural Engineering, University of Wisconsin: Madison, WI, USA, 1973; p. 226.

30. Jacobs, C.O.; Harrell, W.R. *Agricultural Power and Machinery*, 1st ed.; McGraw Hill Book Co.: NewYork, NY, USA 1983; p. 472. Available online: https://books.google.ca/books?id=B2YfAQ AAMAAJ&q=ISBN+0070322104 (accessed on 16 September 2015).

31. A Review of the 1996 Farm Energy Use Survey (FEUS). Available online: http://www.usask.ca/agriculture/caedac/pubs/pindex.html (accessed on 16 September 2015).

32. Vergé, X.P.C.; Maxime, D.; Dyer, J.A.; Desjardins, R.L.; Arcand, Y.; Vanderzaag, A. Carbon footprint of Canadian dairy products: Calculations and issues. *J. Dairy Sci.* **2013**, *96*, 6091–6104, doi:10.3168/jds.2013-6563.

33. Dyer, J.A.; Vergé, X.P.C.; Desjardins, R.L.; McConkey, B.G. Assessment of the carbon and non-carbon footprint interactions of livestock production in Eastern and Western Canada. *Agroecol. Sustain. Food Syst.* **2014**, *38*, 541–572, doi:10.1080/21683565.2013.870631.

34. Koc, A.B.; Mudhafer A.; Fereidouni, M. Soybeans Processing for Biodiesel Production. In *Soybean—Applications and Technology*; Ng, T.-B., Ed.; InTech. Open Access Publisher: Rijeka, Croatia, 2015; pp. 19–37.

35. The George Mateljan Foundation. Is Soybean Oil Considered a Healthy Oil? 2015. Available online: http://www.whfoods.com/genpage.php?tname=dailytip&dbid=187 (accessed on 9 November 2015).

36. Plate, T. Tofu's Carbon Footprint. *The Vegetarian Environmentalist*, 2015. Available online: http://tofuscarbonfootprint.weebly.com/ (accessed on 6 November 2015).

37. The George Mateljan Foundation. What's New and Beneficial about Soybeans? 2015. Available online: http://www.whfoods.com/genpage.php?tname=foodspice&dbid=79 (accessed on 5 November 2015).

38. USDA National Nutrient Database for Standard Reference, Release 26. Available online: http://www.ars.usda.gov/ba/bhnrc/ndl (accessed on 16 September 2015).

39. Rastogi, N. How Green Is Tofu? You'd be Surprised to Know. *The Green Lantern*, 2009. Available online: http://www.slate.com/articles/health_and_science/the_green_lantern/2009/10/how_green_is_tofu.html (accessed on 4 November 2015).

40. Monahan, J. Soybean Fever Transforms Paraguay. *BBC*, 6 June 2005. Available online: http://news.bbc.co.uk/2/hi/business/4603729.stm (accessed on 4 November 2015).

41. Steinfeld, H.; Gerber, P.; Wassenaar, T.; Castel, V.; Rosales, M.; de Haan, C. Livestock's Role in Climate Change and Air Pollution. In *Livestock's Long Shadow. Environmental Issues and Options*; Animal Production and Health Division, Food and Agriculture Organization of the United Nations (FAO): Rome, 2006; Chapter 3, pp. 80–123.

# 14

# Genetic Dissection of Disease Resistance to the Blue Mold Pathogen, *Peronospora tabacina*, in Tobacco

Xia Wu [1,2], Dandan Li [1], Yinguang Bao [1,2], David Zaitlin [3], Robert Miller [1] and Shengming Yang [1,*]

[1] Department of Plant & Soil Sciences, University of Kentucky, Lexington, KY 40546, USA; E-Mails: xwu225@uky.edu (X.W.); dli2@uky.edu (D.L.); ygbao@sdau.edu.cn (Y.B.); rdmiller@uky.edu (R.M.)
[2] Department of Plant Genetics and Breeding, Shandong Agricultural University, Shandong 271018, China
[3] Kentucky Tobacco Research & Development Center, University of Kentucky, Lexington, KY 40546, USA; E-Mail: david.zaitlin@uky.edu

* Author to whom correspondence should be addressed; E-Mail: syang2@uky.edu

Academic Editor: Diego Rubiales

**Abstract:** Tobacco blue mold, caused by the obligately biotrophic oomycete pathogen *Peronospora tabacina* D.B. Adam, is a major foliar disease that results in significant losses in tobacco-growing areas. Natural resistance to *P. tabacina* has not been identified in any variety of common tobacco. Complete resistance, conferred by *RBM1*, was found in *N. debneyi* and was transferred into cultivated tobacco by crossing. In the present study, we characterized the *RBM1*-mediated resistance to blue mold in tobacco and show that the hypersensitive response (HR) plays an important role in the host defense reactions. Genetic mapping indicated that the disease resistance gene locus resides on chromosome 7. The genetic markers linked to this gene and the genetic map we generated will not only benefit tobacco breeders for variety improvement but will also facilitate the positional cloning of *RBM1* for biologists.

**Keywords:** tobacco; blue mold; disease resistance; genetic marker

# 1. Introduction

Common tobacco (*Nicotiana tabacum* L.) is one of the most important non-food crops worldwide, and is also a model plant for biological research [1,2]. Considerable interests have been focused on the molecular mechanisms underlying disease resistance to numerous pathogenic microbes in tobacco [3,4]. Several species of oomycetes, also known as water molds, are among the most devastating plant pathogens that cause notable diseases such as late blight of potato, downy mildew of grape vine, and root and stem rot of soybean. Tobacco blue mold, caused by the obligately biotrophic oomycete pathogen *Peronospora tabacina* D.B. Adam (syn. *P. hyoscyami* de Bary), is a major foliar disease that causes significant crop losses in tobacco-growing areas around the world. Annual losses exceeding $200 million due to blue mold epidemics have been reported in the United States and Canada [5,6].

Chemical treatments have been effective in controlling the spread of blue mold disease but given the economic and environmental costs of fungicide application, harnessing host resistance is the most sustainable strategy for reducing potential crop losses from blue mold. Natural genetic variation in host-pathogen interactions is key to the development of disease-resistant cultivars. Unfortunately, natural resistance to *P. tabacina* is very low in *N. tabacum*, and most commercial varieties are highly susceptible to blue mold disease [7]. A high level of functional resistance to *P. tabacina* infection was identified in both *N. debneyi* and *N. goodspeedii* and was transferred into cultivated tobacco by crossing [8–11]. Resistant tobacco varieties were first released beginning in the 1960s. However, the genes identified in the undomesticated species appear to confer only partial resistance to blue mold infection when incorporated into cultivated tobacco through interspecific hybridizations [8,10,12]. A possible reason for the weakened immunity is that the expression levels of major genes from these undomesticated species are down-regulated by modifier genes in the tobacco genome [7,13]. Alternatively, genetic resistance to blue mold in *N. debeyi* could be determined by multiple factors, but not all of these genes were transferred to tobacco successfully. Nevertheless, both explanations are only speculative at present. Therefore, cloning and characterization of the *Nicotiana* resistance genes directed against blue mold will not only further our understanding of host resistance to oomycete pathogens but also offer new insights into the optimization of genetic resistance to this destructive disease in tobacco.

Molecular markers closely linked to the blue mold resistance locus (*RBM1* hereafter) derived from *N. debneyi* have been developed and used for marker-assisted selection [9,14]. Utilization of these markers greatly facilitates tobacco breeding for blue mold resistance. Host responses to *P. tabacina* are complex and unpredictable under field conditions because multiple factors including plant age, physiological status, and environmental conditions can affect plant reactions to pathogen infection. Consequently, selection solely based on disease phenotype can be misleading. Three sequence characterized amplified region (SCAR) markers, two of which were converted from flanking random amplified polymorphic DNA (RAPD) markers [9] and one that was derived from an amplified fragment length polymorphism (AFLP) marker [14], have become valuable assets to breeding programs worldwide for the improvement of blue mold resistance in tobacco. However, these SCAR markers are dominant, precluding differentiation of plants that are homozygous from those that are heterozygous at the resistance locus, and their genetic locations are unknown. In this study, we conducted genetic mapping of *RBM1* and characterized genetic resistance to *P. tabacina*, providing a robust foundation for map-based cloning of *RBM1* and for engineering *RBM1*-mediated resistance.

## 2. Results

### 2.1. Disease Reaction Assay and Segregation Analysis

Typical symptoms were clearly observed on leaves of TKF (Tennessee Kentucky fertile) 2002 plants six days after spray-inoculation with the pathogen (Supplementary Materials Figure S1). The diseased leaves became spotted with grey lesions that subsequently produced areas of abundant downy sporulation on the lower surface, while leaves of the resistant parent TKF 4321 remained healthy at the same time point (Supplementary Materials Figure S1). Histological analyses were performed to monitor the course of tissue colonization by *P. tabacina*. Inoculated leaves were cleared, and pathogen structures were detected by lactophenol-trypan blue staining (Figure 1). No significant differences were noted between resistant and susceptible lines during the pre-penetration events. Spores germinated on the leaf surface and formed appressoria between 1 and 3 h after inoculation (hpi). Following penetration, pathogen colonization proceeded rapidly in susceptible TKF 2002 cells at 48 hpi, but no colonization was observed during this time in cells of the resistant line (Figure 1A,B). At 120 hpi, the pathogen produced lemon-shaped sporangia (spores) on tree-like branched structures (sporangiophores) that emerged from the leaf stomata in TKF 2002 (Figure 1H). In contrast, pathogen development in TKF 4321 was very restricted. By 72 hpi, only a few hyphae were observed in TKF 4321 leaves (Figure 1C). Although more hyphae were detected at 72 and 96 hpi, sporangiophores were never observed in TKF 4321 leaves. Even if sporangiophores were produced in TKF 4321, they would be very rare and difficult to detect. To test whether the hypersensitive response (HR) was involved in host defense, we inoculated tobacco by injecting *P. tabacina* into the leaves. HR-induced chlorosis was observed at the inoculation sites in TKF 4321, and it limited further development and spread of the pathogen. This was in contrast to the development of a systemic infection beyond the inoculation sites in TKF 2002 (Supplementary Materials Figure S2). Therefore, the HR appears to be an important component of *RBM1*-mediated disease resistance.

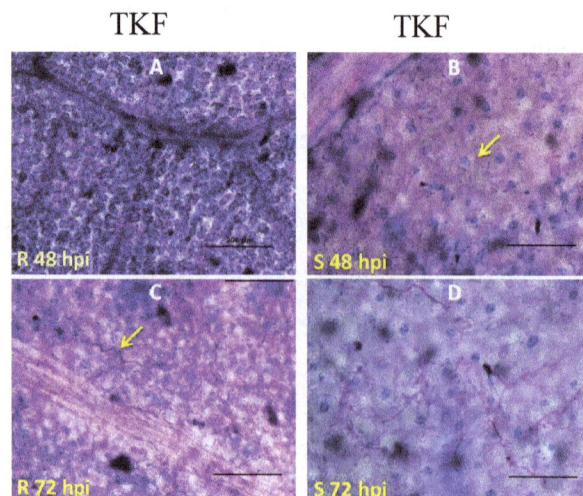

**Figure 1.** *Cont.*

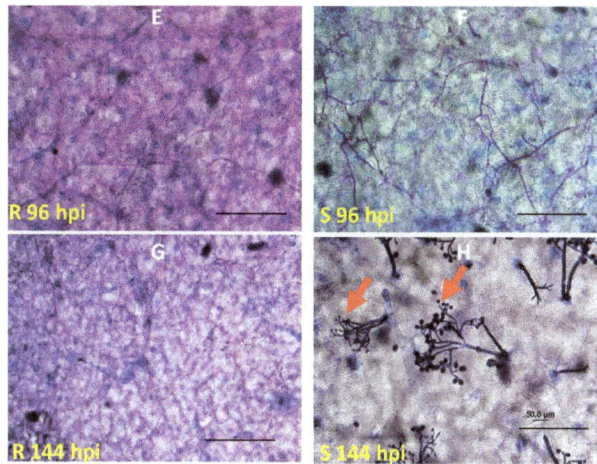

**Figure 1.** Histological analyses to TKF 4321 and TKF 2002 leaves inoculated with blue mold pathogen. The progress of tissue colonization by the pathogen is shown in panels A, C, E, G in the resistant parental line TKF 4321, and in B, D, F, H for the susceptible line TKF 2002. Hyphae were stained with trypan blue and observations were performed at 48 hpi (**A,B**); 72 hpi (**C,D**), 96 hpi (**E,F**), and 144 hpi (**G,H**). The life cycle of *P. tabacina* can be quickly completed on TKF 2002 plants within six days post inoculation, but the spread of pathogen is seriously hampered on TKF 4321 plants with restricted growth of hyphae. The yellow arrows in 1B and 1C indicate hyphae at the early stage of infection. The red arrows in 1H indicate sporangiophores. R, resistant; S, susceptible; hpi, hours post-inoculation.

## 2.2. Genetic Mapping of RBM1

Field experiments showed that the relative disease severity caused by *P. tabacina* infection on $F_1$ plants was intermediate between the two parental lines, suggesting that *RBM1* is a semi-dominant gene (Supplementary Materials Figure S3). To avoid occasional errors in phenotyping that result from incomplete resistance expressed in the heterozygous $F_2$ plants, we selected the susceptible individuals at the first screening and the resistant individuals at the second screening for genetic mapping. Therefore, although we inoculated a total of 862 $F_2$ plants, only 242 resistant and 168 susceptible plants were used in the mapping of the blue mold resistance gene. The first marker linked to *RBM1* we identified is PT61512, which is located on linkage group (LG) 7 and is in repulsion-phase. Taking advantage of the tobacco genetic map constructed by Bindler *et al.* [15], we mapped *RBM1* against polymorphic SSR markers on LG7 and generated a genetic map (Figure 2). As can be seen from this map, the *RBM1* locus is flanked by the two dominant SCAR marker loci developed by Milla *et al.* [9]. After sequencing these two SCAR markers (Supplementary Materials Table S1), we performed BLAST searches, but no high-quality hit was found in any of the sequenced *Nicotiana* genomes. The SSR marker loci that are closely linked to *RBM1*, such as PT61472 and PT51405, are also in repulsion phase, and are present only in blue mold-susceptible TKF 2002. However, heterozygosity of blue mold resistance can be distinguished in a segregating population by using both the coupling SCAR markers and repulsion markers (Supplementary Materials Figure S4).

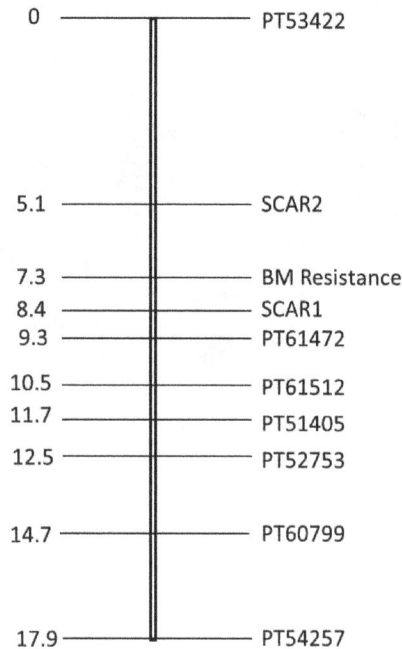

**Figure 2.** Genetic mapping of the *RBM1* locus on the linkage group 7. The genetic distance (cM) for each molecular marker is indicated on the left side of chromosome. The map is drawn to scale.

## 2.3. Quantitative Analysis of Defense Responses to Blue Mold Infection

Mounting an adequate defense response against an invading pathogen is generally dependent on the fine-tuned perception of pathogen infection and the activation of a gene expression network that results in the production of reactive oxygen species (ROS) and synthesis of pathogenesis-related (PR) proteins. To better characterize *RBM1*-mediated resistance to *P. tabacina*, we conducted real time-PCR to quantitatively analyze changes in gene expression for *PR1* and *PR4*, and also for *HSR203J*, a molecular marker of HR cell death [16]. The expression kinetics for all three genes exhibited a similar trend. The genes were expressed at a significantly higher level in the resistant parental line, TKF 4321, than in the susceptible line TKF 2002 starting in the middle of the sampling period, although these genes were also induced gradually to a remarkable level in TKF 2002. The oxidative burst is known to be a hallmark of successful recognition of infection and activation of plant defenses [17]. To test whether ROS plays a role in tobacco resistance to blue mold, we quantified and compared $H_2O_2$ levels in resistant and susceptible plants. TKF 2002 and TKF 4321 had similar basal levels of $H_2O_2$, and pathogen infection induced a comparable increase in $H_2O_2$ levels for both lines at 72 h post inoculation. Interestingly, production of $H_2O_2$ in TKF 2002 appeared to reach a peak at 72 hpi, but $H_2O_2$ production continued to increase in TKF 4321, and the maximum level was observed at 96 hpi. Even at 120 hpi, the level of $H_2O_2$ in TKF 4321 remained high, in contrast with a trend of declining $H_2O_2$ production in TKF 2002 at the same time point.

## 3. Discussion

Development of disease-resistant cultivars is an effective way to control diseases if sufficient genetic variation for host resistance is available. When sources of resistance are limited, breeders must turn to the secondary gene pool for species that can hybridize with the cultivated species. Molecular techniques enable the transfer of resistance genes between much more distantly related species. In the *Solanaceae*, several *R* genes have been shown to confer resistance reactions to pathogens carrying the appropriate *Avr* (*Avirulence*) genes when transferred to other solanaceous species. Transferring tomato *Cf-9* to tobacco and potato, pepper *Bs2* to tomato, tomato *Pto* to tobacco, and the tobacco *N* gene to tomato, demonstrated that *Avr*-dependent *R* protein-triggered signaling cascades are conserved in diverse species in the *Solanaceae* [18–21].

Given the scarcity of genetic resources for resistance to blue mold in common tobacco, breeders introgressed blue mold resistance conferred by *RBM1* from *N. debneyi* into tobacco to reduce the potential for losses from this disease. The two SCAR markers developed by Milla *et al.* [9] provided valuable tools for early selection on breeding for blue mold resistance. However, the SCAR markers are dominant, precluding differentiation of plants that are homozygous from those that are heterozygous at the resistance locus, and their locations in the tobacco genome are unknown. In the present study, genetic mapping indicates that *RBM1* is located on LG 7. In addition, the repulsion markers, in combination with the coupling-phase SCAR markers, make it possible to distinguish heterozygosity of blue mold resistance in a segregating population. As a result, marker-assisted selection for blue mold resistance in tobacco will be achieved with improved efficiency.

Although *RBM1* contributes significantly to the control of blue mold disease, one pitfall is that *RBM1*-mediated immunity in tobacco is not as fully functional as it is in *N. debneyi*. While *P. tabacina* infection was highly restricted in TKF 4321, hyphae were still produced, although with a reduced occurrence (Figure 1C). Thus, while limiting the extent of pathogen spread, *RBM1*-mediated resistance appears to be temporally slower and of lower amplitude than the typical defense responses conferred by plant resistance (*R*) genes such as NBS-LRR genes that encode nucleotide-binding site leucine-rich repeat proteins. *R* gene-mediated disease resistance is often associated with the hypersensitive response (HR), which is characterized by a rapid, localized cell death that serves to suppress pathogen spread at the infection sites. Hand-inoculation of tobacco leaves confirmed that the HR plays a role in defense against *P. tabacina* infection, in agreement with the induced expression of *HSR203J* observed in this host-pathogen interaction (Supplementary Materials Figures S1 and S3). It has been demonstrated that the gene product of *HSR203J* is a serine hydrolase with a potential role in the degradation of harmful compounds [22]. Activation of *HSR203J* is rapid, highly localized, and is correlated with programmed cell death in tobacco in response to various HR-inducing pathogens or elicitors [23]. Therefore, *HSR203J* has been used as a marker gene to identify the triggering of the HR-mediated defense response [24,25]. Activation of this gene is usually observed several hours after pathogen infection. However, transcription of *HSR203J* was induced at 48 hpi and attained its highest level at 120 hpi in TKF 4321 infected with *P. tabacina* (Figure 3). The molecular basis to explain why the HR is delayed or impaired is presently unknown.

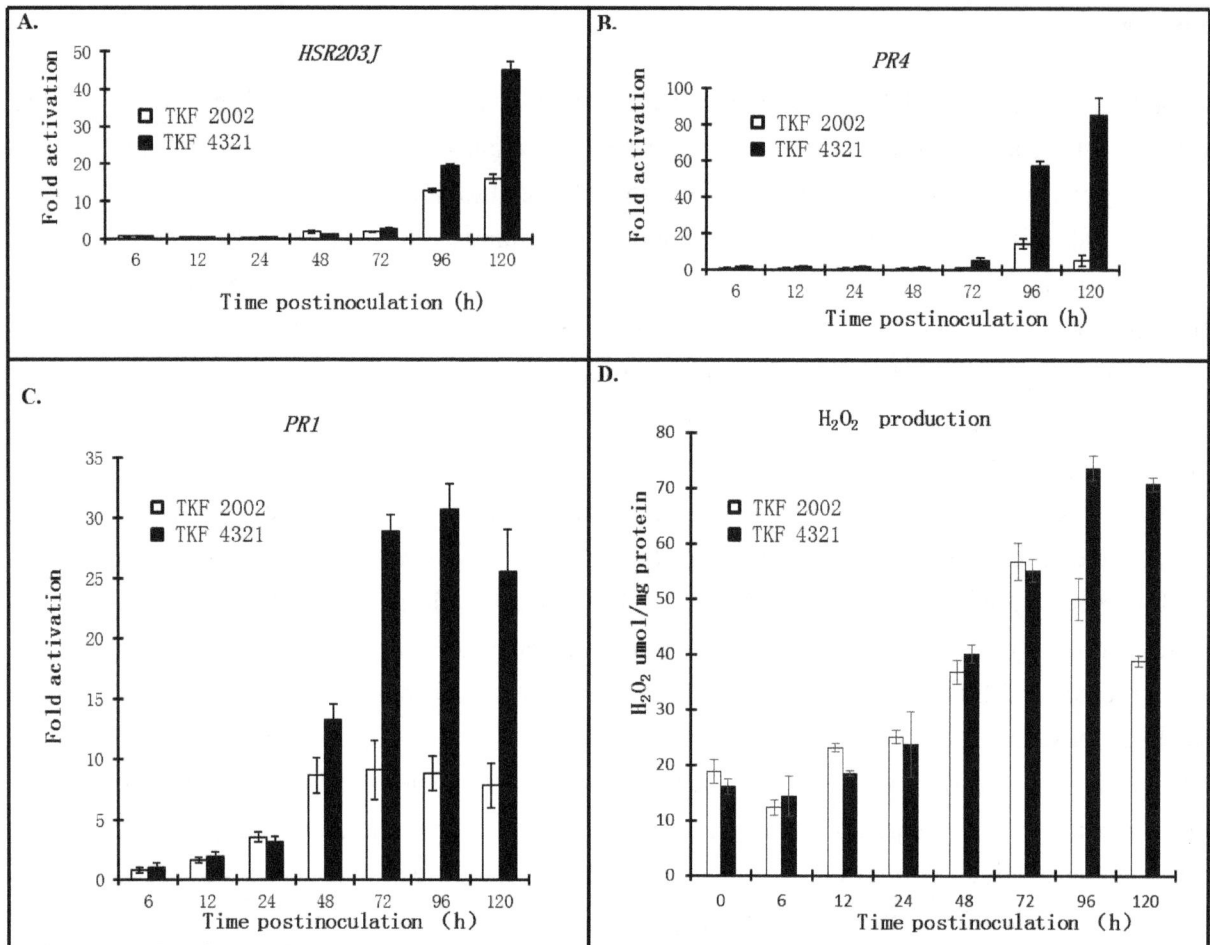

**Figure 3.** Real-time PCR analysis of host responses and H₂O₂ production during blue mold infection of tobacco. Expression profiles for *HSR203J*, *PR1*, and *PR4* were determined at seven time points between six and 120 hpi (**A–C**). Quantification of endogenous H₂O₂ in tobacco leaves is shown in 3D.

The guard hypothesis may provide implications for the transfer of disease resistance. The genetic interaction between R and Avr proteins can be explained by the guard hypothesis [26]. This model seeks to explain how R proteins activate resistance by interacting with another plant protein (a guardee) that is targeted and modified by the pathogen. Defense responses are triggered when the R protein detects an attempt to attack its guardee, which might not necessarily involve direct interaction between the R and Avr proteins. Efforts to transfer *R* genes from model species to crops, or between distantly related crop species, could be hampered by a phenomenon termed "restricted taxonomic functionality (RTF)" [19]. RTF might be caused by variance or absence of an appropriate guardee, rather than the inability of the R protein to recognize pathogen effectors in a different host [27]. If this is the case, *RBM1* alone may not confer full resistance due to the absence of a specific guardee in *N. tabacum*. Therefore, transfer of guard-guardee pairs might extend the range of *R* gene functionality and overcome the RTF limitation.

The production of antimicrobial pathogenesis-related (PR) proteins was first identified to function in defense against tobacco mosaic virus (TMV) infection of tobacco plants [28]. PR proteins include hydrolytic enzymes and defensins, which destroy pathogenic microbes through the hydrolysis of pathogen cell walls and disruption of the pathogen membrane, respectively. Genetic studies in *Arabidopsis thaliana* have shown that distinct sets of PR proteins are induced in response to different pathogens. Dependent on salicylic acid (SA) signaling, *PR1a*, *PR2* (a β-1,3-glucanase), and *PR5* (thaumatin) are responsive to biotrophic pathogens, while *PR3* (a chitinase), *PR4* (a chitinase), and *PR12* (a defensin) are induced by necrotrophic pathogens via the jasmonic acid (JA)-dependent signaling pathway [29]. Increased tolerance to the biotroph *P. tabacina* was demonstrated in transgenic tobacco over-expressing *PR1a* or β-1,3-glucanase [30,31]. In our study, quantitative analysis of *PR1* and *PR4* suggest that both genes are activated in TKF 4321 (Figure 2), indicating that a complicated reaction combining defense responses against both *biotrophic and necrotrophic* pathogens is induced in the host in response to *P. tabacina* infection. The antagonism between the SA and JA signaling pathways in the plant immune network has been well documented [32–34]. Therefore, one question raised here is how these two pathways are conciliated to a synergic mechanisms conferring resistance to blue mold in tobacco.

Production of reactive oxygen species, as well as inducible expression of *PR1a*, are markers of SA accumulation. Elevated $H_2O_2$ levels and *PR1a* gene expression in TKF 4321 (Figure 3) induced by *P. tabacina* infection indicate that SA-dependent signaling pathways are involved in triggering defense reactions. In the natural environment, plants can be infected simultaneously or sequentially by various pathogens with diverse strategies and lifestyles. The antagonistic interaction between the SA and JA signaling pathways has been proposed to be an efficient mechanism to prioritize one over the other, depending on the type of the invading pest or pathogen. We hypothesize that the exceptional defense signaling in blue mold resistance may result from the rapid rate of colonization and the short life cycle of *P. tabacina* because sporangiospores can be produced in as few as five days after the initial infection (Figure 1). If the HR-induced necrosis is unable to keep pace with hyphal development in leaf tissues, the lesions that result from cell death will expand to cover large areas of the leaf, which is also harmful for tobacco growth. In addition, if the initial infection is not controlled in time, the subsequent production and dispersal of infective sporangia can initiate a disease epidemic. Therefore, both the HR reaction, which is specific to biotrophs, and also the defense response against necrotrophs are activated to restrict the rapid colonization of *P. tabacina*. In this scenario, SA signaling-mediated systemic acquired resistance (SAR) protects uninfected parts from further damage, and the JA-dependent immune response suppress hyphal development in living tissues outside the necrotic infection sites. A previous observation of the SA-JA synergistic interactions can shed more light on our understanding of blue mold resistance [35]. A synergistic effect of the JA- and SA-dependent signaling pathways was observed when *Arabidopsis* was treated with low concentrations of JA and SA; however, under higher concentrations the effects were antagonistic, demonstrating that the outcome of the SA-JA interaction is dependent upon the relative abundance of each hormone [35]. We assume that the delayed perception of *P. tabacina* infection fails to induce high concentrations of JA or SA; accordingly, a synergistic SA-JA interaction was achieved. Convincing evidence for involvement of SA, JA, and PR4 in resistance to blue mold in tobacco is lacking at present, and will almost certainly require a study using transgenic plants.

It has been reported that some accessions of *N. langsdorffii*, a wild Brazilian tobacco relative, express resistance to *P. tabacina* infection by developing HR-induced necrotic lesions that eliminate subsequent pathogen colonization and sporulation. This resistance is conferred by a single, dominant gene named *NlRPT* [36,37]. Incompatible interactions with *P. tabacina* have also been identified in *N. obtusifolia* genotypes expressing HR caused by a single, partially dominant gene known as *Rpt1* [38]. Although several sources of genetic resistance to *P. tabacina* are available, we cannot predict whether the genetic effects mediated by these genes will be intact after transfer to tobacco by hybridization. To better exploit blue mold resistance in exotic relatives of tobacco, we need to characterize the mechanism(s) underlying the weakened immunity in tobacco. As for the incomplete *RBM1*-mediated resistance expressed in tobacco, our current hypotheses involve impaired effects caused by tobacco modification genes and the unsuccessful transfer of a complete multi-genic system that is responsible for resistance from *N. debneyi*. Gene cloning and functional analysis of *RBM1* will help explain why the complete resistance to *P. tabacina* infection seen in wild species is reduced in tobacco. Therefore, the genetic mapping of *RBM1* described herein will provide a foundation for molecular cloning of this gene and for engineering of *RBM1*-mediated resistance.

## 4. Experimental Section

### 4.1. The Mapping Population

The $F_2$ mapping population was derived from a cross between the two burley tobacco genotypes TKF 4321 (resistant) and TKF 2002 (susceptible). Blue mold resistance in TKF 4321 was inherited from NC-BMR 90, and the ultimate donor of resistance is believed to be *N. debneyi* [9]. Seedlings of the two parental lines, the $F_1$, and the segregating $F_2$ population were grown in a growth chamber under a 16 h light, 23 °C/8 h dark, 20 °C regime for about six weeks before inoculation with the pathogen.

### 4.2. P. tabacina Culture and Inoculation

*P. tabacina* isolate KY 79 was used for inoculation in the present study [39]. The isolate was continuously maintained and propagated on eight- to 12-week-old plants of *N. tabacum* cv. KY 14 as previously described [37]. The infective sporangia were collected and washed three times by filtration with sterile deionized water, with the final concentration being adjusted to $1 \times 10^5$ spores per ml. Tobacco leaves were inoculated by spraying with the spore suspension, and the inoculated plants were placed in large pre-moistened plastic tubs overnight. Plant reactions to blue mold infection were scored at 7 dpi (days post inoculation) and double-checked at 12 dpi. To investigate whether the hypersensitive response (HR) was involved in *RBM1*-mediated resistance to *P. tabacina*, we used inoculated tobacco leaves manually. The undersides of the leaves were nicked with a syringe needle and the inoculum was forced into the apoplast using a 1-ml disposable syringe with no needle. The inoculated plants were scored four days after injection. Incompatible (hypersensitive) responses were observed as areas of brown sunken tissue at the infiltration sites.

### 4.3. Microscopic Analysis of Inoculated Leaves

Cytological analyses were conducted to monitor the progress of tissue colonization by *P. tabacina*. Inoculated leaves were cleared and fixed in Farmer's fluid (acetic acid/ethanol/chloroform = 1:6:3 *v/v*), and pathogen structures were detected by trypan blue staining [37,40]. Trypan blue was dissolved in a 1:2 mixture of lactophenol/ethanol with a final concentration of 0.03% (*w/v*). Lactophenol was made by adding 10 g of phenol to a mixture of 10 mL of lactic acid, 10 mL of glycerol, and 10 mL of distilled water. Fixed leaves were stained at 100 °C in a water bath for 2 min, followed by de-staining in chloral hydrate solution (2.5 g/mL) at room temperature with gentle shaking prior to being examined with a light microscope.

## 4.4. Quantification of Endogenous Reactive Oxygen Species (ROS) in Tobacco Leaves

The $H_2O_2$ concentration was measured according to Chanda et al. [41]. Small leaf tissue samples (~100 mg) were homogenized in 500 μL of 40 mM Tris-HCl (pH 7.5) and centrifuged (10,000 rpm) at 4 °C for 10 min. A 20 μL aliquot of the supernatant solution was added to 80 μL of a mixture consisting of 77 μL 40 mM Tris-HCl (pH 7.5), 2 μL 1 mM DCFDA (2′, 7′-dichlorofluorescin diacetate; Sigma-Aldrich, St. Louis, MO, USA) and 1 μL 20 mg/mL HRP (horse radish peroxidase; Sigma). The samples were incubated for one hour in the dark, and $H_2O_2$ levels were measured using a spectrophotometer. The concentration of $H_2O_2$ was determined as mmol/mg protein by extrapolating from the standard $H_2O_2$ curve. Total protein was measured using the Bradford Assay which contained 10 μL sample supernatant, 90 μL dd $H_2O$ and 900 μL Coomassie Protein Assay Reagent (Thermo Scientific, Waltham, MA, USA). $H_2O_2$ levels were measured from four independent samples collected from both parental lines at each time point.

## 4.5. Real-Time PCR

Total RNA was extracted with the RNeasy Plant Mini Kit (Qiagen, Valencia, CA, USA) from tobacco leaves collected from TKF 2002 and TKF 4321 plants that had been previously inoculated with P. tabacina at 0, 6, 12, 24, 48, 72, 96, and 120 hpi (hours post inoculation). Three biological replicates were performed for each variety at each time point. First-strand cDNA was synthesized using M-MLV Reverse Transcriptase (Invitrogen) according to the manufacturer's instructions. Fluorescence PCR amplifications were performed in triplicate using the StepOne real-time PCR system (Applied Biosystems, Grand Island, NY, USA). A 2 μL aliquot of each first strand cDNA equivalent of 20 ng of total RNA was amplified using primer pairs specific to the tobacco actin, HSR, and PR genes in a 20 μL reaction containing 2 μL of each primer (2.5 μM), 8.8 μL of water, and 10 μL of iTaq SYBR Green Supermix with ROX (Bio-Rad, Hercules, CA, USA). The names and sequences of the primers used for real-time analysis in this study are: Actin-Forward, 5′-AGGGTTGCTGGAGATGATG-3′, Actin-Reverse, 5′-CGGGTTAAGAGGTGCTTCAG-3′; PR-1aF, 5′-GGATGCCCATAACACAGCTC-3′, PR-1aR, 5′-GCTAGGTTTTCGCCGTATTG-3′; PR-4rtpF, 5′-GGCCAAGATTCCTGTGGTAGAT-3′, PR-4rtpR, 5′-CACTGTTGTTTGAGTTCCTGTTCCT-3′. Amplification conditions were: denaturation at 95 °C for 2 min, followed by 35 cycles of 95 °C for 30 s, 51 °C for 30 s, and 72 °C for 30 s, with a final extension at 72 °C for 5 min.

## 4.6. Genetic Mapping and Marker Design

We initially mapped SSR (simple sequence repeat) markers with known genetic positions to localize the approximate position of RBM1, based on the high-density genetic linkage map of tobacco [15]. Additional markers were then developed from tobacco genome sequence contigs that harbor mapped SSR markers [1]. Only susceptible plants from the initial scoring and resistant plants from the second scoring were used for genetic mapping. The initial mapping population consisted of 93

F2 plants. The size of the mapping population was increased to include 415 susceptible individuals for fine mapping. The genetic linkage map was constructed with Mapmaker version 3.0 [42] (Lander *et al.* 1987). All markers described in this paper are listed in Table 1.

**Table 1.** Molecular markers described in this study.

| Marker Name | Marker Type | Left Primer | Right Primer |
|---|---|---|---|
| PT53422 | SSR | CGCACATACGTACTGAGCATT | GGCTCGAACCCGTAACCTAT |
| PT61472 | SSR | TCCAATACCTTTAATGCATCTCC | GCATGACATGTTGAAGTGGG |
| PT61512Y | SSR | ATCGGACCCAAAGTTTAAGAAACAA | AGGCAAGGATAGGGATAGGAATAGC |
| PT51405 | SSR | AAGTTGGTTATAATCTCGATGCC | AATTCATCTCCAACGCAACTG |
| PT52753 | SSR | TTGGGCCTAGTTTCTACGGA | CAATGCTAACCTGTCACTACCA |
| PT60799 | SSR | GCCGCAGTACTAAAGCTCAGA | TGCACAATCTTCAGGTCAGC |
| PT54257 | SSR | GCAGCACCCAAGTTGCTTA | CCGTCTATTAGCATCAAGGCA |
| SCAR1 | SCAR | CTGAGTTTGGCCGAATAGCAT | CAAACGTCCTAAATGGGGTATAA |
| SCAR2 | SCAR | GTCTACGGCAAGGGGAGATATTA | GTCTACGGCAGCAATCAACATG |

## 5. Conclusions

In the present study, we characterized the *RBM1*-mediated resistance to blue mold, a destructive disease in tobacco. Concomitant with elevated $H_2O_2$ levels and *PR1a* gene expression, hypersensitive response (HR) plays an important role in the host defense reactions. Although *RBM1* confers completed resistance to *P. tabacina* in the original donor, *N. debneyi*, field experiments showed that *RBM1* is a semi-dominant gene with incomplete immunity in tobacco. Genetic mapping indicated that this disease resistance gene resides on chromosome 7. Therefore, our work described herein will enable the molecular cloning of this gene and the engineering of *RBM1*-mediated resistance.

## Acknowledgments

We thank James T. Hall for maintaining the blue mold pathogen and for his help with plant inoculations. We are grateful to the Kentucky Tobacco Research & Development Center (KTRDC) for providing facilities and growth chamber. This work was supported by British American Tobacco (to Shengming Yang).

## Author Contributions

Dandan Li, David Zaitlin, Robert Miller, and Shengming Yang conceived and designed the experiments; Xia Wu, Dandan Li, and Yinguang Bao performed the experiments; Xia Wu, Dandan Li, Yinguang Bao, Robert Miller, and Shengming Yang analyzed the data; Xia Wu, David Zaitlin, and Shengming Yang wrote the manuscript.

## Conflicts of Interest

The authors declare that they have no conflict of interest.

# References

1.  Sierro, N.; Battey, J.N.; Ouadi, S.; Bakaher, N.; Bovet, L.; Willig, A.; Goepfert, S.; Peitsch, M.C.; Ivanov, N.V. The tobacco genome sequence and its comparison with those of tomato and potato. *Nat. Commun.* **2014**, *5*, 3833, doi:10.1038/ncomms4833.

2.  Zhang, J.; Zhang, Y.; Du, Y.; Chen, S.; Tang, H. Dynamic metabonomic responses of tobacco (*Nicotiana tabacum*) plants to salt stress. *J. Proteome Res.* **2011**, *10*, 1904–1914.

3.  Caplan, J.L.; Mamillapalli, P.; Burch-Smith, T.M.; Czymmek, K.; Dinesh-Kumar, S.P. Chloroplastic protein NRIP1 mediates innate immune receptor recognition of a viral effector. *Cell* **2008**, *132*, 449–462.

4.  Whitham, S.; Dinesh-Kumar, S.P.; Choi, D.; Hehl, R.; Corr, C.; Baker, B. The product of the tobacco mosaic virus resistance gene *N*: Similarity to toll and the interleukin-1 receptor. *Cell* **1994**, *78*, 1101–1115.

5.  Heist, E.P.; Nesmith, W.C.; Schardl, C.L. Interactions of *Peronospora tabacina* with Roots of *Nicotiana* spp. in Gnotobiotic Associations. *Phytopathology* **2002**, *92*, 400–405.

6.  Schiltz, P. Downy mildew of tobacco. In *The Downy Mildews*; Spencer, D.M., Ed.; Academic Press: London, UK, 1981; pp. 577–599.

7.  Rufty, R.C. Genetics of host resistance to tobacco blue mold. In *Blue Mold of Tobacco*; McKean, W.E., Ed.; American Phytopathological Society: St. Paul, MN, USA, 1989; pp. 141–164.

8.  Clayton, E.E. The transfer of blue mold resistance from *Nicotiana debneyi*. Part III. Development of a blue mold resistant cigar wrapper variety. *Tob. Sci.* **1967**, *11*, 107–110.

9.  Milla, S.R.; Levin, J.S.; Lewis, R.S.; Rufty, R.C. RAPD and SCAR markers linked to an introgressed gene conditioning resistance to *Peronospora tabacina* D.B. Adam. in tobacco. *Crop Sci.* **2005**, *45*, 2346–2354.

10. Wark, D.C. Nicotiana species as sources of resistance to blue mold (*Peronospora tabacina* Adam) for cultivated tobacco. In Proceedings of the 3rd World Tobacco Science Congress, Salisbury, Southern Rhodesia, 18–26 February 1963, Tobacco Research Board: Harare, Zimbabwe, 1963; pp. 252–259.

11. Wark, D.C. Development of flue-cured tobacco cultivars resistant to a common strain of blue mold. *Tob. Sci.* **1970**, 147–150.

12. Rufty, R.C.; Main, C.E. Components of partial resistance to blue mold in six tobacco genotypes under controlled environmental conditions. *Phytopathology* **1989**, *79*, 606–609.

13. Wuttke, H.H. Blue mould resistance in tobacco. *Aust. Tob. Grow. Bull* **1972**, *20*, 6–10.

14. Julio, E.; Verrier, J.L.; Dorlhac de Borne, F. Development of SCAR markers linked to three disease resistances based on AFLP within *Nicotiana tabacum* L. *Theor. Appl. Genet.* **2006**, *112*, 335–346.

15. Bindler, G.; Plieske, J.; Bakaher, N.; Gunduz, I.; Ivanov, N.; van der Hoeven, R.; Ganal, M.; Donini, P. A high density genetic map of tobacco (*Nicotiana tabacum* L.) obtained from large scale microsatellite marker development. *Theor. Appl. Genet.* **2011**, *123*, 219–230.

16. Pontier, D.; Gan, S.; Amasino, R.M.; Roby, D.; Lam, E. Markers for hypersensitive response and senescence show distinct patterns of expression. *Plant Mol. Biol.* **1999**, *39*, 1243–1255.

17. Tudzynski, P.; Heller, J.; Siegmund, U. Reactive oxygen species generation in fungal development and pathogenesis. *Curr. Opin. Microbiol.* **2012**, *15*, 653–659.

18. Hammond-Kosack, K.E.; Tang, S.; Harrison, K.; Jones, J.D. The tomato *Cf-9* disease resistance gene functions in tobacco and potato to confer responsiveness to the fungal avirulence gene product avr 9. *Plant Cell* **1998**, *10*, 1251–1266.

19. Tai, T.H.; Dahlbeck, D.; Clark, E.T.; Gajiwala, P.; Pasion, R.; Whalen, M.C.; Stall, R.E.; Staskawicz, B.J. Expression of the *Bs2* pepper gene confers resistance to bacterial spot disease in tomato. *Proc. Natl. Acad. Sci. USA* **1999**, *96*, 14153–14158.

20. Thilmony, R.L.; Chen, Z.; Bressan, R.A.; Martin, G.B. Expression of the tomato *Pto* gene in tobacco enhances resistance to *Pseudomonas syringae pv tabaci* expressing *avrPto*. *Plant Cell* **1995**, *7*, 1529–1536.

21. Whitham, S.; McCormick, S.; Baker, B. The N gene of tobacco confers resistance to tobacco mosaic virus in transgenic tomato. *Proc. Natl. Acad. Sci. USA* **1996**, *93*, 8776–8781.

22. Baudouni, E.; Charpenteau, M.; Roby, D.; Marco, Y.; Ranjeva, R.; Ranty, B. Functional expression of a tobacco gene related to the serine hydrolase family—Esterase activity towards short-chain dinitrophenyl acylesters. *Eur. J. Biochem.* **1997**, *248*, 700–706.

23. Pontier, D.; Tronchet, M.; Rogowsky, P.; Lam, E.; Roby, D. Activation of *hsr203*, a plant gene expressed during incompatible plant-pathogen interactions, is correlated with programmed cell death. *Mol. Plant Microbe Interact.* **1998**, *11*, 544–555

24. Pontier, D.; Balague, C.; Bezombes-Marion, I.; Tronchet, M.; Deslandes, L. Identification of a novel pathogen-responsive element in the promoter of the tobacco gene HSR203J, a molecular marker of the hypersensitive response. *Plant J.* **2001**, *26*, 495–507.

25. Takahashi, Y.; Uehara, Y.; Berberich, T.; Ito, A.; Saitoh, H.; Miyazaki, A.; Terauchi, R.; Kusano, T. A subset of hypersensitive response marker genes, including *HSR203J*, is the downstream target of a spermine signal transduction pathway in tobacco. *Plant J.* **2004**, *40*, 586–595.

26. Van der Biezen, E.A.; Jones, J.D. Plant disease-resistance proteins and the gene-for-gene concept. *Trends Biochem. Sci.* **1998**, *23*, 454–456.

27. McDowell, J.M.; Woffenden, B.J. Plant disease resistance genes: Recent insights and potential applications. *Trends Biotechnol.* **2003**, *21*, 178–183.

28. Van Loon, L.C.; van Kammen, A. Polyacrylamide disc electrophoresis of the soluble leaf proteins from *Nicotiana tabacum* var. "Samsun" and "Samsun NN". II. Changes in protein constitution after infection with tobacco mosaic virus. *Virology* **1970**, *40*, 190–211.

29. Thomma, B.P.; Eggermont, K.; Penninckx, I.A.; Mauch-Mani, B.; Vogelsang, R.; Cammue, B.P.; Broekaert, W.F. Separate jasmonate-dependent and salicylate-dependent defense-response pathways in Arabidopsis are essential for resistance to distinct microbial pathogens. *Proc. Natl. Acad. Sci. USA* **1998**, *95*, 15107–15111.

30. Alexander, D.; Goodman, R.M.; Gut-Rella, M.; Glascock, C.; Weymann, K.; Friedrich, L.; Maddox, D.; Ahl-Goy, P.; Luntz, T.; Ward, E.; *et al.* Increased tolerance to two oomycete pathogens in transgenic tobacco expressing pathogenesis-related protein. *Proc. Natl. Acad. Sci. USA* **1993**, *90*, 7327–7331.

31. Lusso, M.; Kuc, J. The effect of sense and antisense expression of the PR-N gene for β-1,3-glucanase on disease resistance of tobacco to fungi and viruses. *Physiol. Mol. Plant Pathol.* **1996**, *49*, 267–283.

32. Gupta, V.; Willits, M.G.; Glazebrook, J. *Arabidopsis thaliana* EDS4 contributes to salicylic acid (SA)-dependent expression of defense responses: Evidence for inhibition of jasmonic acidsignaling by SA. *Mol. Plant Microbe Interact.* **2000**, *13*, 503–511.

33. Koornneef, A.; Leon-Reyes, A.; Ritsema, T.; Verhage, A.; Den Otter, F.C.; van Loon, L.C.; Pieterse, C.M. Kinetics of salicylate-mediated suppression of jasmonate signaling reveal a role for redox modulation. *Plant Physiol.* **2008**, *147*, 1358–1368.

34. Spoel, S.H.; Koornneef, A.; Claessens, S.M.C.; Korzelius, J.P.; van Pelt, J.A.; Dong, X.; Pieterse, C.M. NPR1 modulates cross-talk between salicylate- and jasmonate-dependent defense pathways through a novel function in the cytosol. *Plant Cell* **2003**, *15*, 760–770.

35. Mur, L.A.; Kenton, P.; Atzorn, R.; Miersch, O.; Wasternack, C. The outcomes of concentration-specific interactions between salicylate and jasmonate signaling include synergy, antagonism, and oxidative stress leading to cell death. *Plant Physiol.* **2006**, *140*, 249–262.

36. Zhang, S.; Gao, M.; Zaitlin, D. Molecular linkage mapping and marker-trait associations with *NlRPT*, a downy mildew resistance gene in *Nicotiana langsdorffii. Front. Plant Sci.* **2012**, *3*, 185, doi:10.3389/fpls.2012.00185.

37. Zhang, S.; Zaitlin, D. Genetic resistance to *Peronospora tabacina* in *Nicotiana langsdorffii*, a South American wild tobacco. *Phytopathology* **2008**, *98*,519–528.

38. Heist, E.P.; Zaitlin, D.; Funnell, D.L.; Nesmith, W.C.; Schardl, C.L. Necrotic lesion resistance induced by *Peronospora tabacina* on Leaves of *Nicotiana obtusifolia. Phytopathology* **2004**, *94*, 1178–1188.

39. Reuveni, M.; Nesmith, W.C.; Siegel, M.R. Symptom development and disease severity in *Nicotiana tabacum* and *N. repanda* caused by *Peronospora tabacina. Plant Dis.* **1986**, *70*, 727–729.

40. Keogh, R.C.; Deverall, B.J.; McLeod, S. Comparison of histological and physiological responses to *Phakopsora pachyrhizi* in resistant and susceptible soybean. *Trans. Br. Mycol. Soc.* **1980**, *74*, 329–333.

41. Chanda, B.; Venugopal, S.C.; Kulshrestha, S.; Navarre, D.A.; Downie, B.; Vaillancourt, L.; Kachroo, A.; Kachroo, P. Glycerol-3-phosphate levels are associated with basal resistance to the hemibiotrophic fungus *Colletotrichum higginsianum* in Arabidopsis. *Plant Physiol.* **2008**, *147*, 2017–2029.

42. Lander, E.S.; Green, P.; Abrahamson, J.; Barlow, A.; Daly, M.J.; Lincoln, S.E.; Newburg, L. MAPMAKER: An interactive computer package for constructing primary genetic maps of experimental and natural populations. *Genomics* **1987**, *1*, 174–181.

# Permissions

The contributors of this book come from diverse backgrounds, making this book a truly international effort. This book will bring forth new frontiers with its revolutionizing research information and detailed analysis of the nascent developments around the world.

We would like to thank all the contributing authors for lending their expertise to make the book truly unique. They have played a crucial role in the development of this book. Without their invaluable contributions this book wouldn't have been possible. They have made vital efforts to compile up to date information on the varied aspects of this subject to make this book a valuable addition to the collection of many professionals and students.

This book was conceptualized with the vision of imparting up-to-date information and advanced data in this field. To ensure the same, a matchless editorial board was set up. Every individual on the board went through rigorous rounds of assessment to prove their worth. After which they invested a large part of their time researching and compiling the most relevant data for our readers.

The editorial board has been involved in producing this book since its inception. They have spent rigorous hours researching and exploring the diverse topics which have resulted in the successful publishing of this book. They have passed on their knowledge of decades through this book. To expedite this challenging task, the publisher supported the team at every step. A small team of assistant editors was also appointed to further simplify the editing procedure and attain best results for the readers.

Apart from the editorial board, the designing team has also invested a significant amount of their time in understanding the subject and creating the most relevant covers. They scrutinized every image to scout for the most suitable representation of the subject and create an appropriate cover for the book.

The publishing team has been an ardent support to the editorial, designing and production team. Their endless efforts to recruit the best for this project, has resulted in the accomplishment of this book. They are a veteran in the field of academics and their pool of knowledge is as vast as their experience in printing. Their expertise and guidance has proved useful at every step. Their uncompromising quality standards have made this book an exceptional effort. Their encouragement from time to time has been an inspiration for everyone.

The publisher and the editorial board hope that this book will prove to be a valuable piece of knowledge for researchers, students, practitioners and scholars across the globe.

# List of Contributors

**Abin Sebastian and Majeti Narasimha Vara Prasad**
Department of Plant Sciences, University of Hyderabad, Hyderabad-500046, Telangana, India

**Benjamin Klug**
Aglukon Spezialdünger GmbH & Co. KG, Düsseldorf; Germany

**Thomas W. Kirchner and Walter J. Horst**
Institute of Plant Nutrition, Leibniz Universität Hannover, Germany

**Valérie Page and Urs Feller**
Institute of Plant Sciences, University of Bern, Bern 3012, Switzerland

**Carlos Zúñiga-Espinoza, Raúl Ferreyra and Gabriel Selles**
Instituto de Investigaciones Agropecuarias, INIA La Platina, Santa Rosa 1161, Santiago, Chile

**Cristina Aspillaga**
Roma N° 90, San Esteban, Los Andes 2120000, Chile

**Clyde D. Boyette and Kenneth C. Stetina**
USDA-ARS, Biological Control of Pests Research Unit, Stoneville, MS 38776, USA

**Robert E. Hoagland**
USDA-ARS, Crop Production Systems Research Unit, Stoneville, MS 38776, USA

**Carol Miles and Kelly Ann Atterberry**
Department of Horticulture, Washington State University, Northwestern Washington Research and Extension Center, Mount Vernon, WA 98273, USA

**Brook Brouwer**
Department of Crop and Soil Sciences, Washington State University, Northwestern Washington Research and Extension Center, Mount Vernon, WA 98273, USA

**Elgailani A. Abdalla, Abdelrahman K. Osman, Mahmoud A. Maki, Fadlalmaola M. Nur and Salah B. Ali**
Elobeid Research Station, Elobeid 611, Sudan

**Jens B. Aune**
Department of International Environment and Development Studies, Noragric, Norwegian University of Life Sciences (NMBU) P.O. Box 5003, N-1432 Ås, Norway

**Zhonghui Ou**
School of Mathematics and Computer Science, Fujian Normal University, 350117 Fuzhou, China

**Eleni Siasou and Neil Willey**
Centre for Research in Bioscience, Faculty of Health and Applied Sciences, University of the West of England, Coldharbour Lane, Frenchay, Bristol BS16 1QY, UK

**Arif Hasan Khan Robin, Md. Jasim Uddin and Khandaker Nafiz Bayazid**
Department of Genetics and Plant Breeding, Bangladesh Agricultural University, Mymensingh 02202, Bangladesh

**Muhammad Ijaz**
College of Agriculture, Bahauddin Zakariya University, Bahadur- Sub Campus Layyah 31200, Pakistan
Institute of Agronomy & Plant Breeding I, Biomedical Research Center Seltersberg (BFS), Justus Liebig University Giessen, Schubertstr. 81, Giessen D-35392, Germany

**Khalid Mahmood**
Department of Agro-ecology, Faculty of Science and Technology, Aarhus University, Aarhus C 8000, Denmark

**Bernd Honermeier**
Institute of Agronomy & Plant Breeding I, Biomedical Research Center Seltersberg (BFS), Justus Liebig University Giessen, Schubertstr. 81, Giessen D-35392, Germany

**Heike Bücking and Arjun Kafle**
Biology and Microbiology Department, South Dakota State University, Brookings, SD 57007, USA

**James A. Dyer**
Contract Researcher, 122 Hexam Street, Cambridge, Ontario, N3H 3Z9, Canada

**Xavier P.C. Vergé**
Contract Researcher; 2055, Carling avenue, #1016, Ottawa, Ontario, K2A 1G6, Canada

**Xia Wu**
Department of Plant & Soil Sciences, University of Kentucky, Lexington, KY 40546, USA
Department of Plant Genetics and Breeding, Shandong Agricultural University, Shandong 271018, China

**Dandan Li**
Department of Plant & Soil Sciences, University of Kentucky, Lexington, KY 40546, USA

**Yinguang Bao**
Department of Plant & Soil Sciences, University of Kentucky, Lexington, KY 40546, USA
Department of Plant Genetics and Breeding, Shandong Agricultural University, Shandong 271018, China

**David Zaitlin**
Kentucky Tobacco Research & Development Center, University of Kentucky, Lexington, KY 40546, USA

**Robert Miller and Shengming Yang**
Department of Plant & Soil Sciences, University of Kentucky, Lexington, KY 40546, USA

www.ingramcontent.com/pod-product-compliance
Lightning Source LLC
Chambersburg PA
CBHW080514200326
41458CB00012B/4205